W9-DBF-709

ELEMENTS OF REAL ANALYSIS

DAVID A. SPRECHER
University of California, Santa Barbara

DOVER PUBLICATIONS, INC. *New York*

Published in Canada by General Publishing Company, Ltd., 30 Lesmill Road, Don Mills, Toronto, Ontario.
Published in the United Kingdom by Constable and Company, Ltd., 10 Orange Street, London WC2H 7EG.

This Dover edition, first published in 1987, is an unabridged, corrected republication of the work first published by Academic Press, New York, 1970.

Manufactured in the United States of America
Dover Publications, Inc., 31 East 2nd Street, Mineola, N.Y. 11501

Library of Congress Cataloging-in-Publication Data

Sprecher, David A., 1930–
Elements of real analysis.

Reprint. Originally published: New York : Academic Press, 1970.
Bibliography: p.
Includes index.
1. Functions of real variables. 2. Mathematical analysis. I. Title.
QA331.5.S68 1987 515.8 87-561
ISBN 0-486-65385-4 (pbk.)

CONTENTS

PREFACE

A student's training in calculus is generally related to his everyday experiences, and the symbols he manipulates represent familiar objects. The essence of mathematics, however, is its capacity for abstractions and generalizations. Therefore, the goal of advanced training in mathematics is what Herman Weyl has called the decisive step of mathematical abstraction: forgetting what the symbols stand for. But there must be intermediate steps between calculus and the ultimate stage; this text is intended to serve as a first step. It is designed for a one-year course in introductory analysis, and the material should be accessible to students who have completed a regular calculus course.

Many abstractions and generalizations originate with the real line, which is used as the unifying theme of the text. Both the selected notation and arguments, however, lend themselves to an easy transplant into more abstract settings. Generalizations are indicated at various stages of the discussion.

A most fundamental concept in analysis is that of a Cauchy sequence. It thus seemed natural to construct the real number system from this point of view. This, like every other construction of the real numbers, becomes laborious and very technical, but it is through these agonies that we learn what the real number system is. If desired, however, one can choose Section 9 as a starting point for the course. References to the needed nomenclature and results from Sections 1–8 are included for this reason.

The contents of Part I may be known to some readers, but it has been my experience that this material is not known to many. This part is provided to give a sound footing to the inadequately prepared student.

Each section is followed with exercises. Their total, ranging from routine exercises to very difficult ones, exceeds five hundred. I have found that most students are challenged by some hard exercises, and that they gain insight into the subject matter as a result of their efforts to solve them. But it should be made clear that they are not expected to come up with a solution for every problem tackled.

The author was aided at various stages of writing this manuscript by the criticism and suggestions of Professors Anatole Beck, Richard C. Gilbert, Barnet W. Glickfeld, Stuart M. Newberger, and Max L. Weiss: my thanks go to them. For their care and devotion in producing this book I would like to express my appreciation to my typists, Delores Brannon, Louise Kraus, and Wanda Michalenko, as well as to the production and editorial staff of Academic Press: I could not imagine a happier working relationship.

PART ONE

FUNDAMENTAL CONCEPTS

ONE

SETS AND FUNCTIONS

The satisfactory presentation of any subject matter requires that it be stated in the simplest possible terms. The ultimate language for this is envisioned to possess rather elusive qualities: It must have a minimal vocabulary and a simple grammar, yet admit a perfectly clear and precise listing of all phenomena irrespective of their intuitive appeal. The language, in other words, must be a transparent vehicle which carries ideas without affecting their meaning in any way. Set theory comes closest to being such a language. Introduced in the 1870's by Georg Cantor to cope with infinities in mathematics, the theory eventually became the foundation of most mathematical disciplines. Despite its universality it is quite astounding in its simplicity and elegance.

A systematic and rigorous development of set theory is not the object here: this properly belongs in a course in logic. Relying on the reader's intuition and imagination, we discuss the theory very informally instead. Our primary aim is to get acquainted with the rudiments of set theory. The reader who feels in need of a more detailed account of the theory should consult the bibliography at the end of the book for suggested references.

1 / SETS

The problem of identifying objects and comparing them is certainly basic, but it is an issue we do not want to fuss over here. We assume, therefore, that there is an equality symbol "=" with the property that if x, y, and z are any objects whatsoever, then

$$x = x;$$

$$\text{if} \quad x = y \quad \text{then} \quad y = x;$$

$$\text{if} \quad x = y \quad \text{and} \quad y = z, \quad \text{then} \quad x = z.$$

The statement "$x = y$" means that "x" and "y" are different names of the same thing. The fact that x and y are not equal is expressed by writing $x \neq y$. It is further being assumed that for any objects x and y exactly one of the relations $x = y$ or $x \neq y$ holds. This implies, of course, that any two objects can be compared.

Let us now consider sets. A *set* is imagined to be an aggregate of objects, be these numbers, arbitrary symbols, or entities. It is identified in one of two ways:

by exhibiting its members,
by listing all distinguishing features which members have and non-members do not have.

A most basic notion is that of equality of sets: The sets A and B are *equal*, written

$$A = B,$$

when and only when they contain the same members: otherwise we write

$$A \neq B.$$

Owing to the first prescription, the set consisting of the elements a and b is completely identified by the symbol

$$\{a, b\}.$$

At the same time it is seen that the sets

$$\{b, a\}$$

and

$$\{a, b, b, a\},$$

for example, are likewise determined by the membership of a and b. Because a set is determined by membership alone, we must regard all such sets as one and the same.

To specify sets according to the other scheme we introduce this convenient notation: If p is a property and x an element, then the formula

$$p(x)$$

states "the element x has property p"; the symbolic statement

$$\{x \mid p(x)\}$$

reads "the set of all elements x for which $p(x)$ is a true statement." For example, if p is the property of being a duck, then $p(x)$ means "x is a duck," and $\{x \mid p(x)\}$ represents the set of all ducks; if p is the property of being a or b, then

$$\{x \mid p(x)\} = \{x \mid x = a \quad \text{or} \quad x = b\} = \{a, b\}.$$

For a given set A, the relation

$$a \in A$$

signifies that a is an element of A. Synonymous with this are the statements "a belongs to A" and "a is a member of A". When a does not belong to A we write

$$a \notin A.$$

Consequently, if A is a given set and x an arbitrary element, then one and only one of the relations $x \in A$ or $x \notin A$ is true.

An element must always be carefully distinguished from the set consisting of this element. Thus,

$$a, \quad \{a\}, \quad \{\{a\}\},$$

for instance, are distinct from each other. In the relation $a \in A$, however, a itself may be a set. To be sure, the concept of a set whose members are sets is entirely consistent with the above. As an example we may take the Euclidean plane to be a set of lines which are sets of points. The idea of a set containing all other sets is, on the other hand, not meaningful. This phenomenon is logically quite difficult and no further reference to it will be made in this text.

Consider the situation $A = B$. From the definition of set equality it follows that $x \in A$ implies $x \in B$ and $x \in B$ implies $x \in A$. When part of this symmetry may be false we speak of subsets: When every member of a given set A is concurrently a member of a set B, then A is a *subset* of B, written

$$A \subset B;$$

the statement

$$B \supset A,$$

which reads "B contains A," is equivalent to it. The relation

$$A \not\subset B$$

signifies that A is not a subset of B. These set relations are graphically described in a *Venn Diagram* (Figure 1) in which sets are depicted as figures in the plane.

Figure 1

Having imposed no restrictions on the sets A and B above, we are compelled to admit the relation

$$A \subset A$$

for each set A. Clearly, this is in accord with the concept of a subset. The very important connection existing between the symbols $=$ and $\subset$ is formally summarized in Theorem 1.1:

1.1 / Theorem The following statements are equivalent:
(**1**) $A = B$;
(**2**) $A \subset B$ and $B \subset A$.

In other words, the two statements in the theorem are either both true or both false. This fact is exploited in virtually all proofs in which the equality of sets is in question. When the relations

$$A \subset B \quad \text{and} \quad B \not\subset A$$

are true we say that A is a *proper subset* of B.

A very important concept is that of a set containing no elements. Consider the formula

$$X = \{x \mid x \neq x\}.$$

It specifies a well-defined set, but owing to an earlier observation, there are no elements x such that $x \neq x$. The set X is said to be *empty*. While the set was obtained through a specific formula, its uniqueness is easily established.

1.2 / Theorem There is only one empty set.

Proof We begin with this observation: If X is *not* a subset of A, then there must be an element $x \in X$ such that $x \notin A$. Thus, when A is an arbitrary set and X is an empty set, then $X \subset A$. Similarly, if B is a given set and Y is another empty set, then $Y \subset B$. Since no implication which will restrict the choice of sets A or B was made, we may put $A = Y$ and $B = X$. With this choice we find that the relations $X \subset Y$ and $Y \subset X$ are both true, and the proof follows with an application of Theorem 1.1.

The (unique) empty set is designated with the symbol $\varnothing$. We prove:

1.3 / Theorem If $A \subset B$ and $B \subset C$, then $A \subset C$.

Proof The proof is very simple. If $A = \varnothing$, then the theorem is trivially true, because the empty set is a subset of every set. Otherwise, consider an arbitrary element $x \in A$. Then $x \in B$ because of the inclusion $A \subset B$, and thus $x \in C$ since $B \subset C$. Consequently $A \subset C$.

A point in logic is in order here. Namely, the reader should note that in the case $A = \varnothing$ the conclusion of the above theorem is independent of the hypothesis $B \subset C$. Specifically, the conclusion $\varnothing \subset C$ remains valid even when the assumption $B \subset C$ is replaced by its negation $B \not\subset C$.

The analog of negation in logic in our framework is complementation of sets: If A is an arbitrary set, then the *complement* of A is the set

$$A^c = \{x \mid x \notin A\}.$$

It will always be assumed that the elements in a given discussion belong to some known set X, often called a *universal set.* In particular, associated with each set X is its *power set* $P(X)$ consisting of all subsets of X:

$$P(X) = \{A \mid A \subset X\}.$$

We agree once and for all that all sets in a given discussion belong to the power set of some *a priori* known set X.

1.4 / *Examples* Put $X = \varnothing$. Then

$$P(\varnothing) = \{\varnothing\}$$

and $\varnothing^c = \varnothing$. Note, in particular, that the power set of the empty set contains one element.

When $X = \{1\}$, then $P(X) = \{\varnothing, X\}$. For complements we find that $\varnothing^c = \{1\}$ and $\{1\}^c = \varnothing$.

When $X = \{1, 2\}$ then the power set has four members: $P(X) = \{\varnothing, \{1\}, \{2\}, \{1, 2\}\}$; the complements are these: $\varnothing^c = \{1, 2\}$; $\{1\}^c = \{2\}$; $\{2\}^c = \{1\}$; $\{1, 2\}^c = \varnothing$.

The complementation of sets satisfies the following properties:

1.5 / Theorem Let X be a universal set; let $A \subset X$ be given. Then

(1) $\varnothing^c = X$.

(2) $X^c = \varnothing$.

(3) $(A^c)^c = A$.

Proof We see at once that

$$\varnothing^c = \{x \in X \mid x \notin \varnothing\} = X$$

and

$$X^c = \{x \in X \mid x \notin X\} = \varnothing.$$

This proves the first two assertions in the theorem. To verify the claim made in (3) we first observe that it is true when $A = \varnothing$ for then the preceding cases show that $(\varnothing^c)^c = X^c = \varnothing$. When $x \in A$, then $x \notin A^c$, and this implies that $x \in (A^c)^c$. Hence the inclusion $A \subset (A^c)^c$. Similarly, when $x \in (A^c)^c$ then $x \notin A^c$ so that $x \in A$; that is, $(A^c)^c \subset A$, and the proof follows from Theorem 1.1.

This section is concluded with an important observation.

1.6 / ***Example*** We already observed that order plays no part when a set is described by listing its members. For instance, $\{0, 1\} = \{1, 0\}$. On the other hand, let us form the sets $\{\{0\}, \{0, 1\}\}$ and $\{\{1\}, \{1, 0\}\}$ and put

$$\langle 0, 1\rangle = \{\{0\}, \{0, 1\}\}$$

$$\langle 1, 0\rangle = \{\{1\}, \{1, 0\}\}.$$

Then

$$\langle 0, 1\rangle \neq \langle 1, 0\rangle.$$

Thus, we see that the order in which the symbols 0 and 1 are listed in these sets is relevant to their description. These sets, in other words, are not determined by the membership of 0 and 1. This scheme will be used in Section 3 to define *ordered pairs*.

EXERCISES

1-1 Let $a \in A$ and let B consist of a. Which of the following relations are correct?

$$a \in B, \quad \{a\} \in B, \quad A \subset B, \quad B \subset A, \quad \{a\} \subset B, \quad \{a\} \subset A.$$

1-2 Consider the set $A = \{\varnothing, \{\varnothing\}\}$. Which of the following relations are true?

$$A = \varnothing, \quad \varnothing \in A, \quad \varnothing \subset A, \quad \{\varnothing\} \in A, \quad \{\varnothing\} \subset A.$$

1-3 With the idea suggested in Example 1.6, use parentheses and the symbols 0 and 1 to construct a set containing eight elements.

1-4 Give the symbolic equivalent of the following statement: "$n \in A$

when n is a positive integer which is divisible by 2 or by 5." Symbolically specify this set when *or* is replaced by *and*.

1-5 Show that the set $\mathfrak{N}$ of positive integers has no largest member.

1-6 Consider the set $\mathfrak{M} = \{m \cdot n \mid m \in \mathfrak{N} \text{ and } n \in \mathfrak{N}\}$: Prove that $\mathfrak{M} = \mathfrak{N}$.

1-7 For each positive integer n construct n pairwise disjoint subsets of $\mathfrak{N}$ which have no largest member. The sets are *pairwise disjoint* when no two of them have a common member.

1-8 Every nonempty subset of $\mathfrak{N}$ has a smallest member: Can you prove this statement? If not, see [11].

1-9 Write out the power set $P(X)$ when $X = \{1, 2, 3\}$.

1-10 Let the set X consist of n members. Prove that the set $P(X)$ has exactly

$$\binom{n}{k} = \frac{n!}{k!(n-k)!}$$

subsets of X of k elements each. Hence, show that $P(X)$ contains 2^n members.

Hint: Use the binomial expansion of $(1 + 1)^n$.

1-11 Let the set X consist of n elements. What is the largest number of sets of two elements each that we can select from $P(X)$ if no two sets are to have a common member? Solve this problem also for the case of sets of 3, 4, $\ldots$, $k \leq n$ elements.

1-12 Consider n sets, $A_1, A_2, \ldots, A_n$, which are related as follows:

$$A_1 \subset A_2, \qquad A_2 \subset A_3, \qquad \cdots, \qquad A_{n-1} \subset A_n, \qquad A_n \subset A_1.$$

Show that these sets equal each other.

1-13 Consider an arbitrary set X. Is there an element x such that $x \in X$ and $x \in P(X)$?

1-14 Consider sets X and Y. Can you deduce that $X = Y$ if $P(X) = P(Y)$?

1-15 Illustrate with a Venn diagram the relation between the sets A, B, and C if $A \in P(B)$ and $A \in P(C)$.

1-16 Consider the set

$$\langle x, y, z \rangle = \{\{x\}, \{x, y\}, \{x, z\}, \{y, z\}\}.$$

How many distinct sets can you get by arbitrarily permuting the letters x, y, and z?

2 / THE ALGEBRA OF SETS

The previous section was devoted to the elementary structure of sets and the method for comparing them with each other. Now we consider the means of generating new sets from given ones.

Let A and B be given sets. The collection of all elements belonging to A *or* to B is called the *union* of A and B, expressed as

$$A \cup B.$$

We point out that our usage of the word *or* is in the inclusive sense, which is to say that the statement "p or q" implies "p or q or both p and q."

The *intersection* of the sets, designated as

$$A \cap B,$$

is defined to be the set of all elements which belong concurrently to both A *and* B. The sets are *disjoint* or *mutually exclusive* when their intersection is empty.

The *difference set* of A and B, also known as the *relative complement* of B with respect to A, consists of precisely those members of A which are excluded from B. The notation for this set is

$$A \setminus B.$$

A formal summary of the preceding definitions follows:

2.1 / ***Definition***

(**1**) $A \cup B = \{x \mid x \in A \quad \text{or} \quad x \in B\};$
(**2**) $A \cap B = \{x \mid x \in A \quad \text{and} \quad x \in B\};$
(**3**) $A \setminus B = \{x \mid x \in A \quad \text{and} \quad x \notin B\}.$

As an illustration consider the sets

$$B = \{2k \mid k = 1, 2, 3, \ldots\}$$

$$C = \{5k \mid k = 1, 2, 3, \ldots\}.$$

The set A in Exercise 1-4 is expressible as $A = B \cup C$; the second set in this exercise is $A \cap B$. We also have

$$B \setminus C = \{2k \mid k \neq 5, 10, 15, \ldots\}$$

$$C \setminus B = \{10(k-1) + 5 \mid k = 1, 2, 3, \ldots\}.$$

Some set relations are represented graphically in Figure 2. The student will find the Venn diagram a very important tool in verifying relations between sets.

In analogy to addition and multiplication of numbers, the operations of generating new sets from given ones are endowed with certain algebraic properties. The ensuing statements are readily verified:

2.2 / Theorems

(1) $A \cup A = A$;
(2) $A \cup B = B \cup A$;
(3) $A \cup (B \cup C) = (A \cup B) \cup C$;
(4) $A \cap A = A$;
(5) $A \cap B = B \cap A$;
(6) $A \cap (B \cap C) = (A \cap B) \cap C$;
(7) $A \cup (B \cap C) = (A \cup B) \cap (A \cup C)$;
(8) $A \cap (B \cup C) = (A \cap B) \cup (A \cap C)$;
(9) $A \cup \varnothing = A$;
(10) $A \setminus A = \varnothing$;
(11) $A \setminus \varnothing = A$;
(12) $A \cap \varnothing = \varnothing$;
(13) $C \setminus (A \cup B) = (C \setminus A) \cap (C \setminus B)$;
(14) $C \setminus (A \cap B) = (C \setminus A) \cup (C \setminus B)$.

The last two statements are known as the *De Morgan Laws.*

The reader is urged at this point to discover for himself the analogies and differences between the algebra of set operations and that of addition and multiplication of numbers. Specifically, relating $\cup$ to $+$ and $\cap$ to $\cdot$, which of the above statements has an analog in the realm of numbers?

Two of the facts listed in 2.2 are now proved in order to acquaint the student with the type of argument employed in this part of set theory.

Proof of 2.2(10) Let us consider any set A. If x is an arbitrary element such that $x \in A \setminus A$, then according to Definition 2.1(3) both statements $x \in A$ and $x \notin A$ are true. This, however, is an inadmissible phenomenon in our set-up. Since this dilemma arises whenever we assume the existence of an element in the set $A \setminus A$, we must conclude that this set is empty as asserted.

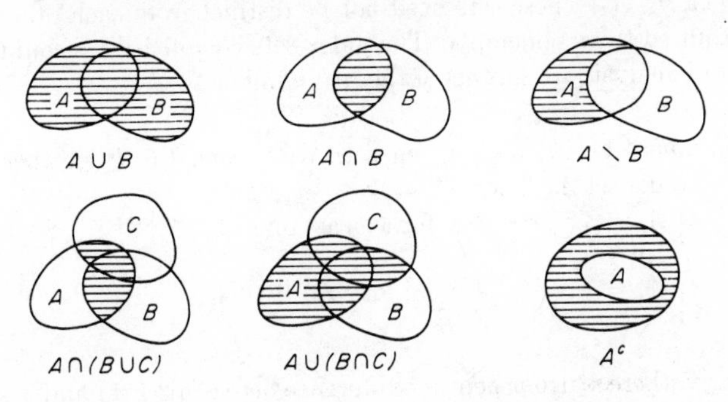

Figure 2

Proof of 2.2(14) In view of Theorem 1.1 it suffices to demonstrate the truth of the two relations

(a) $C \setminus (A \cap B) \subset (C \setminus A) \cup (C \setminus B)$;

(b) $(C \setminus A) \cup (C \setminus B) \subset C \setminus (A \cap B)$.

To verify the inclusion in (a), suppose $x \in C \setminus (A \cap B)$; each of the ensuing steps is easily derived from the preceding one:

$x \in C$ and $x \notin A \cap B$;

$x \in C$ and $x \notin A$ or $x \notin B$;

$x \in C \setminus A$ or $x \in C \setminus B$;

$x \in (C \setminus A) \cup (C \setminus B)$.

The assertion in (a) has thus been proved to be correct.

To dispose of (b) we proceed in a similar fashion. Namely, if

$x \in (C \setminus A) \cup (C \setminus B)$, then

$x \in C \setminus A$ or $x \in C \setminus B$;

$x \in C$ and $x \notin A$ or $x \in C$ and $x \notin B$;

$x \in C$ and $x \notin A \cap B$;

$x \in C \setminus (A \cap B)$.

The proof of the theorem is now complete.

Having gained some degree of familiarity with operations on sets we now extend the definitions of union and intersection to arbitrary families of sets. Specifically, consider a collection of sets A_ι, where t varies over some *index set* T. This simply means that to each element $t \in T$ there corresponds a set A_ι: These sets need not be distinct from each other nor are they required to be nonempty. The index sets customarily encountered in this part of analysis are subsets of the real numbers.

2.3 / ***Definitions*** Let T be a given set. With each $t \in T$ let there be associated a set A_ι. Then

(1) $\bigcup_{\iota \in T} A_\iota = \{x \mid x \in A_\iota \text{ for at least one } t \in T\}$;

(2) $\bigcap_{\iota \in T} A_\iota = \{x \mid x \in A_\iota \text{ for all } t \in T\}$.

Clearly, there is no conceptual difference between 2.1(1) and 2.3(1), and between 2.1(2) and 2.3(2).

2.4 / *Examples*

(1) For each $n \in \mathfrak{N}$, $\mathfrak{N}$ standing for the set of positive integers, let $A_n = \{n, n+1\}$. Then

$$\bigcup_{n \in \mathfrak{N}} A_n = \mathfrak{N},$$

$$\bigcap_{n \in \mathfrak{N}} A_n = \varnothing.$$

(2) Take the sets

$$B_n = \left\{ \frac{n}{m} \;\middle|\; m \geq n, \quad n \in \mathfrak{N} \quad \text{and} \quad m \in \mathfrak{N} \right\}.$$

Then

$$\bigcup_{n \in \mathfrak{N}} B_n = \left\{ \frac{n}{m} \;\middle|\; 0 < \frac{n}{m} \leq 1 \right\},$$

$$\bigcap_{n \in \mathfrak{N}} B_n = B_1.$$

(3) Let $C_n = [-1/n, 1/n]$, where the expression in square brackets stands for a closed interval on the real line. Then the union of the sets C_n is the interval $[-1, 1]$, whereas their intersection consists of the single point 0. It is understood that the index n varies over the set $\mathfrak{N}$.

Important generalizations of the De Morgan laws are given in the following theorems:

2.5 / Theorems For any system of sets A_t $(t \in T)$ we have:

(1) $(\bigcup_{t \in T} A_t)^c = \bigcap_{t \in T} A_t^c$;

(2) $(\bigcap_{t \in T} A_t)^c = \bigcup_{t \in T} A_t^c$.

A very important concept in mathematics is that of an equivalence relation. Consider a set A with a family of subsets A_t $(t \in T)$ such that

$$A = \bigcup_{t \in T} A_t$$

and

$$A_t \cap A_u = \varnothing \quad \text{when} \quad t \neq u.$$

If x and y are in A, let us write $x \sim y$ when both elements belong to the *same* subset A_t, $x \nsim y$ otherwise. We easily convince ourselves that the symbol "$\sim$" satisfies the properties (i)–(iii) of equality (see Section 1), and any such relation is said to be an equivalence relation. Formally we state:

2.6 / *Definition* The relation $\sim$ is an *equivalence relation* if, for arbitrary elements x, y, and z, the following is the case:

$$x \sim x \qquad \text{(reflexivity)}$$
$$\text{if} \quad x \sim y, \quad \text{then} \quad y \sim x \qquad \text{(symmetry)}$$
$$\text{if} \quad x \sim y \quad \text{and} \quad y \sim z, \quad \text{then} \quad x \sim z \qquad \text{(transitivity)}.$$

Set inclusion is an instance of a relation which is not an equivalence relation; so is the order relation "$<$" between real numbers.

We have noted above that the partition of a set into mutually exclusive subsets gives rise to an equivalence relation. Let us show that the converse is also true. For this, let A be a given set, and $\sim$ an equivalence relation. We associate with each element $a \in A$ the subset $E(a)$ of all members of A which are equivalent to it:

$$E(a) = \{x \in A \mid x \sim a\}.$$

The set $E(a)$ is called an *equivalence class*. Clearly

$$A = \bigcup_{a \in A} E(a).$$

We assert that if $E(a)$ and $E(b)$ are arbitrary equivalence classes of A, then either

$$E(a) = E(b),$$

or else

$$E(a) \cap E(b) = \varnothing;$$

that is, two equivalence classes are either identical or disjoint. Indeed, if $x \in E(a) \cap E(b) \neq \varnothing$, then $x \sim a$ and $x \sim b$, and because of Definition 2.6, $a \sim b$. But this implies that each x in $E(b)$ is equivalent to a, and so $E(a) = E(b)$.

EXERCISES

2-1 Verify the facts listed in Theorems 2.2 using Venn diagrams.

2-2 Complete the proof of Theorems 2.2 without Venn diagrams.

2-3 Establish the equivalence of the following facts:

(a) $A \cap B = A$;
(b) $A \cup B = B$;
(c) $B \supset A$;
(d) $B^c \subset A^c$.

2-4 Prove Theorems 2.5.

2-5 Prove or disprove this statement: If $\{A_t\}$ $(t \in T)$ is any family of sets such that

$$\bigcap_{t \in T} A_t = \emptyset,$$

then there are two sets in the family, say A_r and A_s, such that

$$A_r \cap A_s = \emptyset.$$

2-6 If the family of sets $\{A_n\}$ $(n \in \mathfrak{N})$ is such that

$$A_1 \supset A_2 \supset A_3 \supset \cdots,$$

then

$$A_1 = [\bigcup_{n \in \mathfrak{N}} (A_n \setminus A_{n+1})] \cup [\bigcap_{n \in \mathfrak{N}} A_n].$$

Prove this statement. Show also that the sets $A_n \setminus A_{n+1}$ are pairwise disjoint, and that no such set has a common member with the indicated intersection.

2-7 Let $\mathfrak{A}$ and $\mathfrak{B}$ be families of sets such that $\mathfrak{A} \subset \mathfrak{B}$. Show that

$$\bigcap_{A \in \mathfrak{A}} A \supset \bigcap_{B \in \mathfrak{B}} B \quad \text{and} \quad \bigcup_{A \in \mathfrak{A}} A \subset \bigcup_{B \in \mathfrak{B}} B.$$

2-8 The *symmetric difference* of the sets A and B is the set

$$A \triangle B = \{x \mid x \in A \quad \text{or} \quad x \in B \quad \text{and} \quad x \notin A \cap B\}.$$

Verify the equivalence of the following facts:

(a) $C = A \triangle B$;
(b) $C = (A \setminus B) \cup (B \setminus A)$;
(c) $C = (A \cap B^c) \cup (A^c \cap B)$.

2-9 Establish the relations:

(a) $A \triangle (B \triangle C) = (A \triangle B) \triangle C$;
(b) $A \triangle A = \emptyset$;
(c) $A \triangle \emptyset = A$;
(d) $A \cap (B \triangle C) = (A \cap B) \triangle (A \cap C)$;
(e) $(A \cup B) \triangle (C \cup D) \subset (A \triangle C) \cup (B \triangle D)$.

2-10 Prove that for any given sets A and B there is a set X such that $A \triangle X = B$.

2-11 A fruitful concept is that of an *algebra* of sets. Let X be given: let $\mathfrak{A} \subset P(X)$ be fixed, $\mathfrak{A} \neq \emptyset$. Then $\mathfrak{A}$ is an *algebra* if the following conditions are met:

(i) $A \in \mathfrak{A}$ and $B \in \mathfrak{A}$ implies that $A \cup B \in \mathfrak{A}$.
(ii) $A \in \mathfrak{A}$ implies that $A^c \in \mathfrak{A}$.

Assuming $\mathfrak{A}$ to be an algebra, deduce the following theorems:

(a) $A \in \mathfrak{A}$ and $B \in \mathfrak{A}$ implies that $A \cap B \in \mathfrak{A}$ and $A \setminus B \in \mathfrak{A}$.

(b) If $A_1, A_2, \ldots, A_n$ are members of $\mathfrak{A}$, then

$$A_1 \cup A_2 \cup \cdots \cup A_n \in \mathfrak{A}$$

and

$$A_1 \cap A_2 \cap \cdots \cap A_n \in \mathfrak{A}.$$

(c) $\varnothing \in \mathfrak{A}$ and $X \in \mathfrak{A}$.

(d) The power set of an arbitrary set is an algebra.

(e) For any set X, $\{\varnothing, X\}$ is an algebra.

(f) If $X = \{1, 2, 3\}$, find all algebras of $P(X)$.

2-12 Determine which of the following relations are equivalence relations:

(a) In the set $\mathfrak{I}$ of all integers let $m \sim n$ if $m - n$ is divisible by 2 (the numbers m and n are said to be *congruent* modulo 2).

(b) In the set $\mathfrak{N}$ let $m \sim n$ if m and n are relatively prime, which is to say that the integers have no common factor other than 1.

(c) Considering the collection of ordered pairs of positive integers, put $(a, b) \sim (c, d)$ if $a + d = b + c$.

(d) In a given family of sets, $A \sim B$ if $A \cap B = \varnothing$.

(e) In the realm of real numbers $a \sim b$ when $a \cdot b \geq 0$.

(f) The following relations apply to ordered pairs of sets:

$$(A, B) \sim (C, D) \quad \text{if} \quad A \cup D = B \cup C;$$

$$(A, B) \sim (C, D) \quad \text{if} \quad A \cap C = B \cap D.$$

3 / FUNCTIONS

One of the richer and more central ideas in mathematics is that of a function (mapping). Loosely speaking, a function is a relation between sets, its most important feature being *singlevaluedness*. This is to say that a function associates exactly one value (element) with each given element for which it is defined. A function f which relates the members of a set A with values in a set B is said to be a function on A into B, a fact expressed as

$$f: A \rightarrow B.$$

The set A is the *domain* of f. If $y \in B$ is related to $x \in A$ through f, then y is said to be the *image* of x under f: The functional notation

$$y = f(x)$$

is well known (see Figure 3). The *range* of the function f is the set

$$f(A) = \{y \in B \mid y = f(x) \quad \text{and} \quad x \in A\}.$$

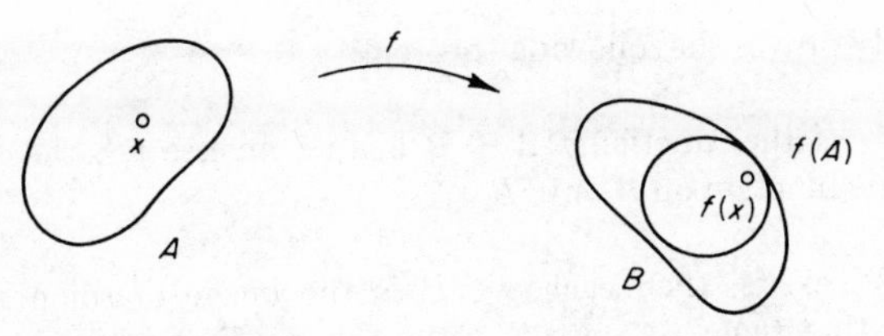

Figure 3

Let a function $f:A \to B$ be given. Then for each subset $C \subset A$ this function gives rise to the function

$$g:C \to B$$

which is defined through the formula

$$g(x) = f(x) \qquad \text{when} \quad x \in C.$$

The function g, which is called the *restriction* of f to C, is uniquely determined; the formula

$$g = f \mid C$$

is often used to represent it.

The function f is said to map A *onto* B if each member of B is the image of some member of A; the function is *one-to-one* if $f(x_1) = f(x_2)$ implies $x_1 = x_2$ for all members of A. If f is a one-to-one function on A onto B, then for each $y \in B$ there is exactly one $x \in A$ such that $y = f(x)$. The function

$$f^{-1}:B \to A,$$

which is defined by the formula

$$f^{-1}(f(x)) = x \qquad \text{for each} \quad x \in A,$$

is called the *inverse* function of f (see Figure 4).

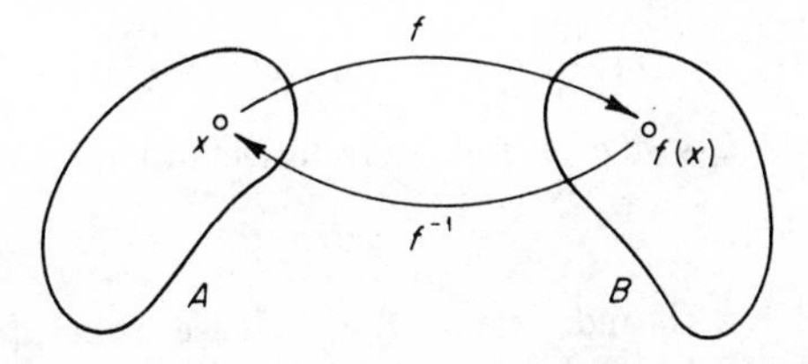

Figure 4

One easily proves the following theorem:

3.1 / Theorem If the function $f:A \to B$ has an inverse f^{-1}, then f is a one-to-one function on A onto B.

Proof When f^{-1} exists, then each $y \in B$ is the image of some $x \in A$, implying that f maps A onto B. From the fact that f^{-1} is a function we deduce that if $f(x_1) = f(x_2)$, then $f^{-1}(f(x_1)) = f^{-1}(f(x_2))$, but the last relation states that $x_1 = x_2$. Hence f is also one-to-one.

The terms *mapping, map, transformation, correspondence,* and *operator* are to be regarded as synonymous with *function.*

A *function* was defined so far only imprecisely, with a strong appeal to intuition. A rigorous definition is obtained with the help of ordered pairs which were hinted at in Example 1.6. This example, in fact, suggests the following:

3.2 / *Definition* Let A and B be given sets. The *ordered pair* $\langle a, b\rangle$, where $a \in A$ and $b \in B$, is the set

$$\langle a, b\rangle = \{\{a\}, \{a, b\}\}.$$

From this we have the following theorem:

3.3 / Theorem The ordered pair $\langle a, b\rangle$ is completely determined by the elements a and b. Furthermore, for any other ordered pair $\langle c, d\rangle$, $\langle a, b\rangle = \langle c, d\rangle$ if and only if $a = c$ and $b = d$.

Proof The first assertion in the theorem is a direct consequence of Definition 3.2 and the fact that a set is completely determined by its membership. The statement that $a = c$ and $b = d$ implies $\langle a, b\rangle = \langle c, d\rangle$ requires no further justification. Hence, it remains to show that the converse statement is true. Thus, let us suppose that $\langle a, b\rangle = \langle c, d\rangle$.

If $c = d$, then

$$\langle c, d\rangle = \{\{c\}, \{c, d\}\} = \{\{c\}, \{c, c\}\} = \{\{c\}, \{c\}\} = \{\{c\}\}.$$

Since the left side equals $\langle a, b\rangle$, it follows from Definition 3.2 that

$$\{\{c\}\} = \{\{a\}, \{a, b\}\}.$$

Consequently, $\{c\} = \{a\}$ and $\{c\} = \{a, b\}$. Hence $\{a\} = \{a, b\}$, and it follows that $a = b$ and $a = c$, and this case follows.

If, on the other hand, $c \neq d$, then both relations

$$\{c\} \neq \{c, d\} \qquad \text{and} \qquad \{a\} \neq \{c, d\}$$

are true. The first of these is clear. The second is verified as follows: If, on the contrary, $\{a\} = \{c, d\}$, then $a = c$ and $a = d$, so that $c = d$. But this violates the assumption that $c \neq d$. Now, the equality $\langle a, b\rangle = \langle c, d\rangle$ can be written in the form

$$\{\{a\}, \{a, b\}\} = \{\{c\}, \{c, d\}\},$$

from which we find that $\{a\} \in \{\{c\}, \{c, d\}\}$, so that $\{a\} = \{c\}$, and consequently $a = c$. But we also see that $\{c, d\} \in \{\{a\}, \{a, b\}\}$. This, together with the inequality $\{a\} \neq \{c, d\}$, shows that $\{c, d\} = \{a, b\}$. The relation $a = c$ being already known, it follows that $\{a, d\} = \{a, b\}$. But $\{a\} \neq \{c, d\} = \{a, b\}$ and hence $a \neq b$, and we are forced to conclude that $b = d$. This, then, completes the proof of the theorem.

Having established the existence and uniqueness of ordered pairs we lay down the next definition:

3.4 / *Definition* Let A and B be given sets. The *product set* $A \times B$ is the aggregate of ordered pairs $\langle a, b\rangle$, $a \in A$ and $b \in B$:

$$A \times B = \{\langle a, b\rangle \mid a \in A \quad \text{and} \quad b \in B\}.$$

Synonymous with *product set* is the term *Cartesian product.*

3.5 / *Definition* The subset $f \subset A \times B$ is a *function* on A into B if for each $x \in A$ there is one and only one $y \in B$ such that $\langle x, y\rangle \in f$.

It will be convenient at times to refer to f as the *graph* of the function. The reader must bear in mind that this is a mere linguistic convenience and not a conceptual distinction between the two. The graph is the function.

Another comment is this: When a function is defined as a set it stands out as an entity, a single object, like a point. Regarding functions in this light is extremely important in modern analysis. This is particularly so when one deals with sets of functions: on occasion one examines the image of a fixed point under different functions. In such a situation the function itself serves as the variable. A case in point is the Weierstrass Approximation Theorem (see Chapter Eight).

Deferring the exploitation of the preceding definitions to later chapters, we present here a variety of examples, which the student should go through carefully.

3.6 / *Example* The set

$$g = \{\langle x^2, x\rangle \mid x^2 \in [0, 1]\}$$

is *not* a function since the correspondence $g: x^2 \to x$ is not single-valued, and hence undefined. We mention this trivial example to

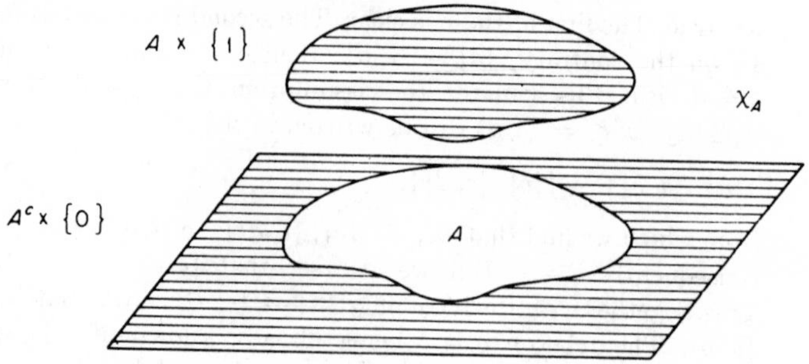

Figure 5

point out that the functional notation is inconsistent here. It would lead to the simultaneous relations

$$g(x^2) = x$$

$$g(x^2) = -x$$

for each admitted value of x. However, $x = -x$ only when $x = 0$.

3.7 / ***Example*** Associated with each set A is the function

$$\chi_A(x) = \begin{cases} 1 & \text{when} \quad x \in A \\ 0 & \text{when} \quad x \in A^c. \end{cases}$$

This function has the equivalent description

$$\chi_A = \{\langle x, 1\rangle \mid x \in A\} \cup \{\langle x, 0\rangle \mid x \in A^c\}$$

in terms of ordered pairs in the product set $X \times \{0, 1\}$ (see Figure 5) and is called the *characteristic function* of A.

3.8 / ***Example*** Addition of positive integers is a function $f : \mathcal{N} \times \mathcal{N} \to \mathcal{N}$. It can be specified via the formula $f(m, n) = m + n$, or as the set $f = \{\langle\langle m, n\rangle, \; m + n\rangle \mid m, n \in \mathcal{N}\}$. The function is presented graphically in Figure 6.

3.9 / ***Example*** Consider the real line $\mathcal{R}$. We recall that the *absolute value* of a number x is defined as follows:

$$|x| = \begin{cases} x & \text{when} \quad x \geq 0 \\ -x & \text{when} \quad x < 0. \end{cases}$$

The *distance* between the points x and y in $\mathcal{R}$ is the number

$$d(x, y) = |x - y|.$$

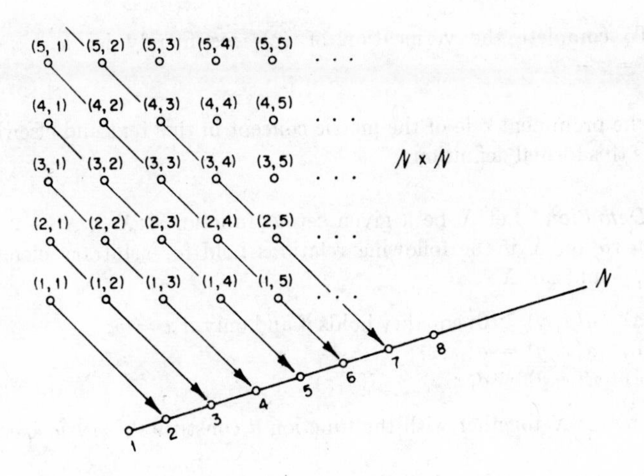

Figure 6

This formula specifies a function $d: \mathcal{R} \times \mathcal{R} \to \mathcal{R}$ which is endowed with three very special properties: For arbitrary numbers x, y, and z:

(a) $d(x, y) \geq 0$, equality holding if and only if $x = y$;
(b) $d(x, y) = d(y, x)$;
(c) $d(x, y) \leq d(x, z) + d(y, z)$.

This function is said to be a *metric*; $\mathcal{R}$ together with d constitute a *metric space.*

Let us show the properties of d to be as claimed. The property in (b) is verified by applying the definition of absolute value to the differences $x - y$ and $y - x$. Regarding (a), it is clear that $d(x, x) = 0$ and if $d(x, y) = 0$, then (b) shows that also $d(y, x) = 0$. But this implies that $x - y = y - x$, and the equality $x = y$ follows.

To establish the inequality in (c) we observe that if u and v are arbitrary members of $\mathcal{R}$, then

$$-|u| \leq u \leq |u|,$$

$$-|v| \leq v \leq |v|,$$

and accordingly

$$-(|u| + |v|) \leq u + v \leq (|u| + |v|).$$

This, however, is equivalent to the statement

$$|u + v| \leq |u| + |v|.$$

To complete the verification of (c), we merely let $u = x - z$, $v = z - y$.

The prominent role of the metric concept in this text and elsewhere warrants this formal definition:

3.10 / *Definition* Let X be a given set. A function $d: X \times X \to \mathcal{R}$ is a *metric* on X if the following relations hold for arbitrary elements x, y, and z of X:

(a) $d(x, y) \geq 0$; equality holds if and only if $x = y$;
(b) $d(x, y) = d(y, x)$;
(c) $d(x, y) \leq d(x, z) + d(y, z)$.

The set X together with the function d constitute a *metric space.*

Properly, one should distinguish notationally between the set X and the corresponding metric space, because the metric is not necessarily unique (see Example 6.4). We do not do so in this book simply because no such ambiguity is possible here: the metric in a given discussion will always be known.

3.11 / *Example* Let $\mathcal{C}$ stand for the set of continuous functions $f: [0, 1] \to \mathcal{R}$. We recall that any function $f \in \mathcal{C}$ is *integrable*, by which we mean that f determines the unique number

$$I(f) = \int_0^1 f(x)\, dx.$$

The subset of ordered pairs $\langle f, I(f) \rangle$ in the product set $\mathcal{C} \times \mathcal{R}$ is clearly a function on $\mathcal{C}$ onto $\mathcal{R}$. This function is endowed with the following linearity properties:

(a) $I(f + g) = I(f) + I(g)$;
(b) $I(af) = a \cdot I(f)$ for each $a \in \mathcal{R}$.

A function such as I is called a *linear functional.*

In contrast to I, the formula

$$J(f) = \int_0^1 | f(x) |\, dx$$

does not specify a linear functional on $\mathcal{C}$. For instance, let $f(x) = x$ and $g(x) = -\frac{1}{2}$. Then

$$J(f + g) = \int_0^1 | x - \tfrac{1}{2} |\, dx = \tfrac{1}{4},$$

whereas

$$J(f) + J(g) = \int_0^1 |x|\,dx + \int_0^1 |-\tfrac{1}{2}|\,dx = 1,$$

and accordingly $J(f+g) \neq J(f) + J(g)$. The functional

$$\rho(f, g) = J(f - g),$$

however, satisfies Definition 3.10. The student should verify this fact.

3.12 / *Example* Consider the differentiation operator $D(f) = f'$. Clearly, D is *not* a function on $\mathcal{C}$ because of functions such as $|x - \frac{1}{2}|$ which are not differentiable. On the other hand, D is a function on the set of polynomials. The function D, which is linear, is called a *linear operator*. The reader should discover for himself the major differences between the functions D and I.

3.13 / *Example* A very important function on $\mathcal{C}$ into $\mathcal{R}$ is the so called *supremum function* which relates to each $f \in \mathcal{C}$ the number

$$\|f\| = \max_{0 \leq x \leq 1} |f(x)|$$

(see Figure 7). Like the function J in Example 3.11, it is a metric on $\mathcal{C}$. The verification of this fact should be at least attempted by the student.

3.14 / *Example* Consider once more the set $\mathfrak{N}$, and the function

$$s(n) = \frac{n}{n+1}.$$

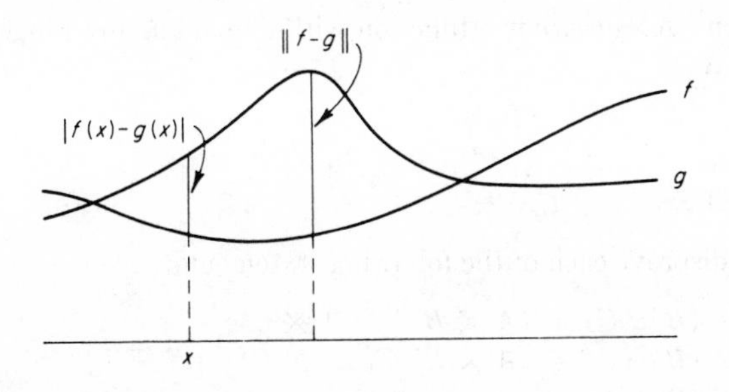

Figure 7

Marking its range by A,

$$A = \left\{\frac{n}{n+1} \;\middle|\; n \in \mathfrak{N}\right\},$$

we can represent the function as

$$s = \left\{\left\langle n, \frac{n}{n+1}\right\rangle \;\middle|\; n \in \mathfrak{N}\right\} \subset \mathfrak{N} \times A.$$

The values which this function takes on can be listed in the order induced by $\mathfrak{N}$:

$$s = (\tfrac{1}{2}, \tfrac{2}{3}, \tfrac{3}{4}, \ldots).$$

We say that this function is a *sequence*.

As a further illustration consider the function $t(n) = 1$ $(n \in \mathfrak{N})$. Its range is the set 1, so t can be written as

$$t = \{\langle n, a_n\rangle \mid a_n = 1, \quad n \in \mathfrak{N}\};$$

that is,

$$t = (1, 1, 1, \ldots).$$

This shows that a sequence must always be carefully distinguished from its range. A sequence, as opposed to a set, is *not* determined by membership alone. It requires for its identification both membership and order.

3.15 / ***Example*** Consider the function $p(n) = x^{n-1}$ $(n \in \mathfrak{N})$. The range of this function consists of the polynomials $1, x, x^2, x^3, \ldots,$ and according to the preceding example we can put

$$p = (1, x, x^2, x^3, \ldots).$$

Thus, also the function p is a sequence.

3.16 / ***Definition*** A *sequence* is a function with domain $\mathfrak{N}$ and range in a given set X.

EXERCISES

3-1 Prove or disprove each of the following statements:

(a) $A \times (B \cup C) = (A \times B) \cup (A \times C)$;
(b) $A \times (B \cap C) = (A \times B) \cap (A \times C)$;
(c) $A \times \varnothing = A$;
(d) $A \times (B \times C) = (A \times B) \times C$.

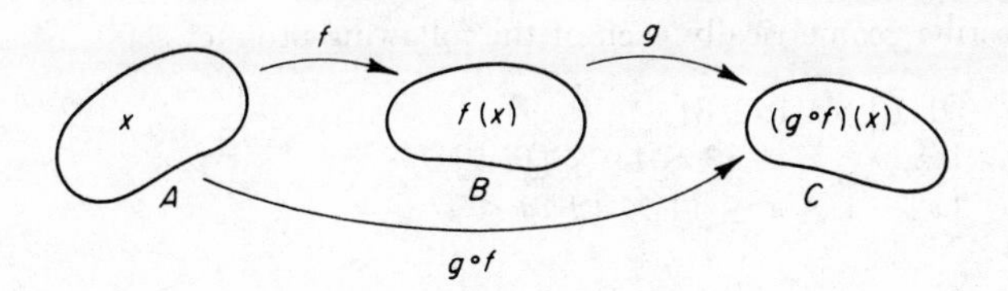

Figure 8

3-2 With the idea used in Definition 3.2 proceed to define an ordered 3-tuple $\langle a_1, a_2, a_3 \rangle$. Verify that your definition is valid. How would you go about defining an ordered n-tuple for $n > 3$?

3-3 Consider the function $f : (A \cup B) \to C$. Is the relation

$$f(A \cup B) = f(A) \cup f(B)$$

correct?

3-4 Let A and B be given sets; let $f: A \to B$ be an arbitrary function. Show that to each fixed element $a \in A$ there corresponds a function $g: A \times A \to B$ such that

$$g(x, a) = f(x).$$

Show that the formula $H(f) = f(0)$ specifies a linear functional on $\mathcal{C}$.

3-5 Let a function $f: \mathcal{R}^* \to \mathcal{R}^*$ be endowed with the properties:

$$f(x + y) = f(x) + f(y) \quad \text{and} \quad f(1) = 2.$$

Show that $f(x) = 2x$.

3-6 Given functions $f: A \to B$ and $g: B \to C$, show that the formula

$$(g \circ f)(x) = g(f(x))$$

determines a function on A into C (see Figure 8). When does this function have an inverse?

3-7 Let $A = \bigcup_{k=1}^{n} A_k$. When are the following relations correct?

(a) $\chi_A = \sum_{k=1}^{n} \chi_{A_k}$.

(b) $\chi_A = \left(\sum_{k=1}^{n} \chi_{A_k}\right) - \chi_{\bigcap_{k=1}^{n} A_k}$.

(c) $\chi_A = \sum_{k=1}^{n-1} \chi_{A_k \cap A_{k+1}}$.

3-8 Describe geometrically each of the following product sets:

(a) $[0, 1] \times \{1, 2, 3\}$;
(b) $\{\langle x, y\rangle \mid x^2 + y^2 = 1\} \times [0, 1]$;
(c) $\{x \mid -1 \leq x \leq 1\} \times \{y \mid y < x\}$.

4 / COUNTABILITY

A primitive classification of sets separates them into finite and infinite sets. To make this statement meaningful we state the following:

4.1 / *Definition* The set A is *finite* if $A = \varnothing$ or if there is a positive integer n and a one-to-one correspondence between the elements of the set A and the set $\{1, 2, \ldots, n\}$. The set A is *infinite* if it is not finite.

This definition tells us how to distinguish quantitatively between finite sets: we count their members. To be able to tell apart quantitatively infinite sets we quite obviously need a much more general concept of "quantity" since statements about the "number of elements" are meaningless already in the case of the set $\mathfrak{N}$. To achieve this we reexamine the counting process. Like Cantor before us, we ask "What do we mean by saying that two finite sets have equally many members?" The answer is, of course, very simple: two finite sets have the same number of elements when there is a one-to-one correspondence between them. The existence of such relations, however, is not confined to finite sets (see below). This immediately suggests a simple yet effective tool for comparing with each other arbitrary sets: this tool is equivalence.

4.2 / *Definition* The sets A and B are *equivalent*, written

$$A \sim B,$$

if and only if there is a one-to-one correspondence between them.

Alternatively stated, equivalence implies the existence of a function $f : A \to B$ of A onto B such that when both $\langle a_1, b_1\rangle$ and $\langle a_2, b_2\rangle$ belong to its graph, then $a_1 = a_2$ if and only if $b_1 = b_2$; conversely, the existence of such a function carries with it the stipulated equivalence. As with equality, this is another instance of an equivalence relation. When A and B are not equivalent we write

$$A \nsim B.$$

The sets A and B in Definition 4.2 are said to have the same *cardinality*. The accepted symbol for the cardinality of the set $\mathfrak{N}$ is $\aleph_0$, that of the cardinality of the continuum $[0, 1]$ is $\aleph$. The symbol $\aleph$ is the letter "aleph" of the Hebrew alphabet.

4.3 / ***Example*** The set $\mathfrak{J}$ of all integers and the set $\mathfrak{N}$ of positive integers are equivalent. In fact, an explicit one-to-one relationship is easily exhibited: The function

$$f(n) = \begin{cases} 2n + 1 & \text{for} \quad n = 0, 1, 2, \ldots \\ -2n & \text{for} \quad n = -1, -2, \ldots \end{cases}$$

maps $\mathfrak{J}$ one-to-one onto $\mathfrak{N}$; its inverse

$$f^{-1}(n) = \begin{cases} -\dfrac{n}{2} & \text{for} \quad \text{even } n \\ \dfrac{n-1}{2} & \text{for} \quad \text{odd } n \end{cases}$$

maps $\mathfrak{N}$ one-to-one onto $\mathfrak{J}$.

4.4 / ***Example*** Consider the intervals

$$[0, 1] = \{x \mid 0 \leq x \leq 1\}$$

and

$$(0, 1) = \{x \mid 0 < x < 1\}.$$

These point sets are asserted to be equivalent. This is verified with the relations:

(a) $(0, 1] \sim (0, 1)$,
(b) $(0, 1) \sim [0, 1)$,
(c) $[0, 1) \sim [0, 1]$,

and the fact that equivalence is transitive.

Clearly, if there is a one-to-one function $f: (0, 1) \leftrightarrow (0, 1]$, then the function

$$g(x) = \begin{cases} f(x) & \text{for} \quad x \in (0, 1) \\ 0 & \text{at} \quad x = 0 \end{cases}$$

establishes an equivalence between the sets in (c). Consequently, it suffices to verify the relations in (a) and (b); let us verify (a).

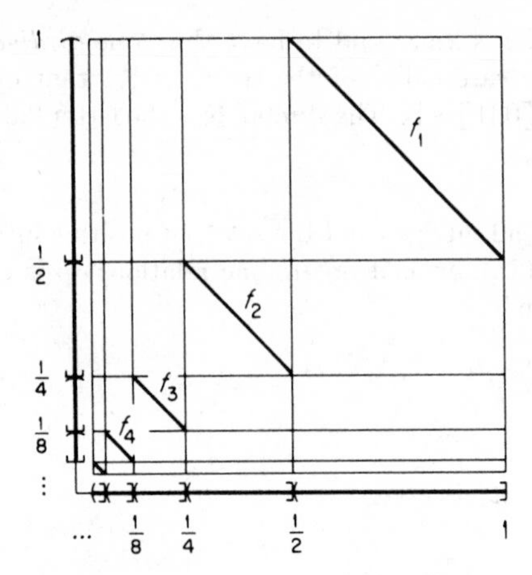

Figure 9

For each positive integer n consider the function

$$f_n(x) = \frac{3}{2^n} - x \qquad \text{for} \quad x \in \left(\frac{1}{2^n}, \frac{1}{2^{n-1}}\right]$$

(see Figure 9). This function maps the specified domain in a one-to-one manner onto the set $[1/2^n, 1/2^{n-1})$. Now define the function f as follows:

$$f(x) = f_n(x) \qquad \text{for} \quad x \in \left(\frac{1}{2^n}, \frac{1}{2^{n-1}}\right].$$

This function maps the interval $(0, 1]$ one-to-one onto the interval $(0, 1)$. The construction of a suitable function for the equivalence in (b) should be carried out by the student.

We observe in the above example that infinite sets may be equivalent to certain of their proper subsets. That this is, indeed, a characterization of infinite sets is shown below. Our Definition 4.1 could be replaced with the statement that a set is infinite when and only when it is equivalent to one of its proper subsets. This latter definition is due to Dedekind.

Examining the counting process a little further we observe that the actual construction of a one-to-one correspondence between the members of a given finite set A and a finite subset of $\mathfrak{N}$ amounts to writing the members of A in a finite sequence. Again, the possibility of arranging the members of a set in a sequence exists also for infinite sets: The elements of $\mathfrak{N}$, for example, can be put in a sequence. This motivates the following generalization of counting:

4.5 / ***Definition*** The set A is *countable* when and only when $A \sim \mathfrak{N}$.

In other words, a set is countable when its members can be arranged in a sequence. Synonymous with countable are the terms *denumerable* and *enumerable*. When a set is neither finite nor countable it is said to be *undenumerable* or *nonenumerable*.

We will do well to pause briefly in order to note that we have taken a very significant step. Namely, we asserted in the above definition that the property of being finite plays no essential role in the quantitative classification of sets.

We proceed to demonstrate now that the class of countable sets is neither confined to $\mathfrak{N}$ nor exhaustive: countability does not extend to all infinite sets. We begin with the set $\mathcal{R}^*$ of rational numbers. This set is distinguished from the set $\mathcal{I}$ of integers in a very important respect: Each member of $\mathcal{I}$ has a definite successor and a definite predecessor, but this property holds for no member of $\mathcal{R}^*$. Yet $\mathcal{R}^*$ is countable:

4.6 / **Theorem** The set $\mathcal{R}^*$ is countable.

Proof The proof devised by Cantor is surprisingly simple. To establish the assertion just made we must produce a scheme for listing all rational numbers in a sequence. Once this is accomplished, their association with the members of $\mathfrak{N}$ can be done routinely. To begin with, let us restrict the process to the positive rational numbers. For this we consider the following scheme:

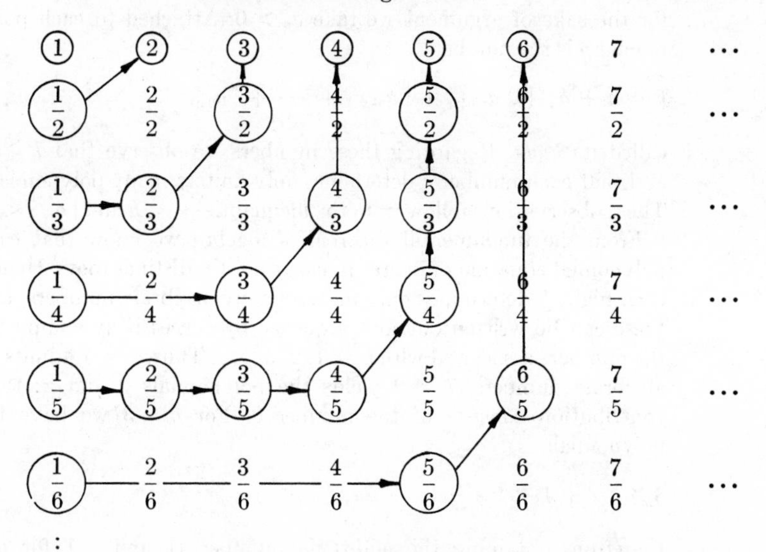

Let us write down in a sequence the rational numbers as they appear in the diagram, omitting those numbers already listed:

$$1, \tfrac{1}{2}, 2, \tfrac{1}{3}, \tfrac{2}{3}, \tfrac{3}{2}, 3, \tfrac{1}{4}, \tfrac{3}{4}, \tfrac{4}{3}, 4, \tfrac{1}{5}, \ldots.$$

It is readily seen that each positive member of $\mathcal{R}^*$ appears in this sequence exactly once. It therefore follows that each member of $\mathcal{R}^*$ is accounted for in the sequence

$$0, 1, -1, \tfrac{1}{2}, -\tfrac{1}{2}, 2, -2, \tfrac{1}{3}, -\tfrac{1}{3}, \ldots$$

and the asserted countability of $\mathcal{R}^*$ ensues.

A yet richer set which is still countable is the set of algebraic numbers: The number b is an *algebraic number* if it is a solution of some polynomial equation

$$p(x) = a_0 + a_1x + a_2x^2 + \cdots + a_nx^n = 0$$

with integral coefficients. In particular, rational numbers are algebraic numbers; so are the numbers $\sqrt{2}$ and $\sqrt{-1}$, for example. Together with the previous result Cantor also proved:

4.7 / Theorem The set of all algebraic numbers is countable.

Proof In the polynomial equation $p(x) = 0$ we may assume that $a_n \neq 0$; for the sake of argument we take $a_n > 0$. Attached to each polynomial p is the number

$$h = n + a_n + |a_{n-1}| + |a_{n-2}| + \cdots + |a_0|,$$

called its *hight*. Regarding these numbers we observe that $h \geq 1$, and that each number h determines only finitely many polynomials. These observations follow from the inequalities $n < h$ and $|a_m| < h$.

From the fundamental theorem of algebra we know that each polynomial equation of degree n has at most n distinct roots. Hence, each hight h determines only finitely many algebraic numbers, and these can be written out in a sequence by successively computing the numbers associated with $h = 1, 2, 3, \ldots$. Thus, $h = 1$ admits no algebraic numbers; $h = 2$ yields the polynomials 2 and x: their contribution consists of the number 0. For $h = 3$ we have the polynomials

$$3, \qquad x + 1, \qquad x - 1, \qquad 2x, \qquad x^2:$$

therefrom obtaining the additional numbers 1 and -1 for our sequence. In the case $h = 4$ we have the following nontrivial polynomials:

$$3x, \quad 2x+1, \quad 2x-1, \quad x+2, \quad x-2, \quad 2x^2, \quad x^2+x, \quad x^2-x, \quad x^2+1, \quad x^2-1, \quad x^3.$$

Roots not already listed are

$$-2, \quad -\tfrac{1}{2}, \quad \tfrac{1}{2}, \quad 2, \quad -\sqrt{-1}, \quad \sqrt{-1}.$$

To complete the proof of the theorem, we merely add the remark that each polynomial has a hight.

The fact that Definition 4.5 does not extend to all sets is the topic of the next theorem.

4.8 / Theorem The set of all real numbers in the interval $0 < x < 1$ is uncountable.

Proof For the purpose of this theorem it suffices to consider normalized decimal expansions $0.a_1a_2a_3\cdots$ $(0 \leq a_k \leq 9)$. We recall that in this set up there is no integer n such that $a_k = 9$ for all values $k \geq n$. The proof consists of showing that no scheme for listing all these decimals in a sequence exists.

Suppose the contrary were true. Then, one by one, we could list these decimals in a table, say,

$$\begin{array}{llllll} 0. & a_{11} & a_{12} & a_{13} & a_{14} & \cdots \\ 0. & a_{21} & a_{22} & a_{23} & a_{24} & \cdots \\ 0. & a_{31} & a_{32} & a_{33} & a_{34} & \cdots \\ 0. & a_{41} & a_{42} & a_{43} & a_{44} & \cdots \\ \cdot & \cdot & \cdot & \cdot & \cdot & \cdots \end{array}$$

in such a way that each decimal occupies a definite position in the scheme. We proceed to show that this is false. Namely from the indicated diagonal we construct the decimal

$$a = 0.a_{11}a_{22}a_{33}a_{44}\cdots;$$

from it we construct a decimal

$$b = 0.b_1b_2b_3b_4\cdots$$

according to the following rule: For each admitted value of k,

$$1 \leq b_k \leq 8 \qquad \text{and} \qquad b_k \neq a_{kk}.$$

In other words, every entry in the expansion of b disagrees with the corresponding entry in the expansion of a. Furthermore, the expansion of b is nonterminating since the digits 0 and 9 are excluded from it. We assert that this decimal is absent from our scheme. This claim is easily verified with the fact that two normalized decimal expansions coincide if and only if they agree in each component. The number b, however, disagrees in at least one component with each number in the table, and the theorem is thus proved.

The technique employed in the above proof has numerous applications in other parts of analysis. It is known as the *Cantor second diagonal method.* It will recur in this text on several occasions.

We are now in a position to establish some elementary yet important facts concerning countability. We begin with the following theorem:

4.9 / Theorem Every infinite set contains a countable subset.

Proof Let A be an arbitrary infinite set. Then it contains some element which may be labeled a_1. The set $A \setminus \{a_1\}$ is not empty. Hence, we can label one of its members a_2: clearly $a_2 \neq a_1$. The set $A \setminus \{a_1, a_2\}$ is not empty, and one of its elements can be designated as a_3. Continuing in this manner, we construct a subset of A, $\{a_1, a_2, a_3, \dots\}$, which has the desired property.

In particular we have:

4.10 / Theorem Every infinite subset of a countable set is countable.

This fact follows at once from the observation that if A is countable, then any one-to-one function $f: A \to \mathfrak{N}$ automatically maps every subset of A onto a subset of $\mathfrak{N}$.

4.11 / Theorem The set A is infinite if and only if it is equivalent to one of its proper subsets.

Proof From Definition 4.2 we know that a finite set A cannot be equivalent to a proper subset. When negated, this statement tells us that when A is equivalent to one of its proper subsets, then A is infinite. Let us consider, therefore, the converse statement.

First, the theorem is true when A is countable, for then its members can be put in a sequence, say, $\{a_1, a_2, a_3, \dots\}$. The function

$f(a_n) = a_{n+1}$ establishes a one-to-one correspondence between the set A and its proper subset $A \setminus \{a_1\}$. We now use such a relation to prove the general case.

If A is uncountable, then it has a denumerable subset $B = \{a_1, a_2, a_3, \ldots\}$ (Theorem 4.9). We assert that $A \sim (A \setminus \{a_1\})$. To verify this, we write $C = A \setminus B$ and express A in terms of two disjoint subsets:

$$A = B \cup C \quad \text{and} \quad B \cap C = \varnothing.$$

Clearly,

$$A \setminus \{a_1\} = (B \setminus \{a_1\}) \cup C.$$

Let $e(x) = x$ be the identity function which associates each $x \in A$ with itself. Let the function f be as above, $f(a_n) = a_{n+1}$, and define the function h as follows:

$$h(x) = \begin{cases} e(x) & \text{when } x \in C \\ f(x) & \text{when } x \in B. \end{cases}$$

The function h has domain A and range $A \setminus \{a_1\}$, and evidently $h(x) = h(y)$ if and only if $x = y$. Hence the theorem is proved.

4.12 / Theorem If the sets A and B are countable, then so is the set $A \cup B$.

Proof Suppose first $A \cap B = \varnothing$. Since the sets A and B are countable we may express them as

$$A = \{a_1, a_2, a_3, \ldots\}$$

$$B = \{b_1, b_2, b_3, \ldots\}.$$

The function

$$f: \begin{cases} a_n \leftrightarrow 2n - 1 \\ b_n \leftrightarrow 2n \end{cases}$$

clearly relates the members of A in a one-to-one manner with the odd positive integers, the members of B with the even ones. Since each member of $A \cup B$ is either an a_n or a b_n this case is proved.

In case $A \cap B \neq \varnothing$ we can write $A \cup B = A \cup (B \setminus A)$, where $A \cap (B \setminus A) = \varnothing$. We distinguish between two cases:

(i) $B \setminus A$ is countable. Here we simply apply the preceding argument to the sets A and $B \setminus A$.

(ii) $B \setminus A$ is finite. Let $B \setminus A = \{c_1, c_2, \ldots, c_m\}$. Then the set $\{c_1, c_2, \ldots, c_m, a_1, a_2, \ldots\}$ is clearly countable, and the proof is complete.

4.13 / Theorem The countable union of countable sets is countable.

In other words, if each of the sets A_n ($n \in \mathfrak{N}$) is countable, then so is the set $A = \bigcup_{n \in \mathfrak{N}} A_n$. The student should be able to prove this theorem with the help of the Cantor diagonal method (see Theorem 4.6).

In concluding this section we should like to mention the following: Consider a collection X of n nonempty sets, say, $X = \{A_1, A_2, \ldots, A_n\}$. For each k, $1 \leq k \leq n$, let a_k be a fixed member of A_k. The formula

$$f(A_k) = a_k \qquad (1 \leq k \leq n)$$

determines a function with domain X: this is all perfectly clear. Is it just as clear when X is an arbitrary collection of nonempty sets? In particular, when X is uncountable, then no scheme for listing its members exists. Is it meaningful under these circumstances to claim that we can assign to each $A \in X$ a member $a \in A$? This, indeed, is a very controversial issue. The assertion that such a correspondence can always be made was stated by Ernst Zermelo in 1904 as an axiom, titled the *axiom of choice* (*Auswahlpostulat* in German). We give here one version thereof.

4.14 / Axiom of Choice Let X be a set whose members are nonempty sets. Then there is a function f with domain X which assigns to each $A \in X$ an element $f(A) \in A$.

In particular, it should be noticed that the element $f(A)$ is uniquely determined for each $A \in X$. The function f is called a *choice function*. The above axiom has numerous applications. A case in point is Theorem 4.8. Another direct application is to be found in the proof of Theorem 4.9.

EXERCISES

4-1 Show that Definition 4.2 specifies an equivalence relation.

4-2 Using a geometric argument, show that the point set of a sphere from which one point has been removed is equivalent to the Euclidean plane $\mathfrak{R} \times \mathfrak{R}$.

4-3 Verify that the union of finitely many finite sets is finite.

4-4 Show that if A is infinite and B is countable, then

$$A \sim A \cup B.$$

4-5 Regarding the real line $\mathcal{R}$, show that an arbitrary collection of pairwise disjoint nondegenerate intervals is at most countable.

4-6 Show that the interval $[0, 1]$ can be expressed as the union of countably many intervals which are mutually exclusive.

4-7 Referring to Theorem 4.7, write down the polynomials corresponding to hight $h = 5$.

4-8 Prove that the set of all polynomials with rational coefficients is countable.

4-9 Prove or disprove: Let A be a countable set of real numbers, and let P be the set of all polynomials with coefficients in A. Then P is countable.

4-10 Show why the Cantor second diagonal method of Theorem 4.8 cannot be used to prove that the repeating decimals (rational numbers) in the interval $0 < x < 1$ are not countable.

4-11 Show that all numbers of the form $0.a_1a_2a_3\cdots$, where $a_n = 0$ or $a_n = 1$, cannot be arranged in a matrix as in Theorem 4.8 in which every element on the main diagonal is 1.

4-12 We know from Theorem 4.8 that all infinite sequences of members of $\mathfrak{N}$ constitute an uncountable set. Prove, however, that the set of all finite (terminating) sequences with members in $\mathfrak{N}$ is countable.

4-13 Suppose $A \neq \varnothing$ and $A \sim B$. Prove that there are functions $f: A \to B$ and $g: B \to A$ such that $g \circ f: A \to A$ and $f \circ g: B \to B$ (see also Exercise 3-6).

4-14 Show that an arbitrary infinite set can be expressed as the countable union of pairwise disjoint infinite sets.

4-15 Let F be the set of all functions $f: \mathfrak{N} \to \{0, 1\}$ (i.e., F consists of all functions with domain $\mathfrak{N}$ and range in $\{0, 1\}$). Show that F is uncountable.

4-16 Let $\{C_n\}$ $(n \in \mathfrak{N})$ be a given family of pairwise disjoint sets such that $C_n \sim C_{n+1}$ for each admitted value of n. Show that

$$\bigcup_{n\in\mathfrak{N}} C_n \sim \bigcup_{n\in\mathfrak{N}} C_{n+1}.$$

TWO

THE RATIONAL NUMBERS

The real line is the source of many generalizations and abstractions in mathematics. The material in this book is closely tied in with the real number system, and a thorough understanding thereof is consequently emphasized. The material in this chapter is by and large known to the reader in one form or another. It is included to ensure an equal footing for readers with an inadequate preparation. Motives for introducing the real number system into analysis abound: ours is measure.

The correspondence between rational numbers and points on a straight line is well known. Distinguishing one point as a fixed origin 0, we assign a measure, or unit length, $\overline{01}$ to the line. The rational number $a = \frac{p}{q}$, where p and q are integers, is then attached to the line segment $\overline{0a}$ which is uniquely determined by the equation

$$q \cdot \overline{0a} = p \cdot \overline{01}.$$

The existence of a major flaw in the structure of the rational number system was already known to the Greeks. They discovered, namely, that each unit of length on the straight line continuum gives rise to line segments which cannot be measured with it: the endpoint of the diagonal of a right triangle with sides of unit length, for example, does not belong to any rational number (see Figure 10). Consequently, not all phenomena on the straight line continuum can be described in the realm of the rational numbers. We say that this system is *incomplete*.

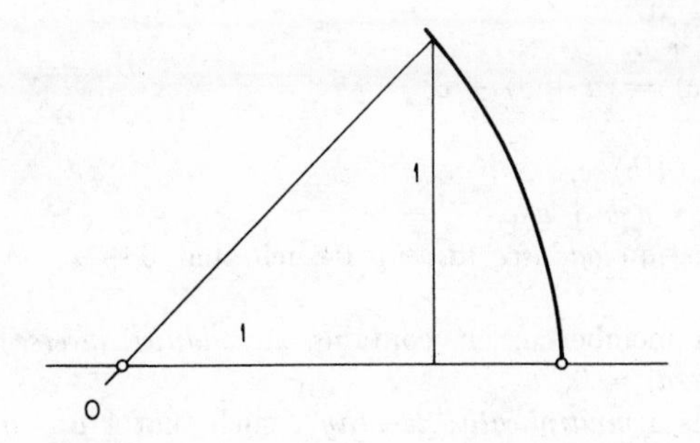

Figure 10

To understand the nature of the indicated defect we must have an intimate knowledge of the anatomy of the system in which we operate. The rational numbers are not derived here from the integers, however, since the strategy required for this sheds no light on the problem at hand. Assuming these numbers to be given, we proceed instead to examine their incompleteness, following, in turn, the line of reasoning of Dedekind and Cantor. Thinking of rational numbers as fractions $\frac{m}{n}$ $(n \neq 0)$, where m and n are integers, suffices initially. The reader is cautioned to keep in mind that the symbol $\frac{m}{n}$ actually represents a whole class of rational numbers namely, it represents the equivalence class of fractions $\frac{p}{q}$ which are related to it through the equation $mq = pn$. The arithmetic of fractions and their ordering are assumed known.

The interested reader is referred to Landau's lucid and elegant text "Foundations of Analysis" for a systematic derivation of the rational numbers from the integers.

5 / ALGEBRAIC PROPERTIES

The family of rational numbers is characterized by two sets of properties: one of these determines their arithmetical structure and the other their order qualities. Specifically, the rational numbers together with the operations of addition and multiplication and ordering constitute an *ordered field* $\mathcal{R}^*$ which is described through the following sets of postulates:

5.1 / Field Axioms

(1) The operations $+$ and $\cdot$ associate with each pair of elements a and b of $\mathcal{R}^*$ the unique elements $a + b$ and $a \cdot b$ which are again in $\mathcal{R}^*$. This association is such that:

(**2**) $a + b = b + a$,
(**3**) $a + (b + c) = (a + b) + c$,
(**4**) $a \cdot b = b \cdot a$,
(**5**) $a \cdot (b \cdot c) = (a \cdot b) \cdot c$,
(**6**) $a \cdot (b + c) = a \cdot b + a \cdot c$,
(**7**) $\mathfrak{R}^*$ contains an *additive identity* 0 such that $0 + a = a$ for all $a \in \mathfrak{R}^*$,
(**8**) With each member a, $\mathfrak{R}^*$ contains an *additive inverse* $-a$ such that $a + (-a) = 0$,
(**9**) $\mathfrak{R}^*$ contains a *multiplicative identity* 1 such that $1 \cdot a = a$ for each $a \in \mathfrak{R}^*$,
(**10**) Together with each nonzero entry a, $\mathfrak{R}^*$ contains the *multiplicative inverse* a^{-1} such that $a \cdot a^{-1} = 1$.

5.2 / Order Axioms

(**1**) For each pair of members $a \in \mathfrak{R}^*$ and $b \in \mathfrak{R}^*$ one and only one of the relations

$$a < b \qquad a = b \qquad a > b$$

holds.
(**2**) If $a < b$ and $b < c$, then $a < c$.
(**3**) If $a < b$, then $a + c < b + c$.
(**4**) If $0 < a$ and $0 < b$, then $0 < ab$.

The first six axioms in 5.1 together with the order axioms are satisfied in the realm of the positive integers $\mathfrak{N}$. Axioms 5.1(7) and 5.1(8) are equivalent to this very elementary result: Each of the equations $a + x = b$ and $x + a = b$ in the indeterminate x has a unique solution.

The resulting system, which is the well-known *ring* of integers, can therefore be regarded as an arithmetic extension of $\mathfrak{N}$ in which subtraction is always meaningful. The last two axioms in 5.1 were designed to accomodate the operation of division. They are equivalent to the statement: Each of the equations $a \cdot x = b$ and $x \cdot a = b$ $(a \neq 0)$ has a unique solution.

Again, $\mathfrak{R}^*$ is an extension of an imperfect system. The process of augmenting sets which are closed under algebraic operations by annexing to them solutions of given algebraic equations which are not already solvable is called an *algebraic extension.*

In the system $\mathfrak{R}^*$ all linear equations $ax + b = 0$ have solutions, but not so for general nonlinear equations. For example, $x^2 - 2 = 0$ has no solution. Geometrically this means that there is no square in the plane with sides of rational length whose area equals 2. Another fact is that the circumference of a circle with rational diameter or the area of the corresponding disk are not measurable with rational numbers. Let us note in passing that there is a significant difference between the problem of measuring the area of squares with rational sides and that of measuring the area (or circumference) of disks with rational diameters. Indeed, the first of these is conveniently disposed of with algebraic extensions of the rational field; the solution of the second problem cannot be so realized: to solve it one has to introduce nonalgebraic (limit) schemes into the extension process. This remarkable fact is difficult to verify, but the existence of numbers which are not solutions of algebraic equations with rational coefficients was established in Section 4. At this point we investigate the first problem in some detail. We state formally:

5.3 / Theorem The equation

$$x^2 - 2 = 0$$

has no solution in $\mathcal{R}^*$.

Proof The proof is by contradiction. Namely, the assumption that $\mathcal{R}^*$ contains an element a such that $a^2 = 2$ is shown to lead to a false (self-contradictory) statement.

Writing $a = \dfrac{p}{q}$, where p and q are integers, $q \neq 0$, we may assume from the start that the fraction has been normalized, meaning thereby that p and q have no common factors. In particular, p and q cannot be both even with this agreement.

From the relation

$$p^2 = 2q^2,$$

which ensues from the supposition $a^2 = 2$, we deduce that p is even, say, $p = 2r$. A substitution for p in the above equation leads to the equation

$$2r^2 = q^2,$$

from which we deduce that q is even. This, however, contradicts the hypothesis that p and q have no common factors. The assertion that a belongs to $\mathcal{R}^*$ was, therefore, false.

From this theorem we see that each member $x \in \mathcal{R}^*$ gives rise to one of the strict inequalities $x^2 < 2$ or $x^2 > 2$. The function $f(x) = | x^2 - 2 |$, however, has no minimum value in $\mathcal{R}^*$. In other words, the area of the square with diagonal 2 can be computed to any desired degree of accuracy with rational numbers. This fact is expressed as follows:

5.4 / Theorem For each element a, such that

$$a^2 < 2,$$

$\mathcal{R}^*$ also contains a member b, $a < b$, such that

$$b^2 < 2.$$

Similarly, for each $a \in \mathcal{R}^*$ which satisfies the inequality

$$2 < a^2$$

$\mathcal{R}^*$ contains a member $b < a$ for which

$$2 < b^2.$$

Proof Without loss of generality we may suppose that $a < 2$: The case $a > 2$ can only occur in the second part of the theorem, but there $b = (a + 2)/2$ is easily seen to be as demanded.

Consider the linear equation

$$x = a + (2 - a)t.$$

As t varies over $\mathcal{R}^*$, x exhausts all rational numbers. Suppose that $a^2 < 2$. To assure the proper relation $a < b < 2$, the variable t will be restricted to the rational values $0 < t < 1$. A direct computation shows that

$$x^2 = a^2 + 2a(2 - a)t + (2 - a)^2t^2$$
$$< a^2 + 2a(2 - a)t + (2 - a)^2t = a^2 + (4 - a^2)t.$$

If we put $x = b$, then $b^2 < 2$ when

$$a^2 + (4 - a^2)t < 2,$$

but this is true whenever t is an arbitrary rational number subject to the inequalities

$$0 < t < \frac{2 - a^2}{4 - a^2}.$$

In case $a^2 > 2$ we have to let $t < 0$. From the above expansion of x^2 we derive the estimate

$$x^2 > a^2 + 2a(2 - a)t,$$

so that the inequality

$$a^2 + 2a(2 - a)t > 2$$

leads to the restriction

$$\frac{2 - a^2}{2a(2 - a)} < t < 0.$$

Owing to the inequalities

$$0 < 2a(2 - a) < 4 - a^2$$

we can, in fact, put

$$\frac{2 - a^2}{4 - a^2} < t < 0.$$

Taking now, for example,

$$t = \frac{2 - a^2}{(2 - a)(n + a)},$$

where n is some integer ≥ 3, and substituting t into $x = a + (2 - a)t$, we obtain the number

$$b = \frac{na + 2}{a + n}.$$

It is easily verified that either $a^2 < 2$ and $b^2 < 2$, or else $a^2 > 2$ and $b^2 > 2$. Clearly, these are the numbers demanded in the proof.

This theorem brings about a very important result:

5.5 / Corollary The number $\sqrt{2}$, which is the formal solution of the equation $x^2 - 2 = 0$, partitions $\mathcal{R}^*$ into two nonempty mutually exclusive classes, $\mathcal{A}$ and $\mathcal{B}$:

$$\mathcal{A} = \{a \in \mathcal{R}^* \mid a < \sqrt{2}\},$$

$$\mathcal{B} = \{b \in \mathcal{R}^* \mid b > \sqrt{2}\},$$

such that $\mathcal{A}$ has no largest member, whereas $\mathcal{B}$ has no smallest member.

The number $\sqrt{2}$ is therefore seen to occupy a definite position in the order hierarchy of $\mathcal{R}^*$. According to Dedekind, an ordered pair of sets $(\mathcal{A}, \mathcal{B})$ such as those specified in the corollary constitutes a *cut*. The cut in 5.5 determines the number $\sqrt{2}$ uniquely, and conversely. Dedekind thus identified the two, thereby defining an irrational number in terms of rational numbers. Pairs of sets $(\mathcal{A}, \mathcal{B})$ in which $\mathcal{A}$ has a largest member *or* $\mathcal{B}$ has a smallest member are also admitted into the family of cuts. They are called rational cuts.† For instance, the sets

$$\mathcal{A} = \{x \in \mathcal{R}^* \mid x \leq 1\}$$

and

$$\mathcal{B} = \{x \in \mathcal{R}^* \mid x > 1\}$$

specify a cut which is attached to the number 1.

† The pairs $(\mathcal{A}, \mathcal{B})$ in which $\mathcal{A}$ has a largest member and $\mathcal{B}$ has a smallest member are not considered to be cuts.

Dedekind has shown that the set $\mathfrak{D}$ of rational cuts is identical in its algebraic structure to $\mathcal{R}^*$. That is, one can define arithmetic operations on rational cuts in such a way that $\mathfrak{D}$ is an ordered field. This, however, applies to the set of all cuts despite the fact that not all cuts belong to rational numbers. This was Dedekind's characterization of the incompleteness of $\mathcal{R}^*$. To realize completeness he regarded all cuts as representing real numbers and showed that when the completion process of forming cuts is repeated in the resulting system of cuts, no new elements are created. This is the *Dedekind Completeness Theorem.*

The method just described is highly ingenious and surprisingly simple. We do not pursue it further here because it has no other applications in the text. Instead we proceed to attack the incompleteness of $\mathcal{R}^*$ with Cantor's strategy. To him incompleteness meant that certain sequences of rational numbers failed to converge.

EXERCISES

5-1 Show that the set of numbers $a + \sqrt{2}b$, where a and b vary over $\mathcal{R}^*$, is a field under ordinary addition and multiplication.

5-2 Show that the field in the above exercise is the *smallest* field in which all linear equations and the equation $x^2 - 2 = 0$ have a solution.

5-3 Consider the numbers

$$r = p_1 + \frac{p_2}{2} + \frac{p_3}{2^2} + \frac{p_4}{2^3} + \frac{p_5}{2^4}$$

in which the p_k take on the values 0 or 1. Find a sequence of digits $(p_1, p_2, \ldots, p_5)$ for which r^2 best approximates 2.

5-4 Demonstrate with axioms in 5.2 that the inequalities $a > 0$ and $b > 0$ imply that $a + b > 0$.

5-5 Let a and b be members of $\mathcal{R}^*$, $a \geq 0$. Show that $|b| \leq a$ if and only if $-a \leq b \leq a$.

5-6 Show that the inequality $|a + b| \leq |a| + |b|$ is universally true in $\mathcal{R}^*$. Hence prove that for an arbitrary finite set $a_1, a_2, \ldots, a_n$ of rational numbers

$$|a_1 + a_2 + \cdots + a_n| \leq |a_1| + |a_2| + \cdots + |a_n|.$$

5-7 Show that no member of $\mathcal{R}^*$ has an immediate successor in $\mathcal{R}^*$.

5-8 How would you define equivalence between cuts?

5-9 How would you characterize the sum of two cuts?

6 / DECIMAL EXPANSIONS

Rational numbers were regarded thus far as fractions, but the reader is already familiar with yet another representation, namely, in the form of decimal expansions. More generally, given an integer $p > 1$, we can associate with each fraction at least one sequence $a.a_1a_2a_3\cdots$, where the integers a_k are confined to the values $0 \le a_k \le p - 1$.

6.1 / *Example* In the familiar case $p = 10$,

$$\tfrac{1}{6} = 0.1\,666666\cdots,$$

$$\tfrac{1}{7} = 0.142857\,142857\cdots;$$

in the case $p = 2$,

$$\tfrac{1}{6} = 0.00\,10\,10\,10\cdots,$$

$$\tfrac{1}{7} = 0.00\,100\,100\,100\cdots.$$

The present discussion is confined for the sake of convenience to fractions $\frac{m}{n}$ in which $0 < m < n$.

The association between fractions and decimal expansions warrants a closer inspection. For this purpose we regard $\mathcal{R}^*$ as a metric space with metric d (see Example 3.9). All the remarks to follow in this section are going to be made relative to this concept.

To say that $\frac{m}{n}$ has a decimal expansion means, of course, that attached to it is a certain *infinite series*

$$s = \sum_{j=1}^{\infty} \frac{a_j}{10^j},$$

where the digits a_k are integers such that $0 \le a_k \le 9$, which completely identifies it. That is, the formula which associates s with $\frac{m}{n}$ cannot be used to associate s with any other fraction which is not equivalent to $\frac{m}{n}$. We should mention that the line of reasoning followed here dates back to Cauchy:

For each $k \in \mathfrak{N}$ we consider the rational number

$$s_k = \sum_{j=1}^{k} \frac{a_j}{10^j}$$

called the *partial sum* of the infinite series s. The sequence

$$\{s_k\} = (s_1, s_2, \ldots, s_k, \ldots)$$

thereby obtained has the property that

$$d\left(s_k, \frac{m}{n}\right) = \sum_{j=k+1}^{\infty} \frac{a_j}{10^j} \leq \frac{9}{10^{k+1}} \sum_{j=0}^{\infty} \frac{1}{10^j} = \frac{1}{10^k}.$$

That is, for each rational number $\epsilon > 0$ the inequality

$$d\left(s_k, \frac{m}{n}\right) < \epsilon$$

holds except possibly for finitely many of the s_k. We already know from calculus that this is the content of the equivalent statement

$$\lim_{k\to\infty} d\left(s_k, \frac{m}{n}\right) = 0.$$

With the properties of the function d we easily prove:

6.2 / Theorem If

$$\lim_{k\to\infty} d\left(s_k, \frac{m}{n}\right) = 0$$

and

$$\lim_{k\to\infty} d\left(s_k, \frac{m'}{n'}\right) = 0,$$

then

$$\frac{m}{n} = \frac{m'}{n'}.$$

Proof Holding k momentarily fixed, we appeal to the properties of d to obtain the inequalities

$$0 \leq d\left(\frac{m}{n}, \frac{m'}{n'}\right) \leq d\left(s_k, \frac{m}{n}\right) + d\left(s_k, \frac{m'}{n'}\right).$$

In view of the hypotheses of the theorem we can fix for each rational number $\epsilon > 0$ an integer $n_\epsilon > 0$, such that

$$\max\left[d\left(s_k, \frac{m}{n}\right), d\left(s_k, \frac{m'}{n'}\right)\right] < \frac{\epsilon}{2}$$

for all integers $k > n_\epsilon$, implying that

$$0 \leq d\left(\frac{m}{n}, \frac{m'}{n'}\right) < \epsilon.$$

Hence

$$d\left(\frac{m}{n}, \frac{m'}{n'}\right) = 0$$

and from Example 3.9 we deduce that $\dfrac{m}{n} = \dfrac{m'}{n'}$.

The uniqueness of decimal expansions comes under scrutiny next.

6.3 / *Example* Consider the decimal expansions

$a = 0.0\,9\,9\,9\,9\cdots$

$b = 0.1\,0\,0\,0\,0\cdots$.

The partial sums of their associated infinite series are

$$s_k = \frac{1}{10} - \frac{1}{10^k}$$

and

$$s_k' = \frac{1}{10},$$

respectively. By the distance between a and b we mean the number

$$d(a, b) = \lim_{k\to\infty} d(s_k, s_k').$$

However, this limit is clearly zero in this case, so that $a = b$.

This nonuniqueness makes the comparison between such expansions difficult. Specifically, we wish to be able to claim that two decimal expansions $a = 0.a_1a_2a_3\cdots$ and $b = 0.b_1b_2b_3\cdots$ coincide when and only when $a_k = b_k$ for all values of k. For this reason the expansions are normalized by stipulating that for no decimal can all the digits equal 9 after a finite number of terms. With this agreement one easily proves the desired uniqueness:

6.4 / *Example* It must be emphasized that the metric d plays a key role in deciding when two rational numbers coincide. To clarify this remark, let us, in fact, change the metric on $\mathcal{R}^*$. Namely, for arbitrary numbers $a = 0.a_1a_2a_3\cdots$ and $b = 0.b_1b_2b_3\cdots$ let

$$\delta(a, b) = \max_{i\in\mathfrak{N}} d(a_i, b_i).$$

The existence of the indicated maximum is guaranteed by the fact that

$$0 \leq d(a_i, b_i) \leq 9$$

for all values $i \in N$. It is clear without further justification that

$$\delta(a, a) = 0$$

and if

$$\delta(a, b) = 0,$$

then

$$d(a_i, b_i) = 0 \qquad (i \in \mathfrak{N})$$

and accordingly $a = b$. Obviously, $\delta(a, b) = \delta(b, a)$.

Finally, consider an arbitrary number $c = 0.c_1c_2c_3\cdots$; for each fixed value of i it follows that

$$d(a_i, b_i) \leq d(a_i, c_i) + d(b_i, c_i),$$

so that the following steps ensue:

$$d(a_i, b_i) \leq d(a_i, c_i) + \max_{i \in \mathfrak{N}} d(b_i, c_i) = d(a_i, c_i) + \delta(b, c)$$

and likewise

$$d(a_i, b_i) \leq \delta(a, c) + \delta(b, c).$$

Since the right side of the last inequality is independent of the choice of i, it follows that

$$\delta(a, b) \leq \delta(a, c) + \delta(b, c).$$

We have just shown that $\mathfrak{R}^*$ together with δ constitute a metric space. How does the foregoing discussion stand in this new structure? The numbers a and b in Example 6.3 are no longer indistinguishable since $\delta(a, b) = 9$.

Rational numbers possess a rather remarkable property: Except for a finite number of terms their expansions are periodic. By this it is meant that every rational number a, $0 < a < 1$, has an expansion of the form

$$a = 0.a_1a_2\cdots a_k\, b_1b_2\cdots b_r\, b_1b_2\cdots b_r\cdots$$

in which the sequence $b_1b_2\cdots b_r$ repeats indefinitely. If we write this expansion in the form

$$a = 0.a_1a_2\cdots a_ka_{k+1}a_{k+2}\cdots,$$

then the digits a_n are determined for $n > k$ as follows: put $n = k + m$. Owing to the Euclidean algorithm there are integers p and q, $p \geq 0$ and $0 \leq q < r$, such that

$$m = pr + q.$$

Hence

$$a_{k+m} = b_q.$$

The following is true:

6.5 / Theorem Let $p > 1$ be an arbitrary but fixed integer. The expansion $0.a_1a_2a_3\cdots$ $(0 \leq a_k \leq p - 1)$ represents a rational number if and only if it is periodic after a finite number of terms.

Proof The proof is outlined for the familiar case $p = 10$, but the general case does not involve any new ideas.

Let the number $\frac{m}{n}$ be given. To show that its decimal expansion is as asserted we make some suppositions which do not detract from the generality of the proof. These are

(a) m and n have no common factors,
(b) n is divisible by neither 2 nor 5.

To justify the last restriction, it suffices to note that if $n = 2^r5^sq$, where r and s are nonnegative integers and q is as demanded in (b), then upon setting $p = 2^s5^rm$ we obtain the number $10^{-r-s}\frac{p}{q}$ in which p and q are related as to satisfy (a). Clearly, the original fraction has a periodic expansion if and only if $\frac{p}{q}$ has one.

With this agreement the Euclidean algorithm guarantees the iteration

$$10\frac{m}{n} = a_1 + \frac{m_1}{n} \qquad (0 \leq m_1 < n)$$

$$10\frac{m_1}{n} = a_2 + \frac{m_2}{n} \qquad (0 \leq m_2 < n)$$

$$10\frac{m_2}{n} = a_3 + \frac{m_3}{n} \qquad (0 \leq m_3 < n)$$

$$\vdots$$

in which the numbers a_k and m_k are uniquely generated. The above restrictions on m and n guarantee that, at each stage, m_k and n have no common factors. In particular, at the kth step we have

$$\frac{m}{n} = \frac{a_1}{10} + \frac{a_2}{10^2} + \cdots + \frac{a_k}{10^k} + \frac{m_k}{10^k n} = 0.a_1a_2\cdots a_k + \frac{m_k}{10^k n}.$$

Since each integer m_k in the iteration is confined to the values $0 \leq m_k < n$, it follows that two of the m_k must coincide after at most n steps, say, $m_p = m_q$, where $1 \leq p \leq q \leq n$. Owing to the uniqueness of the Euclidean algorithm it follows that

$$a_{p+1} = a_{q+1}, \qquad a_{p+2} = a_{q+2}, \cdots, \quad a_{q-1} = a_{2q-p-1},$$

and hence we have established the existence of a period.

To prove the converse, let $a = 0.a_1a_2\cdots a_k\, b_1b_2\cdots b_r\, b_1b_2\cdots b_r\cdots$; let s_n be the partial sum consisting of the first $k + nr$ terms of the expansion of a. Then the following computations are correct:

$$s_n = \sum_{j=1}^{k} \frac{a_j}{10^j} + \sum_{j=1}^{r} \frac{b_j}{10^{k+j}} + \sum_{j=1}^{r} \frac{b_j}{10^{k+r+j}} + \cdots + \sum_{j=1}^{r} \frac{b_j}{10^{k+(n-1)r+j}}$$

$$= \sum_{j=1}^{k} \frac{a_j}{10^j} + \left[1 + \frac{1}{10^r} + \frac{1}{(10^r)^2} + \cdots + \frac{1}{(10^r)^{n-1}}\right] \cdot \sum_{j=1}^{r} \frac{b_j}{10^{k+j}}$$

$$= \sum_{j=1}^{k} \frac{a_j}{10^j} + \frac{1 - \dfrac{1}{10^{nr}}}{1 - \dfrac{1}{10^r}} \cdot \sum_{j=1}^{r} \frac{b_j}{10^{k+j}}.$$

Hence,

$$a = \lim_{n\to\infty} s_n = \sum_{j=1}^{k} \frac{a_j}{10^j} + \frac{10^r}{10^r - 1} \sum_{j=1}^{r} \frac{b_j}{10^{k+j}}:$$

being a finite sum of rational numbers, it follows that a is rational.

Nonrepeating decimals are easily constructed.

6.6 / ***Example*** The decimal

$0.10\,100\,1000\,10000\cdots$

is clearly not periodic. Let us, in fact, derive a formula for generating nonrepeating decimal expansions: The equation

$$n = \frac{m(m+1)}{2} + t \qquad (0 \leq t \leq m)$$

determines a one-to-one correspondence between $\mathfrak{N}$ and the set of ordered pairs (m, t) in which the entries are related by the above equation.

Consider the decimal

$$0.a_{(1,0)}a_{(1,1)}a_{(2,0)}a_{(2,1)}a_{(2,2)}a_{(3,0)}\cdots;$$

the decimal in $0.10\ 100\ 1000\ 10000\cdots$ can now be specified through the requirement

$$a_{(m,t)} = \begin{cases} 1 & \text{when} \quad t = 0 \\ 0 & \text{when} \quad 0 < t \leq m \end{cases} \qquad m \in \mathfrak{N}.$$

With similar conditions on t one readily obtains decimals which have no finite period.

The preceding considerations motivate the following:

6.7 / *Definition* Consider the formal series $\sum_{j=1}^{\infty} \frac{a_j}{10^j}$ with partial sums $s_k = \sum_{j=1}^{k} \frac{a_j}{10^j}$. The sequence $\{s_k\}$ of partial sums *converges* if there is a number $a \in \mathfrak{R}^*$ such that for each rational number $\epsilon > 0$ there is an $n_\epsilon \in \mathfrak{N}$ such that $d(s_k, a) < \epsilon$ whenever $k > n_\epsilon$.

As an immediate consequence of this definition we have:

6.8 / Corollary If the sequence $\{s_k\}$ converges, then for each $\epsilon > 0$ there is an integer n_ϵ such that

$$d(s_k, s_m) < \epsilon$$

for all integers $k > n_\epsilon$ and $m > n_\epsilon$.

A sequence which satisfies the inequality in the corollary is called a *Cauchy sequence* or *fundamental sequence.* As a convergence criterion the last inequality has an advantage over that in Definition 6.7 in that the limit element is not displayed and, in fact, need not even be known. Yet this test cannot be adopted because not every Cauchy sequence converges.

6.9 / *Example* The sequence of partial sums corresponding to the decimal in Example 6.6 constitutes a Cauchy sequence which does not converge. Its members are seen to be

$$s_k = \sum_{i=1}^{k} 10^{-i(i+1)/2}$$

and a simple computation shows the powers in $s_m - s_{k-1}$ $(m \geq k)$ to be expressible as

$$\frac{k(k+1)}{2}, \quad \frac{k(k+1)}{2} + k + 1, \quad \frac{k(k+1)}{2} + 2k + 3, \quad \frac{k(k+1)}{2} + 3k + 6, \quad \cdots$$

Hence, given integers $k \leq m$ we find that

$$\begin{aligned} d(s_k, s_m) &= \sum_{\nu=k}^{m} 10^{-\nu(\nu+1)/2} \\ &< 10^{-k(k+1)/2}(1 + 10^{-k} + 10^{-2k} + \cdots + 10^{-(m-k+1)k}) \\ &< 2 \cdot 10^{-k(k+1)/2}. \end{aligned}$$

This shows that the sequence $\{s_k\}$ is a Cauchy sequence.

Now, regarding the numbers s_k as points on the straight line continuum tells us that there is a point on it which cannot be distinguished with the function d from the sequence $\{s_k\}$. It is therefore natural to attach a symbol to this sequence and admit it into the family of numbers. This, indeed, is our procedure in the next chapter.

EXERCISES

6-1 Construct a nonperiodic expansion to the base 2.

6-2 Construct a nonperiodic decimal using at least four different digits.

6-3 Show that with the appropriate normalization each rational number has a unique expansion in any given base.

6-4 Characterize all terminating expansions to the base $p > 1$, p being an arbitrary integer.

6-5 Characterize all expansions to the base $p > 1$ whose period has length m.

6-6 Show that the partial sums of an arbitrary expansion form a Cauchy sequence.

6-7 Suppose $\lim_{k\to\infty} d(s_k, a) = 0$ and $\lim_{k\to\infty} d(s'_k, a') = 0$ $a, a' \in \mathfrak{R}^*)$. Show that

(a) $\lim_{k\to\infty} d(s_k + s'_k, a + a') = 0$

(b) $\lim_{k\to\infty} d(s_k \cdot s'_k, a \cdot a') = 0$

(c) $a < a'$ if and only if there is a rational number $\rho > 0$ such that $s'_k > s_k + \rho$ except possibly for finitely many values of k.

(d) $a = a'$ if and only if $\lim_{k \to \infty} d(s_k, s'_k) = 0$.

6-8 Show that if the sequence $\{s_k\}$ of partial sums converges to a nonzero limit, and if $s_k \neq 0$ for all $k \in \mathfrak{N}$, then also the sequence $\left\{\frac{1}{s_k}\right\}$ converges.

6-9 Show that for arbitrary numbers $a, b, c, d \in \mathfrak{R}^*$,

(a) $d(a \cdot b, a \cdot c) = |a| \cdot d(b, c)$

(b) $d(a + b, c + d) \leq d(a, c) + d(b, d)$.

6-10 Let $\{a_k\}$ be a Cauchy sequence. Show that if

$$\lim_{k \to \infty} d(a_{2k}, a) = 0,$$

then also

$$\lim_{k \to \infty} d(a_k, a) = 0.$$

PART TWO

THE REAL LINE

THREE

THE REAL NUMBER SYSTEM

With the ground work laid in Chapter Two we are well on our way in constructing the real numbers. The springboard for the developments to follow is the convergence criterion in Definition 6.7, which is extended now to encompass arbitrary sequences of rational numbers. The abbreviated notation $\{a_k\}$ $(k \in \mathfrak{N})$ is used throughout this chapter in lieu of the designation $(a_1, a_2, a_3, \ldots)$ for a sequence.

7 / CAUCHY SEQUENCES AND THEIR EQUIVALENCE CLASSES

Without further ado we set down the following definitions:

7.1 / ***Definition*** The sequence $\{a_k\}$ of rational numbers is a *Cauchy sequence* if, for each rational number $\epsilon > 0$, an integer $n_\epsilon > 0$ can be found such that

$$d(a_k, a_m) < \epsilon$$

for all integers $k, m > n_\epsilon$.

7.2 / ***Definition*** The Cauchy sequences $\{a_k\}$ and $\{b_k\}$ are *equivalent*, written

$$\{a_k\} \sim \{b_k\},$$

if, for each positive $\epsilon \in \mathcal{R}^*$, there is a number $n_\epsilon \in \mathfrak{N}$ such that

$$d(a_k, b_k) < \epsilon$$

for all integers $k > n_\epsilon$.

The information embodied in the above definitions is also conveyed by the respective statements

$$\lim_{k,m \to \infty} d(a_k, a_m) = 0$$

and

$$\lim_{k \to \infty} d(a_k, b_k) = 0.$$

The relation "$\sim$" specified in the last definition affords a partition of the set of Cauchy sequences of rational numbers into mutually exclusive classes, as confirmed in the theorem:

7.3 / Theorem The relation "$\sim$" characterized in Definition 7.2 is an equivalence relation.

The proof, which follows at once from the properties of the function d, is left as an exercise. The resulting system of equivalence classes is marked with the letter $\mathcal{R}$; Greek letters $\alpha, \beta, \gamma, \ldots$ will designate members of $\mathcal{R}$; the notation

$$\alpha = \{a_k\}$$

will signify the fact that $\{a_k\}$ is an arbitrary representative of the class α, the justification for this being that an equivalence class is completely determined by any one of its representatives.

We briefly interrupt the sequence of events at this point to anticipate the main program of this chapter. Our immediate objective is to define operations of addition and multiplication on $\mathcal{R}$ under which $\mathcal{R}$ will be a field. Following this we introduce an order relation into $\mathcal{R}$ which extends the axioms in 5.2 to the new system. This, in turn, is followed by the imbedding of $\mathcal{R}^*$ in $\mathcal{R}$ in the sense that the members of $\mathcal{R}^*$ are identified with equivalence classes of Cauchy sequences. Defining real numbers to be equivalence classes of Cauchy sequences of rational numbers, we culminate this phase of the program with an argument proving that no new equivalence classes are created when Cauchy sequences are redefined to envelop arbitrary Cauchy sequences as members. This will show that $\mathcal{R}$ is complete. With this in mind we formulate:

7.4 / Theorem Let $\{a_k\}$ and $\{b_k\}$ be given Cauchy sequences of rational numbers. Then

(1) $\{a_k + b_k\}$,
and
(2) $\{a_k \cdot b_k\}$

are Cauchy sequences. Furthermore, if $\{a_k\} \sim \{a_k'\}$ and $\{b_k\} \sim \{b_k'\}$, then

(3) $\{a_k + b_k\} \sim \{a_k' + b_k'\}$,
(4) $\{a_k \cdot b_k\} \sim \{a_k' \cdot b_k'\}$.

The proof of (2) and (4) depends on the boundedness of the sequences in question as described in the next lemma:

7.5 / Lemma If $\{a_k\}$ $(a_k \in \mathcal{R}^*)$ is a Cauchy sequence, then there is an integer $r > 0$ such that $|a_k| \leq r$ for all $k \in \mathfrak{N}$.

Proof The hypothesis in the lemma implies that we can find an integer $n > 0$ corresponding to $\epsilon = 1$ such that $d(a_n, a_k) \leq 1$ for all values $k > n$. For each such integer k, however,

$$|a_k| = d(a_k, 0) \leq d(a_n, 0) + d(a_n, a_k) \leq d(a_n, 0) + 1 = |a_n| + 1.$$

Hence, fixing r to be

$$r \geq \max(|a_1|, |a_2|, \ldots, |a_n|) + 1,$$

then this number is as demanded in the lemma.

Proof of 7.4 Directing our attention back to the theorem, let k and m be momentarily fixed. Then

$$d(a_k + b_k, a_m + b_m) \leq d(a_k, a_m) + d(b_k, b_m).$$

The sequences $\{a_k\}$ and $\{b_k\}$ being Cauchy sequences assures us that for each rational number $\epsilon > 0$ there are integers n_1 and n_2 such that

$d(a_k, a_m) < \frac{\epsilon}{2}$ for all indices $k, m > n_1$, and $d(b_k, b_m) < \frac{\epsilon}{2}$ whenever $k, m > n_2$. Let $n = \max(n_1, n_2)$. Then we find that

$$d(a_k + b_k, a_m + b_m) < \frac{\epsilon}{2} + \frac{\epsilon}{2} = \epsilon$$

whenever $k, m > n$, thereby establishing the first claim made in the statement of the theorem.

Regarding the assertion in (2), we refer to Exercise 6-9: When k and m are fixed, then

$$\begin{aligned} d(a_k \cdot b_k, a_m \cdot b_m) &= d(a_k \cdot b_k + a_m \cdot b_k, a_m \cdot b_k + a_m \cdot b_m) \\ &\leq d(a_k \cdot b_k, a_m \cdot b_k) + d(a_m \cdot b_k, a_m \cdot b_m) \\ &= |b_k| \cdot d(a_k, a_m) + |a_m| \cdot d(b_k, b_m). \end{aligned}$$

Appealing to Lemma 7.5, we can find integers r and s such that $|a_k| \leq r$ and $|b_k| \leq s$ for all $k \in \mathfrak{N}$. Moreover, given a rational number $\epsilon > 0$, we can reason as in the proof of (1) to obtain an integer n such that $d(a_k, a_m) < \frac{\epsilon}{2s}$ and $d(b_k, b_m) < \frac{\epsilon}{2r}$ for all $k, m > n$. These estimates show that

$$|b_k| \cdot d(a_k, a_m) < s \cdot \frac{\epsilon}{2s} = \frac{\epsilon}{2}$$

and

$$|a_m| \cdot d(b_k, b_m) < r \frac{\epsilon}{2r} = \frac{\epsilon}{2}$$

whenever $k, m > n$, so that

$$d(a_k b_k, a_m b_m) < \frac{\epsilon}{2} + \frac{\epsilon}{2} = \epsilon$$

when $k, m > n$, as was to be shown. The proof of (3) is fashioned after the proof of (1), that of (4) after (2). The student should be able to construct these proofs.

With this theorem we can define addition and multiplication in $\mathfrak{R}$ as follows:

7.6 / *Definition* If $\alpha \in \mathfrak{R}$, $\beta \in \mathfrak{R}$, $\alpha = \{a_k\}$, and $\beta = \{b_k\}$, then

(1) $\alpha + \beta = \{a_k + b_k\}$.

(2) $\alpha \cdot \beta = \{a_k \cdot b_k\}$.

We thus arrive at the first major result of this section:

7.7 / Theorem With the operations specified in Definition 7.6, $\mathfrak{R}$ is a field.

Proof We begin with the important observation that addition and multiplication in $\mathfrak{R}$ are defined in terms of the corresponding operations in $\mathcal{R}^*$. For this reason the following premises hold in $\mathcal{R}$ (see 5.1):

(1) $\mathcal{R}$ is closed, as evident from Theorem 7.4.
(2) $\alpha + \beta = \beta + \alpha$.
(3) $\alpha + (\beta + \gamma) = (\alpha + \beta) + \gamma$.
(4) $\alpha \cdot \beta = \beta \cdot \alpha$.
(5) $\alpha \cdot (\beta \cdot \gamma) = (\alpha \cdot \beta) \cdot \gamma$.
(6) $\alpha \cdot (\beta + \gamma) = \alpha \cdot \beta + \alpha \cdot \gamma$.
(7) $\mathcal{R}$ contains among its members the equivalence class $\theta = \{0\}$ which is endowed with the property $\theta + \alpha = \alpha$ for each $\alpha \in \mathcal{R}$.
(8) With each $\alpha = \{a_n\}$, $\mathcal{R}$ also contains the class $-\alpha = \{-a_n\}$ for which $\alpha + (-\alpha) = \theta$.
(9) $\mathcal{R}$ contains the class $i = \{1\}$ which is such that $i \cdot \alpha = \alpha$ for each $\alpha \in \mathcal{R}$.
(10) For each $\alpha \neq \theta$, $\mathcal{R}$ contains a member α^{-1} such that $\alpha \cdot \alpha^{-1} = i$.

Proof of 7.7(10) This last premise requires proof (see Exercise 6–8). Specifically, if $\alpha \neq \theta$ then it has a representative $\{a_k\}$ such that $a_k \neq 0$ for all $k \in \mathfrak{N}$. Hence, consider the sequence $\{a_k^{-1}\}$; let its class be α^{-1}. Then

$$\alpha \cdot \alpha^{-1} = \{a_k \cdot a_k^{-1}\} \sim \{1\} = i.$$

According to this, (10) will be proved once we show that the sequence $\{a_k^{-1}\}$ is a Cauchy sequence. This argument runs as follows:

The inequality $\lim_{k \to \infty} |a_k| > 0$ implies the existence of a rational number r and an integer n_r such that $|a_k| > r$ for all values $k > n_r$. Holding k and m fixed, $k, m > n_r$, we deduce that

$$d(a_k^{-1}, a_m^{-1}) = |a_k^{-1} \cdot a_m^{-1}| \cdot d(a_k, a_m) = |a_k^{-1}| \cdot |a_m^{-1}| \cdot d(a_k, a_m)$$
$$\leq r^{-2} \cdot d(a_k, a_m).$$

Since the sequence $\{a_k\}$ is a Cauchy sequence, we can assign to each rational number $\epsilon > 0$ an integer n_ϵ such that $d(a_k, a_m) < r^2 \cdot \epsilon$ for $k, m > n_\epsilon$. Setting $n = \max(n_r, n_\epsilon)$, we find that

$$d(a_k^{-1}, a_m^{-1}) < \epsilon$$

for $k, m > n$, thereby concluding that the sequence $\{a_k^{-1}\}$ is as claimed.

Faithful to our plan, we proceed to define an order relation in $\mathcal{R}$. Again this is based on the relations already existing in $\mathcal{R}^*$. Motivated by the rational case (see Exercise 6-7(c)) we formulate:

7.8 / *Definition* Let α and β be given numbers of $\mathcal{R}$, $\alpha = \{a_k\}$, $\beta = \{b_k\}$.

Then

$$\alpha < \beta$$

if a rational number $\rho > 0$ and an integer n_ρ can be found such that

$$b_k - a_k > \rho \quad \text{for} \quad \text{all } k > n_\rho.$$

It is tacitly implied in this definition that the order relation so specified is independent of the choice of representatives of the respective equivalence classes. This fact is confirmed in Lemma 7.10 below, but first we state

7.9 / Theorem For every pair of numbers $\alpha, \beta \in \mathcal{R}$ one and only one of the relations

$$\alpha < \beta, \qquad \alpha = \beta, \qquad \alpha > \beta$$

holds.

This theorem is a consequence of the next lemma:

7.10 / Lemma Let $\{a_k\}$ and $\{b_k\}$ be given Cauchy sequences of rational numbers. Then exactly one of the following relations is true:

(**1**) $\{a_k\} \sim \{b_k\}$;

(**2**) there are positive numbers $n \in \mathfrak{N}$ and $\rho \in \mathcal{R}^*$ such that

$$b_k - a_k > \rho \quad \text{for} \quad \text{all } k > n;$$

(**3**) there are positive numbers $n \in \mathfrak{N}$ and $\rho \in \mathcal{R}^*$ such that

$$a_k - b_k > \rho \quad \text{for} \quad \text{all } k > n.$$

Furthermore, if $\{a'_k\} \sim \{a_k\}$ and $\{b'_k\} \sim \{b_k\}$, then $\{a'_k\}$ and $\{b'_k\}$ stand in the same relation in which $\{a_k\}$ stands to $\{b_k\}$.

Proof Owing to Definition 7.2 and the transitivity of equivalence this lemma is trivially true when (1) is the case. We therefore assume that $\{a_k\} \not\sim \{b_k\}$. With this we are assured the existence of a rational number $\rho > 0$ and a corresponding integer n_ρ such that

$$d(a_k, b_k) > 3 \cdot \rho \qquad (k > n_\rho).$$

For this ρ we can find positive integers n'_ρ and n''_ρ such that

$$d(a_k, a_m) < \rho \qquad (k, m > n'_\rho)$$

$$d(b_k, b_m) < \rho \qquad (k, m > n''_\rho).$$

Thus, when we set

$$n = \max(n_\rho, n'_\rho, n''_\rho)$$

then the following inequalities emerge:

$$d(a_n, b_n) > 3 \cdot \rho \tag{i}$$

$$d(a_n, a_k) < \rho \qquad (k > n) \tag{ii}$$

$$d(b_n, b_k) < \rho \qquad (k > n). \tag{iii}$$

The fixed numbers a_n and b_n satisfy one and only one of the inequalities $a_n < b_n$ or $b_n < a_n$. In the first case we deduce from (i) that $b_n - a_n > 3 \cdot \rho$. When applied to the inequalities in (ii) and (iii) this shows that

$$b_k - a_k > b_n - a_n - 2\rho > 3\rho - 2\rho = \rho \qquad (k > n).$$

The second alternative is handled in the same manner.

Now imagine sequences $\{a'_k\}$ and $\{b'_k\}$ which are as claimed in the statement of the lemma. With the number ρ still held fixed we can find integers r and s such that

$$d(a_k, a'_k) < \frac{\rho}{3} \qquad (k > r) \tag{iv}$$

$$d(b_k, b'_k) < \frac{\rho}{3} \qquad (k > s). \tag{v}$$

Setting

$$q = \max(n, r, s)$$

it is clear that all statements involving n remain valid when n is replaced by q. Hence, when (2) is the case, then (iv) and (v) can be used to derive the relations

$$b'_k - a'_k > b_k - a_k - \frac{2\rho}{3} > \rho - \frac{2\rho}{3} = \frac{\rho}{3} \qquad (k > q),$$

whereas when (3) holds the same inequalities show that $a'_k - b'_k > \rho/3$ for all $k > q$. This completes the proof of the lemma.

Concerning the remaining postulates in 5.2 we state:

7.11 / Theorem If $\alpha < \beta$ and $\beta < \gamma$, then $\alpha < \gamma$.

Proof Suppose $\alpha = \{a_k\}$, $\beta = \{b_k\}$, and $\gamma = \{c_k\}$. Owing to Definition 7.8 there are positive numbers $\rho', \rho'' \in \mathfrak{R}^*$ and $n', n'' \in \mathfrak{N}$ such that

$$b_k - a_k > \rho' \qquad (k > n')$$

$$c_k - b_k > \rho'' \qquad (k > n'').$$

Hence, for all values of $k > \max(n', n'')$,

$$c_k - a_k > \rho' + \rho''$$

and this inequality shows that $\alpha < \gamma$.

7.12 / Theorem If $\alpha < \beta$, then $\alpha + \gamma < \beta + \gamma$.

7.13 / Theorem If $\alpha > \theta$ and $\beta > \theta$, then also $\alpha \cdot \beta > \theta$.

Proofs of these theorems should be furnished by the reader. $\mathfrak{R}$ is now known to be an ordered field.

EXERCISES

7-1 Prove Theorem 7.3.
7-2 In Theorem 7.7, verify statements (2)–(6).
7-3 Demonstrate that $\theta \cdot \alpha = \theta$ for each $\alpha \in \mathfrak{R}$.
7-4 Prove the uniqueness of the real numbers θ and i.
7-5 Show that every decimal expansion represents a number in $\mathfrak{R}$.
7-6 Prove the identity $\alpha^m \cdot \alpha^n = \alpha^{m+n}$ for arbitrary $m, n \in \mathfrak{N}$.
7-7 Prove that $\alpha = \beta$ implies $\alpha^n = \beta^n$ for each $n \in \mathfrak{N}$.
7-8 Show that $\alpha^2 > \theta$ for each $\alpha \in \mathfrak{R}$ unless $\alpha = \theta$.
7-9 Show that $\alpha > \theta$ implies that $-\alpha < \theta$.

8 / THE REAL NUMBER SYSTEM

The next phase of our program consists of imbedding $\mathfrak{R}^*$ in $\mathfrak{R}$. For this purpose we introduce:

8.1 / *Definition* The number $a \in \mathfrak{R}^*$ is a *limit* of the Cauchy sequence $\{a_k\}$ if to each rational number $\epsilon > 0$ there corresponds an integer $n_\epsilon > 0$ such that

$$d(a_k, a) < \epsilon$$

whenever $k > n_\epsilon$.

The following facts now ensue:

8.2 / Theorem Every Cauchy sequence of rational numbers has at most one limit in $\mathcal{R}^*$.

In fact, we know from Example 6.9 that not all such sequences have a limit in $\mathcal{R}^*$. The proof of the uniqueness of the limit (when it exists) is modeled after the proof of Theorem 6.2.

8.3 / Theorem Let $\{a_k\}$ and $\{b_k\}$ be given Cauchy sequences of rational numbers. If

(a) $\{a_k\} \sim \{b_k\}$,

and

(b) $\lim_{k\to\infty} d(a_k, a) = 0$ for some $a \in \mathcal{R}^*$,

then a is also the limit of $\{b_k\}$.

Proof Let $a \in \mathcal{R}^*$ be as demanded. Given a rational number $\epsilon > 0$, then owing to Definitions 7.1 and 7.2 there are integers $k_\epsilon > 0$ and $m_\epsilon > 0$ such that $d(a_k, a) < \frac{1}{2}\epsilon$ for all values $k > k_\epsilon$, and $d(a_k, b_k) < \frac{1}{2}\epsilon$ whenever $k > m_\epsilon$. Setting $n_\epsilon = \max(k_\epsilon, m_\epsilon)$ we find that for each $k > n_\epsilon$

$$d(b_k, a) \leq d(a_k, a) + d(a_k, b_k) < \frac{\epsilon}{2} + \frac{\epsilon}{2} = \epsilon,$$

and the theorem follows.

The following is a formal summary of previously obtained results:

8.4 / Theorem To each $a \in \mathcal{R}^*$ there corresponds one and only one equivalence class $\alpha \in \mathcal{R}$ such that for any representative $\{a_k\}$ thereof $\lim_{k\to\infty} d(a_k, a) = 0$.

8.5 / *Remarks* Let the equivalence class so associated with a be marked by the symbol $\varphi(a)$. Then it is clear that $\varphi: \mathcal{R}^* \to \mathcal{R}$ is a function which maps $\mathcal{R}^*$ in a one-to-one manner onto its range $\varphi(\mathcal{R}^*)$. This function has the properties

$$\varphi(a) + \varphi(b) = \varphi(a + b)$$

$$\varphi(a) \cdot \varphi(b) = \varphi(a \cdot b)$$

and is said to be an *isomorphism.* The set $\varphi(\mathcal{R}^*)$ with these operations of addition and multiplication is a field. It is an ordered field under the stipulation that $\varphi(a) < \varphi(b)$ if $a < b$. Having thus imbedded $\mathcal{R}^*$ in $\mathcal{R}$ we now redefine the property of being a rational number while saying at the same time what we understand real numbers to be.

8.6 / *Definition* *Real numbers* are equivalence classes of Cauchy sequences of rational numbers. The number $\alpha \in \mathcal{R}$ is *rational* if it has a representative with a limit in $\mathcal{R}^*$, *irrational* otherwise.

This definition should be singled out for its very special significance. In it we committed ourselves for the first time by specifying what real numbers actually are to us, and it is only now, after the set $\mathcal{R}$ has been painstakingly identified and isolated from all other sets not isomorphic with it, that a meaningful formulation of the completeness property can be given. In striving to accomplish this feat we again augment the class of Cauchy sequences, this time to include sequences of real numbers (equivalence classes) as members.

By way of motivation, consider a sequence $\{\alpha_k\}$ of real numbers whose members have representatives $\alpha_k = \{a_{km}\}$. Let us write the sequences in a matrix:

$$
\begin{array}{cccc|ccccc}
a_{11} & a_{12} & a_{13} & \cdots & a_{1r} & \cdots & a_{1m} & \cdots \\
a_{21} & a_{22} & a_{23} & \cdots & a_{2r} & \cdots & a_{2m} & \cdots \\
\vdots & & & & & & & \\
\hline
a_{p1} & a_{p2} & a_{p3} & \cdots & a_{pr} & \cdots & a_{pm} & \cdots \\
\vdots & & & & & & & \\
a_{q1} & a_{q2} & a_{q3} & \cdots & a_{qr} & \cdots & a_{qm} & \cdots \\
\vdots & & & & & & &
\end{array}
$$

A statement such as "$d(\alpha_p, \alpha_q) < \epsilon$" requires that inequalities $d(a_{pm}, a_{qm}) < \epsilon$ hold for suitable large values of m and for every choice of representatives in the equivalence classes. We therefore demand the following:

8.7 / *Definition* The sequence $\{\alpha_k\}$ $(k \in \mathfrak{N})$ of real numbers is a *Cauchy sequence* if, for each rational number $\epsilon > 0$ and choice of representatives $\alpha_k = \{a_{km}\}$ $(m \in \mathfrak{N})$, an integer $n_\epsilon > 0$ can be found such that the following is true:

For each pair of fixed values $p, q > n_\epsilon$ there is a number $r = r(p, q)$ such that

$$d(a_{pm}, a_{qm}) < \epsilon$$

for all values $m > r$.

We prove at once the following lemma:

8.8 / Lemma If Definition 8.7 is true for a particular choice of representatives $\alpha_k = \{a_{km}\}$ then it is true for every choice of representatives of the given class.

Proof For each k, suppose $\{a_{km}\} \sim \{b_{km}\}$ (all sequences are assumed, of course, to be Cauchy sequences of rational numbers); imagine the various numbers in Definition 8.7 to be already known for the representatives $\{a_{km}\}$. We wish to show that under these conditions there is also a number $s = s(p, q)$ such that

$$d(b_{pm}, b_{qm}) < \epsilon$$

for all $m > s$.

Let $r' = r'(p, q)$ be such that

$$d(a_{pm}, a_{qm}) < \frac{\epsilon}{3} \qquad (m > r').$$

The equivalence between the sequences in question tells us that for each $k \in \mathfrak{N}$ there is a positive integer $\mu(k)$ such that

$$d(a_{km}, b_{km}) < \frac{\epsilon}{3} \qquad (m > \mu(k));$$

for each pair of indices p and q we set

$$s(p, q) = \max(r'(p, q), \mu(p), \mu(q)).$$

Thus, when $p, q > r'$,

$$d(b_{pm}, b_{qm}) \leq d(b_{pm}, a_{pm}) + d(a_{pm}, a_{qm})$$
$$+ d(a_{qm}, b_{qm}) < \frac{\epsilon}{3} + \frac{\epsilon}{3} + \frac{\epsilon}{3} = \epsilon$$

for all $m > s$, and the lemma follows.

A motivation similar to that leading to Definition 8.7 bridges the gap between Definition 8.1 and the next definition:

8.9 / *Definition* Let $\{\alpha_k\}$ $(k \in \mathfrak{N})$ be a given Cauchy sequence of real numbers. The number $\alpha \in \mathfrak{R}$ is a *limit* of the sequence if the following holds:

Let $\alpha = \{a_m\}$ and $\alpha_k = \{a_{km}\}$ be arbitrary representatives of the respective classes; let a rational number $\epsilon > 0$ be given. Then an integer $n_\epsilon > 0$ can be found such that for each $k > n_\epsilon$ there is an $r = r(k)$ for which the inequality $m > r$ implies

$$d(a_{km}, a_m) < \epsilon.$$

The completeness of $\mathfrak{R}$ is formulated in the following:

8.10 / Completeness Theorem Every Cauchy sequence of real numbers has a unique limit.

Since $\lim_{k\to\infty} \alpha_k = \alpha$ implies that $\{\alpha_k\}$ is a Cauchy sequence, we are justified in using interchangeably the terms *Cauchy sequence* and *convergent sequence*. The existence and uniqueness asserted in this theorem are proved with the following two lemmas:

8.11 / Lemma Each fixed Cauchy sequence $\{\{a_{km}\}\}$ of Cauchy sequences $\{a_{km}\}(k \in \mathfrak{N})$ of rational numbers has such a sequence as a limit.

8.12 / Lemma Let $\{\{a_{km}\}\}$ and $\{\{b_{km}\}\}$ $(k \in \mathfrak{N})$ be sequences as in 8.11; let $\{a_m\}$ and $\{b_m\}$ be respective limits. Then the equivalence relations

$$\{a_{km}\} \sim \{b_{km}\} \qquad (k \in \mathfrak{N})$$

imply

$$\{a_m\} \sim \{b_m\}.$$

Proof of 8.11 Let $\sum_{i=1}^{\infty} \epsilon_i$ be a convergent series of positive rational numbers (see Section 4). The following weaker version of 8.7 is adopted in the proof: If the sequence $\{\{a_{km}\}\}$ is a Cauchy sequence, then corresponding to each ϵ_n there is an index $k_n > 0$ having the property that with each $k > k_n$ a number $m_n = m_n(k_m, k)$ can be associated such that

$$d(a_{k_n m_n}, a_{km}) < \epsilon_n \qquad (m > m_n).$$

A sequence $\{a_{k_m m_n}\}$ is constructed by induction on n; it is subsequently shown to have the desired properties.

Consider the case $n = 1$. Since $\{a_{k_1 m}\}$ is a Cauchy sequence, we may select m_1 such that concurrently with

$$d(a_{k_1 m}, a_{km}) < \epsilon_1 \qquad (m > m_1); \tag{i}$$

also the inequalities

$$d(a_{k_1 m_1}, a_{k_1 m}) < \epsilon_1 \qquad (m > m_1) \tag{ii}$$

ensue. We observe at this point that for each fixed index $k_2 > k_1$, there is an $m_2 > m_1$, $m_2 = m_2(k_1, m_1)$, such that

$$d(a_{k_2 m_2}, a_{k_1 m_1}) \leq d(a_{k_2 m_2}, a_{k_1 m_2}) + d(a_{k_1 m_2}, a_{k_1 m_1}) < \epsilon_1 + \epsilon_1 = 2\epsilon_1. \tag{iii}$$

Again appealing to the above version of 8.7 we can fix the index k_2 in such a way that a number $m_2 = m_2(k_2, k)$ can be attached to each fixed value $k > k_2$ such that

$$d(a_{k_2m_2}, a_{km}) < \epsilon_2 \qquad (m > m_2) \tag{iv}$$

and

$$d(a_{k_2m_2}, a_{k_2m}) < \epsilon_2 \qquad (m > m_2) \tag{v}$$

and at the same time we see that (iii) is not being violated by so fixing the indices. As in (iii),

$$d(a_{k_3m_3}, a_{k_2m_2}) < 2\epsilon_2$$

for every $k_3 > k_2$ with suitable numbers m_3.

The procedure we are following here is illustrated in the scheme below in which the sequences are written in a matrix:

$$\begin{array}{lllllllll}
a_{k_1m_1} & \cdots & a_{k_1m_2} & \cdots & & & & & \\
\vdots & & & & & & & & \\
a_{k_2m_1} & \cdots & a_{k_2m_2} & \cdots & a_{k_2m_3} & & & & \\
\vdots & & \vdots & & & & & & \\
 & & a_{k_3m_2} & \cdots & a_{k_3m_3} & & & & \\
 & & & & & \ddots & & & \\
 & & & & & & a_{k_pm_p} & \cdots & a_{k_pm_{p+1}} \\
\vdots & & & & & & \vdots & & \\
 & & & & & & a_{k_{p+1}m_p} & \cdots & a_{k_{p+1}m_{p+1}} \\
 & & & & & & & & \ddots
\end{array}$$

Using an inductive hypothesis, we suppose the rational numbers $a_{k_nm_n}$ are known for $n = 1, 2, \ldots, p$ $(p \geq 2)$. For each $k_{p+1} > k_p$ we can find an $m_{p+1} = m_{p+1}(k_p, k_{p+1})$ such that

$$d(a_{k_{p+1}m_{p+1}}, a_{k_pm_p}) < 2\epsilon_p \tag{vi}$$

Furthermore, the preceding argument can again be applied to deduce the existence of a number k_{p+1} and a corresponding $m_{p+1} = m_{p+1}(k_{p+1}, k)$ such that

$$d(a_{k_{p+1}m_{p+1}}, a_{km}) < \epsilon_{p+1} \qquad (m > m_{p+1})$$

and

$$d(a_{k_{p+1}m_{p+1}}, a_{k_{p+1}m}) < \epsilon_{p+1} \qquad (m > m_{p+1}),$$

without conflicting with (vi). We thus arrive at the sequence

$$\{a_{k_nm_n}\} \qquad (k \in \mathfrak{N}).$$

To demonstrate that the sequence is a Cauchy sequence we recall that the series $\sum \epsilon_i$ converges. This means that for each $\epsilon > 0$ there is an integer $n_\epsilon > 0$ such that $\epsilon_q + \epsilon_{q+1} + \cdots + \epsilon_{q+r} < \frac{1}{2}\epsilon$ for $q > n_\epsilon$ and $r > 0$. The inequalities in (vi) being valid for each $p \in \mathfrak{N}$, we can therefore deduce that

$$d(a_{k_{q+r}m_{q+r}}, a_{k_q m_q}) \leq \sum_{i=1}^{r} d(a_{k_{q+i-1}m_{q+i-1}}, a_{k_{q+i}m_{q+i}}) < \sum_{i=1}^{r} 2\epsilon_{q+i-1} < \epsilon.$$

Hence, the sequence is a Cauchy sequence.

To verify that the sequence is a limit, we simply consult Definition 8.9. Given $\epsilon > 0$, we can select an integer q so large, that $\epsilon_q + \epsilon_{q+1} + \cdots + \epsilon_{q+r} < \frac{1}{4}\epsilon$ for every $r \in \mathfrak{N}$. Referring to our inductive process, suppose $m > m_q$ and consider the corresponding fundamental sequence $\{a_{km}\}$ in the theorem. For this sequence we can find an integer $n_\epsilon > 0$ such that $d(a_{km}, a_{kt}) < \frac{1}{4}\epsilon$ for all k, $t > n_\epsilon$, and such that $d(a_{k_q m_q}, a_{ks}) < 2\epsilon_q$ when $s > n_\epsilon$. Consequently, whenever $t > \max(q, k_q)$, then

$$d(a_{k_t m_q}, a_{kt}) \leq d(a_{k_t m_q}, a_{k_q m_q}) + d(a_{k_q m_q}, a_{km})$$
$$+ d(a_{km}, a_{kt}) < \frac{\epsilon}{4} + \frac{2\epsilon}{4} + \frac{\epsilon}{4} = \epsilon$$

thereby completing the proof of the theorem.

The student should be able to construct a proof for Lemma 8.12.

EXERCISES

8-1 Referring to Definition 8.6, show that if $\alpha \neq 0$ is rational and β is irrational, then both $\alpha + \beta$ and $\alpha \cdot \beta$ are irrational.

8-2 Construct irrational numbers α and β such that $\alpha + \beta$ is rational.

8-3 Prove that there is a rational number between any two irrational numbers.

8-4 Show that the preceding exercise remains true when "rational" and "irrational" are interchanged.

8-5 Prove that $\mathcal{R}$ is *Archimedian*, which is to say that if $\alpha > \theta$ and β are given members of $\mathcal{R}$, then a positive integer n can be found such that $n\alpha > \beta$.

8-6 Prove that if $\{\alpha_k\}$ is unbounded then the sequence is *not* a Cauchy sequence.

8-7 The concept of a subsequence is very important in analysis. Consider the sequence $\{k\}$ of positive integers. Any nonterminating sequence $\{k_n\}$ such that $k_1 < k_2 < k_3 < \cdots$ is a *subsequence* of $\{k\}$; given a sequence $\{\alpha_k\}$, then $\{\alpha_{k_n}\}$ is again a subsequence. Prove that a Cauchy sequence is equivalent to each of its subsequences.

8-8 A sequence $\{\alpha_k\}$ of real numbers is *nonincreasing if* $\alpha_1 \geq \alpha_2 \geq \alpha_3 \geq \cdots$; it is *nondecreasing* if $\alpha_1 \leq \alpha_2 \leq \alpha_3 \leq \cdots$. Prove that every sequence of rational numbers contains a nonincreasing or a nondecreasing subsequence.

8-9 Show that if $\{\alpha_k\}$ is a Cauchy sequence, then so is $\{| \alpha_k |\}$, where the absolute value is defined as in Example 3.9. Prove or disprove the converse of this statement.

8-10 Let $\{\alpha_k\}$ be a given Cauchy sequence. If $\alpha \neq \theta$ is the limit of the sequence $\{| \alpha_k |\}$, show that $+\alpha$ or $-\alpha$ is the limit of $\{\alpha_k\}$.

8-11 A *null sequence* is a sequence having zero as its limit. If $\{a_k\}$ and $\{b_k\}$ are null sequences, show that also $\{a_k + b_k\}$ and $\{a_k \cdot b_k\}$ are null sequences.

9 / COMPLETENESS PROPERTIES OF $\mathcal{R}$

In the previous two sections we constructed a *sequentially complete* system $\mathcal{R}$ whose members were identified as being real numbers. Before formalizing the completeness of $\mathcal{R}$ we will do well to transcribe into the new setting the key concepts which led to Theorem 8.10.

Having fixed the metric d on $\mathcal{R}^*$ we treated $\mathcal{R}^*$ as a metric space and singled out a certain class of sequences, those called Cauchy sequences, and demonstrated that every such sequence has a limit (in the sense of Definition 8.9). Moreover, we have seen that all members of $\mathcal{R}$ can be obtained as limits of Cauchy sequences of rational numbers, and this fact permits us to extend to $\mathcal{R}$ the metric d as follows:

9.1 / *Definition* Let α and β be arbitrary members of $\mathcal{R}$ with respective representatives $\{a_k\}$ and $\{b_k\}$. Then

$$d(\alpha, \beta) = \lim_{k \to \infty} d(a_k, b_k).$$

We know, of course, that this definition is independent of the choice of representatives of α and β. Definitions 8.7 and 8.9 assume the following form here:

9.2 / *Definition* The sequence $\{\alpha_k\}$ of real numbers is a *Cauchy sequence* if to each $\epsilon > 0$ there corresponds a number $n_\epsilon \in \mathfrak{N}$ such that

$$d(\alpha_k, \alpha_m) < \epsilon$$

for all integers $k, m > n_\epsilon$.

Synonymous with *Cauchy sequence* is the term *convergent sequence.* A nonconvergent sequence is said to be *divergent.*

9.3 / ***Definition*** The number $\alpha \in \mathcal{R}$ is the *limit* of the sequence $\{\alpha_k\}$ $(\alpha_k \in \mathcal{R})$ if for each $\epsilon > 0$ an integer n_ϵ can be found such that

$$d(\alpha_k, \alpha) < \epsilon$$

whenever $k > n_\epsilon$.

Equivalent statements are

$$\lim_{k \to \infty} d(\alpha_k, \alpha) = 0$$

and

$$\lim_{k \to \infty} \alpha_k = \alpha;$$

the notation

$$\alpha_k \longrightarrow \alpha$$

is also used.

Having never formally defined completeness, it is appropriate to do so now:

9.4 / ***Definition*** The system of numbers $\mathcal{R}$ is *complete* if every Cauchy sequence in $\mathcal{R}$ has a limit in $\mathcal{R}$.

Finally, we note that the algebraic and order qualities of $\mathcal{R}^*$ induced the similar relations in $\mathcal{R}$. All this leads us to the following theorem:

9.5 / **Theorem** The system $\mathcal{R}$ is a complete ordered field.

According to Exercise 8-5, $\mathcal{R}$ is Archimedian. Theorem 9.5 tells us that all the gaps previously existing in the rational numbers are now filled. This statement is equivalent to the Dedekind completeness which was mentioned in Section 5, the *least upper bound* property of the real numbers, as well as their *nested interval* property. While our main concern is to explore the latter two versions of completeness, we shall elaborate briefly on Dedekind's idea.

9.6 / ***Definition*** If $\mathcal{A}$ and $\mathcal{B}$ are nonempty subsets of $\mathcal{R}$, then the ordered pair $(\mathcal{A}, \mathcal{B})$ is a *cut* of $\mathcal{R}$ if

(a) $\mathcal{R} = \mathcal{A} \cup \mathcal{B}$
(b) $a \in \mathcal{A}$ and $b \in \mathcal{B}$ implies $a < b$.

9.7 / Theorem The system $\mathcal{R}$ satisfies the *Dedekind cut* condition: If $(\mathcal{A}, \mathcal{B})$ is a cut of $\mathcal{R}$, then either $\mathcal{A}$ has a largest member or $\mathcal{B}$ has a smallest member.

Strictly speaking, this theorem is unacceptable in its present setting since the concepts "largest member" and "smallest member" were not defined. Appealing to the ingenuity of the reader we leave it to him to come up with a valid definition of the troublesome terms. A fundamental difference between the Cauchy and the Dedekind completeness is worth mentioning. Namely, while the Cauchy completeness is built around the metric of $\mathcal{R}$, Dedekind in his version exploited the ordering of $\mathcal{R}$.

Three further definitions now stand between us and the main result of this section:

9.8 / *Definitions* Let A be a given set of real numbers.

(1) If there is a number τ such that $\alpha \leq \tau$ for all $\alpha \in A$, then A is said to be *bounded above.* τ is an *upper bound* of A.

(2) If there is a number σ such that $\alpha \geq \sigma$ for all $\alpha \in A$, then A is *bounded below* and σ is a *lower bound* of A.

(3) The set A is *bounded* when it is bounded above and below.

Evidently a set has infinitely many upper bounds when bounded above, infinitely many lower bounds when bounded below.

9.9 / *Definitions*

(1) Let A be bounded above. Then τ is a *least upper bound* of A if

(a) τ is an upper bound of A, and
(b) no $\xi < \tau$ is an upper bound of A.

(2) Let A be bounded below. Then σ is a *greatest lower bound* of A if

(c) σ is a lower bound of A, and
(d) no $\xi > \sigma$ is a lower bound of A.

The abbreviations "lub" and "glb" are used to designate least upper bounds and greatest lower bounds, respectively. Common in the literature are also the terms "supremum" in lieu of least upper bound and "infimum" in place of greatest lower bound. The designations "sup" and "inf" are self-explanatory. We state without proof:

9.10 / Lemma A set of real numbers has at most one lub and one glb.

9.11 / *Examples*

(1) The set $\mathfrak{N}$ has no upper bound, but every real number $\xi \leq 1$ is a lower bound, 1 being, in fact, the glb of $\mathfrak{N}$.

(2) Consider the set $A = \left\{\frac{1}{n} \,\middle|\, n \in \mathfrak{N}\right\}$. This set is bounded, has lub 1 and glb 0. It should be noted that a glb or lub may or may not belong to the set in question.

A very important family of sets consists of the so-called intervals. Associated with every pair of real numbers, $\alpha < \beta$, are the sets

$$[\alpha, \beta] = \{\xi \in \mathfrak{R} \mid \alpha \leq \xi \leq \beta\}$$

and

$$(\alpha, \beta) = \{\xi \in \mathfrak{R} \mid \alpha < \xi < \beta\},$$

called, respectively, *closed interval* and *open interval*. The sets

$$(\alpha, \beta] = \{\xi \in \mathfrak{R} \mid \alpha < \xi \leq \beta\}$$

and

$$[\alpha, \beta) = \{\xi \in \mathfrak{R} \mid \alpha \leq \xi < \beta\}$$

are intervals which are neither open nor closed.

9.12 / *Definition* If I stands for any of the intervals displayed above, then the unique number

$$m(I) = d(\alpha, \beta)$$

associated with it is called the *measure* or *length* of I.

It should be observed that the symbol $[\alpha, \alpha]$, where $\alpha \in \mathfrak{R}$ is arbitrary, is admitted into the family of intervals: consisting of the single element α, it is called a *singleton*. The sequence $\{I_k\}$ of intervals is *nested* when $I_1 \supset I_2 \supset I_3 \supset \cdots$.

9.13 / Theorem The following statements are true in $\mathfrak{R}$:

(1) Every nonempty set which is bounded above has a least upper bound.

(2) Every nested sequence of closed intervals has a nonempty intersection.

Proof To prove the first statement, let A be a given nonempty set of real numbers; let γ be a fixed upper bound thereof and let β stand for a real number which is not an upper bound of A. Consider the numbers

$$\xi = \beta + (\gamma - \beta)t_{km},$$

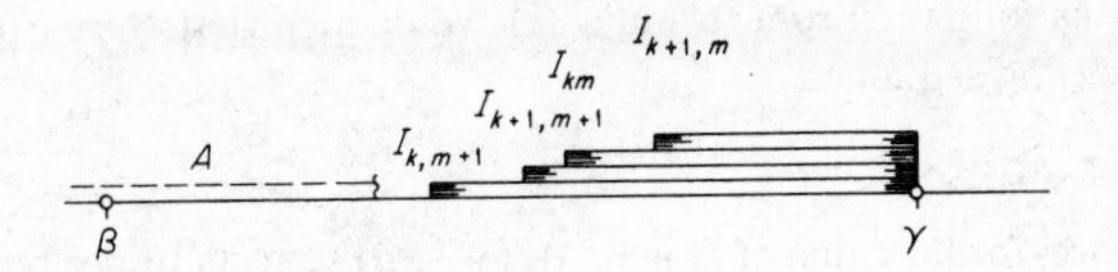

Figure 11

where

$$t_{km} = k \cdot 2^{-m} \qquad (k = 1, 2, \ldots, 2^m; m \in \mathfrak{N}).$$

Holding m momentarily fixed, consider the intervals

$$I_{km} = (\beta + (\gamma - \beta)t_{km}, \gamma]$$

(see Figure 11). Let k_m be the smallest admitted value of k for which

$$A \cap I_{km} = \emptyset.$$

Setting

$$\tau_m = \beta + (\gamma - \beta)t_{k_m m} \tag{i}$$

it is clear that this τ_m is an upper bound of A. When carried out for each value of m this process results in a sequence $\{\tau_m\}$ of upper bounds of A. This sequence is asserted to be a Cauchy sequence. To see this, we note that the τ_m are such that

$$d(\tau_m, \tau_{m+1}) \leq (\gamma - \beta) \cdot 2^{-m-1}$$

for all $m \in \mathfrak{N}$. Thus, when r and s are fixed positive integers, we can conclude that

$$d(\tau_r, \tau_{r+s}) \leq \sum_{i=1}^{s} d(\tau_{r+i-1}, \tau_{r+i}) \leq \sum_{i=1}^{s} (\gamma - \beta) \cdot 2^{-r-i}$$
$$= (\gamma - \beta)(1 - 2^{-s})2^{-r}.$$

Let a real number $\epsilon > 0$ be given. Since $1 - 2^{-s} < 1$, we can select r so large that, irrespective of the value of s,

$$d(\tau_m, \tau_{m+s}) < \epsilon \qquad (m > r),$$

thereby proving that the sequence is as claimed. Let τ be its limit.

To complete the demonstration of (1) it remains to show that τ is the desired least upper bound. That it is an upper bound is quite clear, for otherwise there would be an $\alpha \in A$ such that $\tau < \alpha$. In view of the fact that $\tau_m \geq \alpha$ for each $m \in \mathfrak{N}$, this would imply that $\lim_{m \to \infty} \tau_m \geq \alpha > \tau$, thereby contradicting the fact that τ is the limit of the sequence $\{\tau_m\}$. That τ is, indeed, a least upper bound can now be argued as follows: From formula (i) we deduce that the τ_m form a nonincreasing sequence,

$$\tau_1 \geq \tau_2 \geq \tau_3 \geq \cdots \geq \tau.$$

Furthermore, for no value of m is $\tau_m - (\gamma - \beta) \cdot 2^{-m}$ an upper bound of A. Consequently neither is any of the numbers $\tau - (\gamma - \beta) \cdot 2^{-m}$ an upper bound of A. Since m is arbitrary, we can conclude that τ is as asserted.

We shall use the fact just established to prove the statement in (2). Let $\{I_k\}$ $(k \in \mathfrak{N})$ be a given nested sequence of closed intervals. The statement in (2) is trivially true when there is a number $\mu > 0$ for which $m(I_k) > \mu$ for all $k \in \mathfrak{N}$; it is also true *a priori* when one of the intervals is a singleton. We therefore prove the following:

9.14 / Lemma If $\{I_k\}$ is a given nested sequence of closed intervals and if

$$\lim_{k \to \infty} m(I_k) = 0,$$

then $\bigcap_{k \in \mathfrak{N}} I_k$ consists of a single real number.

Proof For each k, let $I_k = [\alpha_k, \beta_k]$. By supposition the sequence $\{\alpha_k\}$ is nondecreasing whereas the sequence $\{\beta_k\}$ is nonincreasing. Deducing that the range of the sequence $\{\alpha_k\}$ has a least upper bound α we claim that this number is the common intersection of the I_k. Suppose, on the contrary, that $\alpha \notin I_n$ for some n. These circumstances necessitate the relation $\alpha > \beta_j$ for some j since $\alpha_1 \leq \alpha_2 \leq \cdots \leq \alpha$ (see Figure 12). Each β_j, however, is an upper bound of the range of $\{\alpha_k\}$, and this implies that α is not the claimed least upper bound. Hence $\alpha \in \bigcap_{k \in \mathfrak{N}} I_k$.

To prove the uniqueness assertion made in the lemma, suppose that also β belongs to the intersection in question. Then for each value of k,

$$0 \leq d(\alpha, \beta) \leq d(\alpha_k, \beta_k) = m(I_k).$$

Since $m(I_k) \to 0$ as $k \to \infty$, it follows that $d(\alpha, \beta) = 0$, i.e., $\alpha = \beta$.

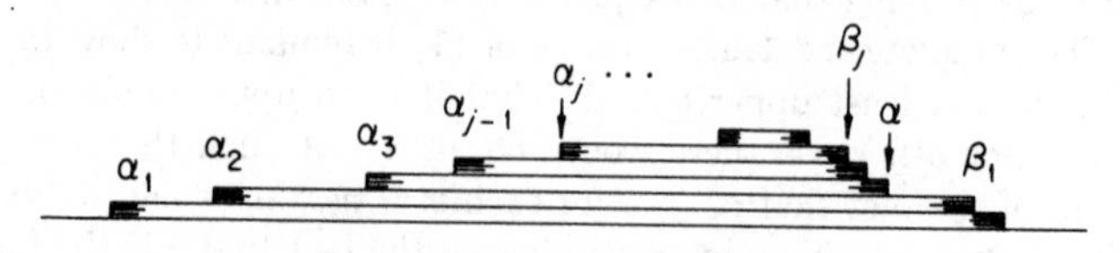

Figure 12

EXERCISES

9-1 Prove Lemma 9.10.

9-2 If they exist, find the lub and glb of the following sets:

(a) $A = \{\xi \mid \ |\xi| > 1\}$

(b) $B = \{\eta \mid \eta^2 + 2^2 \leq 3^2\}$

(c) $C = \left\{\dfrac{\nu}{\nu - 1} \;\middle|\; 0 < \nu < 1\right\}$

(d) $D = A \cap B$

(e) $E = A \setminus B$

(f) $F = B \setminus A$

(g) $G = C \setminus A$.

9-3 Show that if the nonempty set A is bounded below, then it has a greatest lower bound. (*Hint*: The set $A' = \{\alpha \mid -\alpha \in A\}$ is bounded above.)

9-4 Show that τ is a lub of A if and only if the following is true:

(a) $\tau \geq \alpha$ for each $\alpha \in A$,

(b) for each $\epsilon > 0$ the set A contains a member α such that $\alpha > \tau - \epsilon$.

Formulate and verify the corresponding characterization of a glb.

9-5 Consider a set $X \subset \mathcal{R} \times \{1\}$. How would you generalize Definitions 9.8 and 9.9 to apply to X? How would you proceed for sets $Y \subset \mathcal{R} \times \mathcal{R}$?

9-6 Prove that each of the intervals (α, α), $(\alpha, \alpha]$, and $[\alpha, \alpha)$ is empty.

9-7 Prove or disprove: Lemma 9.14 remains correct when the closed intervals are replaced with nonempty open intervals.

9-8 What can be said about the intersection of finitely many nested intervals which are neither open nor closed?

9-9 Consider closed intervals $I_1, I_2, \ldots, I_n$. Show that the set

$$I = \bigcap_{k=1}^{n} I_k$$

is either empty, a singleton, or a closed interval.

9-10 Let the intervals I_k in the above exercise be open. Show that I is either empty or an open interval.

9-11 A subset of $\mathcal{R}$ is *elementary* when it is the finite union of pairwise disjoint intervals. Show that

$$A \cup B, \qquad A \cap B, \qquad A \setminus B, \qquad A \triangle B$$

are elementary sets when A and B are.

9-12 Show that every bounded infinite sequence of real numbers contains a convergent subsequence.

9-13 The nested interval property of the real numbers offers a nice proof of Theorem 4.8. Namely, suppose the real numbers are countable. Let them be arranged in a sequence, say, $a_1, a_2, a_3, \ldots$. Then closed intervals I_k can be constructed iteratively as follows: there is a closed interval I_1 such that $m(I_1) < 1$ and $a_1 \notin I_1$. For each $k > 1$ there is an interval I_k such that $I_k \subset I_{k-1}$, $m(I_k) < 1/k$ and $a_k \notin I_k$. Complete the proof with the help of Lemma 9.14.

9-14 Show that for each Cauchy sequence of rational numbers there are divergent sequences of rational numbers with the given sequence as a subsequence.

9-15 What is the cardinality of the set of all divergent sequences of rational numbers?

10 / THE EXTENDED REAL LINE

On occasion it is convenient to extend $\mathcal{R}$ by adjoining to it two "ideal" elements, designated as ∞ and $-\infty$, respectively. One reason for doing so is to guarantee that *every* subset of the extended system has a least upper bound and a greatest lower bound, a property which is reserved in $\mathcal{R}$ for the bounded sets only. To have a meaningful extension of $\mathcal{R}$, the following properties are being attributed to the new symbols:

10.1 / *Definitions* For each $\xi \in \mathcal{R}$,

(1) $\xi + \infty = \infty$.

(2) $\xi - \infty = -\infty$.

(3) $\dfrac{\xi}{\infty} = 0$ and $\dfrac{\xi}{-\infty} = 0$.

(4) $\xi \cdot \infty = \infty$ and $\xi \cdot (-\infty) = -\infty$ when $\xi > 0$.

(5) $\xi \cdot \infty = -\infty$ and $\xi \cdot (-\infty) = \infty$ when $\xi < 0$.

(6) $-\infty < \xi < \infty$.

10.2 / *Definition* Let the set A belong to the extended real number system. Then

(1) The lub ∞ is assigned to A when A is not bounded above;

(2) The glb $-\infty$ is assigned to A when the set is not bounded below.

The extended real number system is designated with the symbol $\mathcal{R}^E$.

While alleviating one problem, the new symbols create other enigmas since, for example, the extended system is no longer a field: no meaning is given in the system to the sum $\infty + (-\infty)$. It would be quite instructive for the student to figure out why.

FOUR

SEQUENCES AND SERIES OF NUMBERS

In the system $\mathcal{R}$ every Cauchy sequence has a limit. By now we have amassed a substantial fund of knowledge concerning the convergence of sequences, notably being the proposition that a sequence of real numbers converges if and only if it is a Cauchy sequence (Theorem 8.10). To decide whether or not a given sequence converges, however, is quite difficult in the general case and a variety of convergence criteria is indispensable. Once a sequence is known to possess a limit, there remains the problem of its evaluation. This is in general a very difficult problem because irrational numbers were noted to be numerically nonconstructible in the sense that, whatever the base, no iteration exists which will produce their expansion (see Section 6): The actual limit of a sequence can, therefore, be at best approximated in most cases.

A very important class of sequences is that associated with infinite series. The student already encountered such in connection with decimal expansions and in calculus. For motivation the student is invited to examine the following sequences:

Examples

(0.1) $\{n\} = (1, 2, 3, 4, 5, \ldots);$

(0.2) $\{n^{(-1)^n}\} = \left(1, 2, \frac{1}{3}, 4, \frac{1}{5}, 6, \frac{1}{7}, \ldots\right);$

(0.3) $\{(-1)^n\} = (-1, 1, -1, 1, \ldots);$

(0.4) $\{\sum_{k=1}^{n} (-1)^k \cdot k^{-p}\}$

$$= \left(-1, -1 + \frac{1}{2^p}, -1 + \frac{1}{2^p} - \frac{1}{3^p}, \ldots\right), \quad p \text{ fixed};$$

(0.5) $$\left\{\left(1 + \frac{1}{n}\right)^n\right\} = \left(2, 2 + \frac{1}{4}, 2 + \frac{10}{27}, 2 + \frac{113}{256}, \ldots\right);$$

(0.6) $$\left\{\sum_{\nu=0}^{n} \frac{1}{\nu!}\right\} = \left(1, 2, 2 + \frac{1}{1\cdot 2}, 2 + \frac{1}{1\cdot 2} + \frac{1}{1\cdot 2\cdot 3}, \ldots\right),$$

where $0! = 1$ and $n! = 1\cdot 2\cdot 3\cdot \cdots \cdot n$ for $n \geq 1$: What properties set each sequence apart from the others?

From Chapter Three we require here the nomenclature and results of the last two sections, but we will no longer adhere to the convention of designating real numbers with Greek letters.

11 / SEQUENCES: BASIC LIMIT THEOREMS

A crude classification of sequences is obtained when we attribute to a sequence, as to any other function, the boundedness properties of its range.

11.1 / ***Definition*** A sequence is said to be *bounded* when its range is bounded, *unbounded* otherwise.

We also speak of sequences which are bounded above or below. Of those depicted above, the first two sequences are clearly unbounded. (What about the remaining ones?). We observe that every subsequence in (0.1) fails to be bounded whereas the sequence in (0.2) does contain bounded subsequences. Knowing that all unbounded sequences diverge, we can next partition all bounded sequences into convergent and divergent ones. Here one is invariably concerned with phrases such as "arbitrarily close" and "sufficiently near" which are expressed in terms of certain inequalities. Such phrases are conveniently stated in a geometric language which, in addition to its intuitive appeal, will subsequently be seen to be easily adaptable to more abstract situations. The fundamental elements of this language are the open intervals:

11.2 / *Definition* The *neighborhood* $N(a; r)$ $(r > 0)$ of a number $a \in \mathcal{R}$ is the open interval

$$N(a; r) = \{x \in \mathcal{R} \mid d(a, x) < r\}.$$

The number r is called the *radius* of the neighborhood.

We note, in particular, that the empty set is excluded from membership in the family of neighborhoods, and prove the following theorem:

11.3 / Theorem The sequence $\{a_k\}$ converges to the limit a if and only if every neighborhood of a contains all but finitely many of the a_k.

Proof Suppose $\lim_{k\to\infty} a_k = a$. Then for every $\epsilon > 0$ there is an integer $n_\epsilon > 0$ such that $d(a_k, a) < \epsilon$ for all values $k > n_\epsilon$. On the other hand, when $N(a; r)$ is any neighborhood of a, then $x \in N(a; r)$ whenever x satisfies the relation $d(x, a) < r$. Selecting $\epsilon = r$, we guarantee that $a_k \in N(a; r)$ for all values $k > n_r$, as was to be shown.

Conversely, if every neighborhood of a contains all but finitely many of the a_k, then for each $\epsilon > 0$ there is an integer $n_\epsilon > 0$ such that $a_k \in N(a; \epsilon)$ for all values $k > n_\epsilon$, implying that $d(a_k, a) < \epsilon$ for these values of k.

The sequence $\{k\}$ is said to diverge to infinity, a statement expressed symbolically as

$$\lim_{k\to\infty} k = \infty.$$

In the extended real number system $\mathcal{R}^E$ (see Section 10) such a sequence would be classified as convergent, the justification being this: Defining for each $r \in \mathcal{R}$ the set

$$N(r; \infty) = \{x \in \mathcal{R} \mid x > r\}$$

to be a *neighborhood of* ∞, we find that every such neighborhood contains all but finitely many of the members of the sequence in question. We introduce:

11.4 / *Definition* If the sequence $\{a_k\}$ is endowed with the property that for each integer n there is an integer m such that $k > m$ implies that $a_k > n$, then we write

$$\lim_{k\to\infty} a_k = \infty.$$

When an integer m is associated with each n such that $a_k < -n$ for all values $k > m$ we write

$$\lim_{k\to\infty} a_k = -\infty.$$

Regarded as sequences in $\mathcal{R}^E$, they are said to converge to ∞ and $-\infty$, respectively.

The sequence in (0.2) diverges also in $\mathcal{R}^E$ since every neighborhood of ∞ corresponding to $r > 1$ excludes infinitely many members of this sequence: it has, in fact, a subsequence which converges to zero. This leads us to the general question "What, if any, is the connection between the behavior of a sequence and that of its subsequences?" In (0.3) we have an oscillating sequence which contains two convergent subsequences: The sequence itself diverges since the distance between infinitely many of its members exceeds a fixed constant. However, not every subsequence of $\{(-1)^n\}$ converges. The following is, in fact, a rephrasing of Exercise 8-7:

11.5 / Theorem A sequence of real numbers converges when and only when each of its subsequences converges. Moreover, all subsequences of a convergent sequence have the same limit.

The basic arithmetic of sequences should be known from calculus. It is nevertheless treated very extensively in the exercises.

A useful class of bounded convergent sequences is that of monotone sequences: Referring to Exercise 8-8 we call a sequence *monotone* when it is nonincreasing or nondecreasing. Clearly, a nonincreasing sequence is bounded above whereas a nondecreasing sequence is bounded below. The following is an important application of Theorem 9.13:

11.6 / Theorem Every bounded monotone sequence converges.

Proof Imagine, for the sake of argument, the sequence $\{a_k\}$ to be nondecreasing: $a_1 \leq a_2 \leq a_3 \leq \cdots$. Being bounded, its range A has a least upper bound, say, a. We assert that $\lim_{k\to\infty} a_k = a$. To verify this we note that associated with each $\epsilon > 0$ is a member a_n such that

$$a - \epsilon < a_n \leq a < a + \epsilon,$$

i.e.,

$$d(a_n, a) < \epsilon$$

(If this were false for some $\epsilon = \epsilon'$, then $a - \epsilon'$ would be an upper bound). Because of the monotone character of the sequence it follows that

$$d(a_k, a) < \epsilon$$

for all values $k > n$, and the case under discussion is hereby proved. A similar argument establishes the case of a nonincreasing sequence.

We mention in passing that monotone sequences always converge in the extended real number system.

Null sequences (see Exercise 8-11) constitute another prominent class of convergent sequences. We begin with the very basic lemma:

11.7 / Lemma Imagine sequences $\{a_k\}$ and $\{b_k\}$ and an integer n such that $0 \leq a_k \leq b_k$ for all values $k > n$. If $\{b_k\}$ is a null sequence, then so is $\{a_k\}$.

Proof Every neighborhood $N(0; r)$ of zero contains all but finitely many of the b_k: The relation $0 \leq a_k \leq b_k$ $(k > n)$ guarantees that $a_k \in N(0; r)$ whenever $b_k \in N(0; r)$ and $k > n$. Hence, the statement concerning the b_k's applies to the a_k's and the lemma follows from Theorem 11.3.

11.8 / Lemma If $\{a_k\}$ is a null sequence, then so is the sequence $\{a_k{}^r\}$ for each fixed positive r.

Proof The hypothesis of the lemma permits us to state that, for each $\epsilon > 0$, $d(a_k, 0) < \epsilon^{1/r}$ except possibly for finitely many of the a_k, so that $d(a_k{}^r, 0) < \epsilon$ holds for the same values of k.

Considered next are some often occurring null sequences:

11.9 / Theorems

(1) For each $r > 0$, $\{k^{-r}\}$ is a null sequence.

(2) If $0 < a < 1$ and $r > 0$ is arbitrary, then $\{k^r \cdot a^k\}$ is a null sequence.

(3) The sequence $\{\sqrt[k]{k} - 1\}$ is a null sequence.

(4) For arbitrary $r > 0$ and $s > 0$, $\{k^{-r} \ln^s k\}$ is a null sequence.

Proofs

(1) Selecting for each $\epsilon > 0$ an integer $n_\epsilon > \epsilon^{-1/r}$, we find that the inequality $d(k^{-r}, 0) < \epsilon$ holds for all values $k > n_\epsilon$.

(2) We verify this assertion in two steps. First, suppose $r = 1$. Fix a number b such that $0 < b < \dfrac{1}{a} - 1$: then $a < \dfrac{1}{1+b}$. Let k be fixed. Appealing to the binomial theorem, we have the expansion

$$(1 + b)^k = 1 + kb + \frac{k(k-1)}{2} b^2 + \cdots + b^k$$

where all terms are positive. Rejecting all but the third term on the right side, we conclude that

$$(1 + b)^k > \frac{k(k-1)}{2} b^2,$$

leading to the estimate

$$ka^k < \frac{k}{(1+b)^k} < \frac{2}{b^2(k-1)} \qquad (k \geq 2).$$

If for each $\epsilon > 0$ we now demand that $\frac{2}{(k-1)b^2} < \epsilon$, then the inequality $k > \frac{2}{\epsilon b^2} + 1$ results and for these values of k clearly $0 < ka^k < \epsilon$, thereby establishing the case $r = 1$.

Next, we observe that $0 < a^{1/r} < 1$, and therefore the case just proved tells us that $\{ka^{k/r}\}$ is a null sequence. Finally, realizing that for each k, $k^r a^k = (ka^{k/r})^r$, we conclude from Lemma 11.8 that (2) is correct.

(**3**) For each value of k we set $a_k = \sqrt[k]{k} - 1$: clearly $a_k \geq 0$. Again appealing to the binomial theorem, we find that for each $k \geq 2$

$$k = (1 + a_k)^k > \frac{k(k-1)}{2} a_k^2,$$

and a_k is seen to be subject to the inequalities

$$0 < a_k < \frac{\sqrt{2}}{\sqrt{k-1}}.$$

The proof now follows from case (1) above.

(**4**) Let $b > 1$ be a given integer. Corresponding to each k there is an integer m such that

$$b^m \leq k < b^{m+1},$$

implying that

$$m \leq \frac{\ln k}{\ln b} < m + 1.$$

Applied to the problem at hand, these inequalities show that

$$k^{-r} \ln^s k \leq b^{-mr}(m+1)^s \ln^s b = b^r(b^{-r})^{m+1}(m+1)^s \ln^s b.$$

Since $0 < b^{-r} < 1$, we conclude from (2) above that the original sequence is as asserted.

EXERCISES

11-1 Prove: If $\lim_{k\to\infty} a_k = a$ and $\lim_{k\to\infty} b_k = b$, then

(**1**) $\lim_{k\to\infty} (c + a_k) = c + a$ for each $c \in \mathcal{R}$

(2) $\lim_{k\to\infty} c \cdot a_k = c \cdot a$ for each $c \in \mathcal{R}$

(3) $\lim_{k\to\infty} (a_k + b_k) = a + b$

(4) $\lim_{k\to\infty} (a_k \cdot b_k) = a \cdot b$

(5) $\lim_{k\to\infty} \dfrac{a_k}{b_k} = \dfrac{a}{b}$, provided that $b \neq 0$ and $b_k \neq 0$ for all $k \in \mathfrak{N}$.

11-2 If the sequence $\{a_k\}$ converges whereas $\{b_k\}$ diverges, then the sequence $\{a_k + b_k\}$ diverges. On the other hand, no conclusions regarding the behavior of the sequences $\{a_k\}$ and $\{b_k\}$ can be derived from the convergence of either $\{a_k + b_k\}$ or $\{a_k \cdot b_k\}$. Give at least two examples of divergent sequences $\{a_k\}$ and $\{b_k\}$ such that $\{a_k + b_k\}$ converges.

11-3 Prove: If $\lim_{k\to\infty} (a_k + b_k) = c$ and $\lim_{k\to\infty} (a_k - b_k) = d$, then $\lim_{k\to\infty} a_k \cdot b_k = \frac{1}{4}(c^2 - d^2)$.

11-4 Show that the convergence of the sequences $\{a_k\}$ and $\{a_k + b_k\}$ implies the convergence of $\{b_k\}$.

11-5 Suppose the sequences $\{a_k + b_k\}$ and $\{a_k \cdot b_k\}$ converge. Can you conclude that the sequences $\{a_k\}$ and $\{b_k\}$ must also converge?

11-6 Suppose $a_k \to a$ as $k \to \infty$. Prove:

(1) If there is an integer n such that $a_k \geq 0$ for all values $k \geq n$, then $a \geq 0$.

(2) If $a > 0$, then for some integer n, $a_k > 0$ whenever $k > n$.

(3) If the sequence contains infinitely many positive and infinitely many negative members, then necessarily $a = 0$.

11-7 Generalizing Theorem 11.9(2), show that if $|a| < 1$ and $r > 0$ is arbitrary, then $\{k^r \cdot a^k\}$ is a null sequence.

11-8 Demonstrate that $\lim_{k\to\infty} \sqrt[k]{a} = 1$ when $a > 0$.

11-9 Let $\{a_k\}$ and $\{b_k\}$ be convergent sequences. Show that:

(1) $\lim_{k\to\infty} |a_k + b_k| \leq \lim_{k\to\infty} |a_k| + \lim_{k\to\infty} |b_k|$.

(2) $\lim_{k\to\infty} |a_k \cdot b_k| = \lim_{k\to\infty} |a_k| \cdot \lim_{k\to\infty} |b_k|$.

(3) If $P(k) = a_0 + a_1k + \cdots + a_mk^m$, $Q(k) = b_0 + b_1k + \cdots + b_nk^n$ $(b_n \neq 0)$, then

$$\lim_{k\to\infty} \frac{P(k)}{Q(k)} = \begin{cases} 0 & \text{when } m < n \\ \dfrac{a_n}{b_n} & \text{when } m = n \\ \pm\infty & \text{when } m > n. \end{cases}$$

11-10 Verify that the following are null sequences:

(a) $\left\{\frac{a^k}{k!}\right\}$;

(b) $\left\{\frac{k!}{k^k}\right\}$;

(c) $\{\sqrt[k]{k+1} - 1\}$.

11-11 Prove that for any null sequence $\{a_k\}$ and number $b > 0$ $\{b^{a_k} - 1\}$ is likewise a null sequence.

11-12 Show that $\{a_k\}$ is a null sequence if and only if $\{(-1)^k a_k\}$ is a null sequence.

11-13 Prove: If $\{a_k\}$ is a null sequence and $\{b_k\}$ is bounded, then $\{a_k \cdot b_k\}$ is a null sequence.

11-14 Verify that $\frac{1}{k}(1 + \sqrt{2} + \sqrt[3]{3} + \cdots + \sqrt[k]{k}) \to 1$ as $k \to \infty$.

11-15 Assuming the sequences to converge, verify the following estimates:

(a) $\lim_{k\to\infty}\left(1 + \frac{1}{k}\right)^k < 3$;

(b) $\lim_{k\to\infty}\frac{k}{\sqrt[k]{k!}} < 3$;

(c) $\lim_{m\to\infty}\sum_{k=0}^{m}\frac{k}{k!} < 3$.

Hint: One way to go about (b) is the following:

$$a_{k+1} = \left[\frac{(k+1)^{k+1}}{(k+1)!}\right]^{1/k+1} = \left[\frac{k^k}{k!}\frac{(k+1)^k}{k^k}\right]^{1/k+1}$$

$$= \left[a_k{}^k\left(1 + \frac{1}{k}\right)^k\right]^{1/k+1} < (3a_k{}^k)^{1/k+1}$$

according to the estimate in (a). Alternatively, you could make use of *Stirling's formula*, which asserts that

$$\sqrt{2\pi}n^{n+1/2}e^{-n} < n! < \sqrt{2\pi}n^{n+1/2}e^{-n}\left(1 + \frac{1}{4n}\right).$$

11-16 Show that $\lim_{k\to\infty}\left(1 + \frac{1}{k}\right)^k \le \lim_{k\to\infty}\sum_{m=0}^{k}\frac{1}{m!}$.

11-17 Show that the sequences $\left\{\left(1 + \frac{1}{k}\right)^k\right\}$ and $\left\{\left(1 + \frac{1}{k+1}\right)^k\right\}$ are equivalent. If m is an arbitrary positive integer, are these equivalent to the sequence $\left\{\left(1 + \frac{1}{m+k}\right)^k\right\}$?

11-18 Are the sequences $\left\{\left(1 + \frac{2}{k}\right)^k\right\}$ and $\left\{\left(1 + \frac{1}{k}\right)^{2k}\right\}$ equivalent?

11-19 Evaluate the following limits:

(1) $\lim\limits_{k\to\infty} \frac{\sin k}{k}$.

(2) $\lim\limits_{k\to\infty} (\sqrt{k^2 + k} - k)$.

(3) $\lim\limits_{k\to\infty} \sum\limits_{m=1}^{k} \frac{1}{m}$.

(4) $\lim\limits_{k\to\infty} \left(\frac{k}{k+1}\right)^k$.

Which of the above sequences is monotone?

11-20 Imagine equivalent sequences $\{a_k\}$ and $\{b_k\}$ with limit, say, x. Suppose the sequence $\{c_k\}$ is such that, for some integer n, either $a_k \le c_k \le b_k$ or $b_k \le c_k \le a_k$ whenever $k > n$. Show that $c_k \to x$ as $k \to \infty$.

11-21 Generalizing Exercise 11-11, suppose $\lim\limits_{k\to\infty} a_k = a$. Show that for each $b > 0$, $\lim\limits_{k\to\infty} b^{a_k} = b^a$.

11-22 Let $\{a_k\}$ and $\{b_k\}$ be convergent sequences having the same range. Prove or disprove the claim $\lim\limits_{k\to\infty} a_k = \lim\limits_{k\to\infty} b_k$.

11-23 For each n, let

$$\gamma_n = \sum_{k=1}^{n} \frac{1}{k} - \ln n.$$

Show that the sequence $\{\gamma_n\}$ is decreasing and bounded. Hence, show that its limit γ lies in the interval $(\frac{1}{2}, 1)$. The number γ is known as *Euler's constant.* It is not known if γ is rational.

12 / UPPER AND LOWER LIMITS

Associated with every bounded sequence is the set of limits of its convergent subsequences: according to Exercise 9-12 this set is never empty. We know, in fact, that *every* sequence has a subsequential limit in the extended system $\mathfrak{R}^E$.

12.1 / ***Definition*** The number a is a *subsequential limit* of the sequence $\{a_k\}$ if there is a subsequence $\{a_{k_\nu}\}$ thereof such that

$$\lim_{\nu \to \infty} a_{k_\nu} = a.$$

We find it convenient at this time to dispense with boundedness and we consequently formulate the results of this section in $\mathcal{R}^E$ (see Section 10).

The set of subsequential limits of a convergent sequence is of no interest here, for Theorem 11.5 tells us that such a set is a singleton. On the other hand, it is natural to inquire if general sets of subsequential limits have distinguishing features (such as cardinality, etc.). For instance, the subsequential limits in (0.2) are 0 and ∞, those in (0.3) are 1 and -1. The following additional examples will supply a partial answer to this inquiry.

12.2 / ***Examples***

(**1**) Consider the sequence

$$\{a_n\} = (\tfrac{1}{2}, \tfrac{1}{3}, \tfrac{2}{3}, \tfrac{1}{4}, \tfrac{3}{4}, \tfrac{1}{5}, \tfrac{2}{5}, \ldots)$$

in which each rational number in the interval $(0, 1)$ occurs exactly once. We assert that the subsequential limit set of this sequence is the interval $[0, 1]$.

Let $x \in [0, 1]$ be arbitrary. This number is the limit of a sequence $\{b_n\}$ of rational numbers of $(0, 1)$ which are all different. Owing to Theorem 11.5 this sequence is equivalent to each of its subsequences. Hence, to demonstrate that $\{a_n\}$ contains a subsequence which converges to x we have but to show that it contains a subsequence of $\{b_n\}$. This is done as follows:

Clearly $b_1 = a_{n_1}$ for some integer $n = n_1$. If any of the numbers $a_1, \ldots, a_{n_1-1}$ are members of the sequence $\{b_n\}$ delete them from the latter, thereby forming a sequence, say, $(b_1, b_2', b_3', b_4', \ldots)$. Obviously now there is a value $n = n_2$ for which $b_2' = a_{n_2}$. Again rejecting those a_n for $n < n_2$ which are to be found among the b_n' $(n \geq 3)$, we proceed as above to produce a sequence which is simultaneously a subsequence of $\{a_n\}$ and of $\{b_n\}$, as was to be shown.

(**2**) Consider the sequence $(1, 1, 2, 1, 2, 3, 1, 2, 3, 4, 1, \ldots)$. Here every positive integer appears infinitely often, and the subsequential limit set of this sequence is seen to consist of precisely the positive integers and infinity.

(**3**) Let $\{b_n\}$ be the sequence of all rational numbers as constructed, for example, in the proof of Theorem 4.6, but without the deletion of repeating elements. Here we have a sequence which has *every sequence of rational numbers as a subsequence.* Every member of $\mathcal{R}^E$ is evidently a subsequential limit thereof.

From the last example we deduce:

12.3 / Corollary Every subset of $\mathcal{R}^E$ is *contained* in the subsequential limit set of some sequence of rational numbers.

We are thus led to ask this question: "Is every set of real numbers actually *equal* to the subsequential limit set of some sequence?" The answer to this question is negative. While the full significance of this fact will be brought out in the next chapter, it will be observed already in Theorem 12.7 that there is a valid characterization of sets having this property. We first prove:

12.4 / Lemma Let A be the subsequential limit set of a given sequence $\{a_k\}$. Let $\{b_m\}$ be any convergent sequence with range in A. If $\lim_{m\to\infty} b_m = b$, then $\{a_k\}$ contains a subsequence $\{a_{k_\nu}\}$ such that also $\lim_{\nu\to\infty} a_{k_\nu} = b$.

Proof It is easy to see that there is an increasing sequence, $k_1 < k_2 < k_3 < \cdots$, of positive integers, such that

$$d(a_{k_\nu}, b_\nu) < \frac{1}{\nu}$$

for each value of ν. Now, given an arbitrary number $\epsilon > 0$, let n' and n'' be integers such that

$$\frac{1}{n'} < \frac{\epsilon}{2}$$

and

$$d(b_\nu, b) < \frac{\epsilon}{2} \qquad \text{when} \quad \nu > n''.$$

If we put $n = \max(n', n'')$, then it follows that

$$d(a_{k_\nu}, b) \leq d(a_{k_\nu}, b_\nu) + d(b_\nu, b) < \frac{\epsilon}{2} + \frac{\epsilon}{2} = \epsilon$$

whenever $\nu > n$, and the theorem follows.

The above arguments lead us to introduce:

12.5 / *Definition* A set $A \subset \mathcal{R}^E$ is *closed* when it contains the limit of every convergent sequence with range in A.

12.6 / *Examples*

(**1**) Closed intervals and finite unions of such are closed sets. A singleton, in particular, is a closed set as is $\mathcal{R}$; an open interval is not a closed set.

(**2**) Let A be the range of the convergent sequence $\{a_k\}$ and a its limit. The set $A \cup \{a\}$ is closed.

With the language of Definition 12.5 we now formulate the result we are after:

12.7 / Theorem Every subsequential limit set of a sequence of real numbers is closed.

The converse of this theorem is proved in Chapter Five. Now we consider:

12.8 / *Definition* Let $\{a_k\}$ be a given sequence; let its subsequential limit set be A. Setting

$$a^* = \text{lub of } A$$

$$a_* = \text{glb of } A$$

we call a^* the *upper limit* of $\{a_k\}$, a_* the *lower limit* of $\{a_k\}$, and write

$$\limsup_{k\to\infty} a_k = a^*$$

$$\liminf_{k\to\infty} a_k = a_*.$$

Synonymous with upper and lower limits are *limit superior* and *limit inferior*, respectively.

Thus, for instance, in (0.1) $\limsup_{n\to\infty} n = \liminf_{n\to\infty} n = \infty$; the respective limits in (0.2) are ∞ and 0, those in (0.3) are 1 and -1. For the sequence in 12.2(3) we have $\limsup_{k\to\infty} a_k = \infty$, $\liminf_{k\to\infty} a_k = -\infty$.

From Theorem 12.7 we deduce the important corollary:

12.9 / Corollary The numbers a^* and a_* in Definition 12.8 always belong to A.

In addition, these numbers are endowed with the following property:

12.10 / Theorem Let the notation be that of 12.8. For each $\epsilon > 0$ there is an integer n such that

$$a_* - \epsilon < a_k < a^* + \epsilon$$

for all values $k > n$.

Proof Consider the inequality $a_k < a^* + \epsilon$. If there were infinitely many $a_k > a^* + \epsilon$, then a subsequence of these would converge (in $\mathcal{R}^E$) to a limit $x \geq a^* + \epsilon$, as provided by Exercise 9-12. Owing to Theorem 12.7 this x would be a member of A, thereby contradicting the supposition that a^* was the lub of A.

The other inequality is verified in an entirely analogous manner.

Exercises 11-1, dealing with the arithmetic properties of limits, have no analogs in the realm of upper and lower limits. This is seen from the following examples:

12.11 / ***Examples***

(1) To show that, in general,

$$\limsup_{k\to\infty} (a_k + b_k) \neq \limsup_{k\to\infty} a_k + \limsup_{k\to\infty} b_k,$$

consider the sequences $\{a_k\} = \{(-1)^k\}$ and $\{b_k\} = \{(-1)^{k+1}\}$. Then $\limsup_{k\to\infty} a_k + \limsup_{k\to\infty} b_k = 1 + 1 = 2$ whereas $\limsup_{k\to\infty}(a_k + b_k) = 0$ since $a_k + b_k = 0$ for each value of k. Since the lower limit of each of the sequences $\{a_k\}$ and $\{b_k\}$ is -1, it follows that also

$$\liminf_{k\to\infty}(a_k + b_k) \neq \liminf_{k\to\infty} a_k + \liminf_{k\to\infty} b_k.$$

(2) With the same sequences as above we note that, for each k, $a_k \cdot b_k = (-1)^k \cdot (-1)^{k+1} = -1$. Hence, $\limsup_{k\to\infty} a_k \cdot b_k = -1$ but $\limsup_{k\to\infty} a_k \cdot \limsup_{k\to\infty} b_k = 1 \cdot 1 = 1$, with the same holding for the product of the lower limits.

The following are some simple yet very useful properties of upper and lower limits:

12.12 / **Theorem** Let $\{a_k\}$ and $\{b_k\}$ be arbitrary sequences of real numbers. Then:

(1) If $a_k \leq b_k$ $(k \in \mathfrak{N})$ then

$$\limsup_{k\to\infty} a_k \leq \limsup_{k\to\infty} b_k$$

and

$$\liminf_{k\to\infty} a_k \leq \liminf_{k\to\infty} b_k.$$

(2) $\limsup_{k\to\infty}(a_k + b_k) \leq \limsup_{k\to\infty} a_k + \limsup_{k\to\infty} b_k$

whereas

$$\liminf_{k\to\infty}(a_k + b_k) \geq \liminf_{k\to\infty} a_k + \liminf_{k\to\infty} b_k.$$

Proof (**1**) This case is an immediate consequence of Theorem 12.10. Consider the first inequality: The cited theorem and the relation $a_k \leq b_k$ permit us to state that, for each $\epsilon > 0$, $a_k \leq b^* + \epsilon$ for all but finitely many of the a_k, where $b^* = \limsup_{k\to\infty} b_k$. We can therefore conclude that such inequalities are true for the limit of every convergent subsequence of the a_k. In particular, it follows that $a^* = \limsup_{k\to\infty} a_k \leq b^* + \epsilon$: Being true for each $\epsilon > 0$, the inequality tells us that $a^* \leq b^*$. The second assertion in (1) is disposed of in a like manner, this time working with the inequalities $a_* - \epsilon \leq b_k$ which, according to Theorem 12.10 and the above, are true for each $\epsilon > 0$ and $k > k_\epsilon$, k_ϵ being some integer associated with each given ϵ.
(**2**) Again appealing to Theorem 12.10, we conclude that, for each $\epsilon > 0$, there are integers m and n such that $a_k \leq a^* + \frac{1}{2}\epsilon$ $(k > m)$ and $b_k \leq b^* + \frac{1}{2}\epsilon$ $(k > n)$: Hence, $a_k + b_k \leq a^* + b^* + \epsilon$ whenever $k > \max(m, n)$. Emulating the above procedure, the student should be able to bring the proof of (2) to its completion.

To illustrate the utility of upper and lower limits let us prove:

12.13 / **Theorem** Let $\{a_k\}$ be a given sequence; for each value of k let

$$s_k = \frac{a_1 + a_2 + \cdots + a_k}{k}.$$

If $a_k \to a$, then also $s_k \to a$.

Proof This theorem is proved by showing that

$$a_* \leq s_* \leq s^* \leq a^*,$$

the notation being self-explanatory.

Given an $\epsilon > 0$, there is an integer n_ϵ such that

$$a_* - \epsilon < a_k < a^* + \epsilon \qquad (k > n_\epsilon).$$

Let us hold such an integer k fixed. Then for each $m > k$

$$s_m = \frac{(a_1 + \cdots + a_k) + (a_{k+1} + \cdots + a_m)}{m}$$

$$\leq \frac{a_1 + \cdots + a_k}{m} + \left(1 - \frac{k}{m}\right)(a^* + \epsilon).$$

Hence we may take the upper limit with respect to m to find that

$$s^* = \limsup_{m\to\infty} s_m \le a^* + \epsilon.$$

Similarly we show that

$$s_m \ge \frac{a_1 + \cdots + a_k}{m} + \left(1 - \frac{k}{m}\right)(a_* - \epsilon)$$

for each $m > k$ and the proof is easily completed.

The sequence $\{s_m\}$ is the sequence of *arithmetic means* of the numbers a_k. A sequence $\{a_k\}$ whose sequence of arithmetic means converges is said to be $(C, 1)$ *summable*. It is interesting to note that a $(C, 1)$ summable sequence may diverge (i.e., $(C, 1)$ summability does not imply convergence). For example, consider the divergent sequence $\{(-1)^k\}$: regarding this sequence $s_{2k} = 0$ and $s_{2k+1} = -1/(2k + 1)$ for each k, so that $\{s_k\}$ is a null sequence.

As a further application of the tools introduced here we shall prove that the sequences in Examples (0.5) and (0.6) have a common limit. We first note that the latter sequence is nondecreasing: Owing to Exercise 11-15 it is bounded, and Theorem 11.6 lets us conclude that the sequence converges. Its limit is designated with the letter "e" and we formally set

$$e = \sum_{n=0}^{\infty} \frac{1}{n!},$$

meaning, of course, that

$$e = \lim_{k\to\infty} \sum_{n=0}^{k} \frac{1}{n!}.$$

We prove:

12.14 / Theorem

$$\lim_{k\to\infty} \left(1 + \frac{1}{k}\right)^k = e.$$

Proof To establish this theorem it will be shown that

$$\limsup_{k\to\infty} \left(1 + \frac{1}{k}\right)^k \le e \le \liminf_{k\to\infty} \left(1 + \frac{1}{k}\right)^k.$$

The first of these inequalities is obtained when we apply the binomial theorem to the expression $\left(1 + \frac{1}{k}\right)^k$, k being held fixed:

$$\left(1 + \frac{1}{k}\right)^k = 1 + k\frac{1}{k} + \frac{k(k-1)}{2!k^2} + \frac{k(k-1)(k-2)}{3!k^3} + \cdots + \frac{1}{k^k}$$

$$= 1 + 1 + \frac{1}{2!}\left(1 - \frac{1}{k}\right) + \frac{1}{3!}\left(1 - \frac{1}{k}\right)\left(1 - \frac{2}{k}\right) + \cdots$$

$$+ \frac{1}{k!}\left(1 - \frac{1}{k}\right)\left(1 - \frac{2}{k}\right)\cdots\left(1 - \frac{k-1}{k}\right)$$

$$\leq 1 + 1 + \frac{1}{2!} + \frac{1}{3!} + \cdots + \frac{1}{k!} = \sum_{n=0}^{k} \frac{1}{n!}.$$

Consequently

$$\left(1 + \frac{1}{k}\right)^k \leq \sum_{n=0}^{k} \frac{1}{n!} \leq e$$

(see in this connection Exercise 11-16) and the first inequality follows an application of Theorem 12.12.

To arrive at the second inequality we observe that, for fixed k,

$$\left(1 + \frac{1}{k}\right)^k \geq 1 + 1 + \frac{1}{2!}\left(1 - \frac{1}{k}\right) + \cdots$$

$$+ \frac{1}{m!}\left(1 - \frac{1}{k}\right)\left(1 - \frac{2}{k}\right)\cdots\left(1 - \frac{m-1}{k}\right)$$

whenever $m \leq k$. Holding m fixed we find that the right side converges. Hence, in particular,

$$\liminf_{k \to \infty} \left(1 + \frac{1}{k}\right)^k \geq \sum_{n=0}^{m} \frac{1}{n!}.$$

Since the left side in the last inequality is independent of m, we may let $m \to \infty$ to arrive at the desired result.

In concluding this section we observe that we have been able to classify the sequences listed at the beginning of this chapter with the exception of (0.4). The relevant properties of the sequences can be seen in the tabulation.

	Bounded	Monotone	Upper limit	Lower limit	Convergent
(0.1)	no	yes	∞	∞	no
(0.2)	no	no	∞	0	no
(0.3)	yes	no	1	-1	no
(0.5)	yes	yes	e	e	yes
(0.6)	yes	yes	e	e	yes

EXERCISES

12-1 Let $\{a_k\}$ and $\{b_k\}$ be two sequences with a common range A. If S and T are the respective subsequential limit sets of the sequences in question, is it true that $S = T$? (in this connection see Exercise 11-22).

12-2 Referring to Definition 12.5, can the set $\mathcal{R}$ of real numbers be expressed as the union of two nonempty closed sets having no common members?

12-3 Consider a sequence of closed intervals I_k such that $\bigcap_{k \in \mathfrak{N}} I_k \neq \varnothing$: is this intersection a closed set?

12-4 Is the countable union of closed intervals a closed set?

12-5 An alternative characterization of upper and lower limits is this: Let $\{a_k\}$ be a given sequence. For each fixed m let

$$A_m = \{a_{m+1}, a_{m+2}, \ldots\};$$

let

$$\mu_m = \text{lub of } A_m$$

$$\nu_m = \text{glb of } A_m.$$

Prove:

(a) the sequence $\{\mu_m\}$ is nonincreasing.

(b) The sequence $\{\nu_m\}$ is nondecreasing.

(c) $\limsup_{m\to\infty} a_m = \lim_{m\to\infty} \mu_m$ and $\liminf_{m\to\infty} a_m = \lim_{m\to\infty} \nu_m$.

12-6 Find the upper and lower limits of the following sequences:

(1) $\left\{\left(1 + \dfrac{1}{k}\right)^k\right\}$;

(2) $\left\{\sum_{k=1}^{m} \frac{1}{k}\right\}$;

(3) $\left\{\frac{1}{k} \sin k + k \sin \frac{1}{k}\right\}$.

12-7 Suppose $a_k \to a$ and $\limsup_{k\to\infty} b_k = b^*$. Prove or disprove:

$$\limsup_{k\to\infty} a_k b_k = a \cdot b^*.$$

12-8 Prove or disprove: If the sequence $\{a_k\}$ converges and the sequence $\{b_k\}$ is monotone, then the sequence $\{a_k b_k\}$ converges (in $\mathcal{R}^E$).

12-9 Associated with each sequence $\{a_k\}$ of nonnegative terms is the sequence of geometric means $\{(a_1 \cdot a_2 \cdots a_k)^{1/k}\}$: Prove that if $a_k \to a$ then also $(a_1 \cdots a_k)^{1/k} \to a$.

12-10 Verify the following generalization of Theorem 12.13: Let $\{r_k\}$ be a sequence of positive numbers; let the associated sequence $\{\sum_{j=1}^{k} r_j\}$ diverge to ∞. If $\lim_{k\to\infty} a_k = a$ then also

$$\lim_{k\to\infty} \frac{r_1 a_1 + \cdots + r_k a_k}{r_1 + \cdots + r_k} = a.$$

12-11 Prove that the insertion or deletion of finitely many members of a sequence does not affect its convergence behavior.

12-12 Suppose $\lim_{k\to\infty} a_k = a$ and $\limsup_{k\to\infty} a_k b_k = c$. Under what circumstances is it true that

$$\limsup_{k\to\infty} a_k b_k = \lim_{k\to\infty} a_k \limsup_{k\to\infty} b_k?$$

13 / BASIC PROPERTIES OF SERIES

Series are not new to the reader. They are formally introduced therefore without much fuss. Yet to make absolutely certain that no misconceptions prevail in this vital area we shall repeat some basic facts.

13.1 / *Definition* Let $\{a_k\}$ be a given sequence. The symbol

$$\sum_{i=0}^{\infty} a_i$$

is called an *infinite series* or simply *series*. For each n

$$s_n = \sum_{j=0}^{n} a_j$$

is the nth *partial sum* of the infinite series. The series *converges* to the limit s, written

$$s = \sum_{i=0}^{\infty} a_i,$$

when

$$\lim_{n \to \infty} s_n = s.$$

The series diverges otherwise; it diverges to ∞ when $s_n \to \infty$, to $-\infty$ when $s_n \to -\infty$.

For convenience in notation we let the symbol $\sum a_n$ represent the infinite series above on those occasions for which no ambiguity is likely to arise. The student is cautioned once more to bear in mind that symbols such as these do *not* indicate a sum but the limit of a certain sequence: To add together more than finitely many terms is meaningless. To exemplify this, regard the series

$$\sum_{n=0}^{\infty} (-1)^n = 1 - 1 + 1 - 1 + \cdots.$$

How are we to "add" the members of the series? Surely an inductive procedure will not do for, among other difficulties, the "sum" would fluctuate between the values 0 and 1. Also the insertion of parentheses (i.e., applying the associativity of addition) is not permitted since this would lead to possible "sums"

$$\sum (-1)^n = (1 - 1) + (1 - 1) + (1 - 1) + \cdots$$
$$= 0 + 0 + 0 + \cdots = 0$$

and

$$\sum (-1)^n = 1 - (1 - 1) - (1 - 1) - \cdots$$
$$= 1 - 0 - 0 - \cdots = 1.$$

We already know that the sequence of partial sums of this series diverges.

It should be reemphasized that the sequences of partial sums as such are not the subject of study here: sequences were dealt with in the foregoing sections. What is of interest, however, is the correlation of the behavior of the sequences $\{a_i\}$ and $\{s_i\}$. In particular, we wish to predict the behavior of one from that of the other. In this connection we point out that these sequences must always be carefully distinguished from each other: To stress their different roles one might define the associated infinite series to be the ordered pair $(\{a_i\}, \{s_i\})$ but this refinement will not be dwelt upon here.

Let us, then, proceed to examine the basic properties governing the behavior of infinite series. To begin with we have:

13.2 / Theorem If the series $\sum a_i$ converges, then the sequence $\{a_i\}$ is a null sequence.

Proof Suppose $s_n = s$, where $s_n = \sum_{i=0}^{n} a_i$. Then also $s_{n-1} \to s$ (by 11.5). Hence, the relations $a_n = s_n - s_{n-1}$ lead to

$$\lim_{n\to\infty} a_n = \lim_{n\to\infty} (s_n - s_{n-1}) = s - s = 0,$$

as was to be shown.

The condition in the theorem is necessary for a series to converge but not sufficient: the divergent series $\sum_{n=1}^{\infty} \frac{1}{n}$ clearly meets the requirement $\frac{1}{n} \to 0$ (see Exercise 12-6). As in the case of sequences (see Exercise 12-11) we have the following important theorem:

13.3 / Theorem Changing finitely many members of a series does not affect its convergence behavior.

In particular, convergence is preserved when finitely many members are added or deleted, so that the series $\sum_{i=n}^{\infty} a_i$ converges (n being held fixed) whenever $\sum_{i=0}^{\infty} a_i$ converges.

The first major convergence criterion for infinite series follows from the equivalence of convergence of sequences with the Cauchy property. Specifically, we have:

13.4 / Cauchy Convergence Test The series $\sum a_i$ converges if and only if an integer k_ϵ can be associated with each $\epsilon > 0$ such that

$$\left| \sum_{i=m}^{n} a_i \right| < \epsilon$$

for all values $n > m > k_\epsilon$.

This is the most general test and the most difficult one to apply to arbitrary series. A very useful test is the following:

13.5 / Comparison Test Imagine sequences $\sum a_i$ and $\sum b_i$ ($b_i \geq 0$) such that, for some integer k, $|a_i| \leq b_i$ for all values $i > k$: If $\sum b_i$ converges, then also $\sum a_i$ converges.

Proof Let k be as demanded in the theorem. Appealing to Theorem 13.4, if $\epsilon > 0$ is given, then there is an integer k_ϵ such that

$$\sum_{i=m}^{n} b_i < \epsilon$$

for $n > m > k_\epsilon$. Hence

$$\left| \sum_{i=m}^{n} a_i \right| \leq \sum_{i=m}^{n} |a_i| \leq \sum_{i=m}^{n} b_i < \epsilon$$

for all values $n > m > \max(k, k_\epsilon)$.

It is clear that the last criterion cannot be sharpened without further restricting the series involved. Thus, for example, the test is inconclusive when $\sum b_i$ diverges or when $|a_i| > b_i$ $(i > k)$.

Also useful is the following:

13.6 / Dirichlet Test Let $\sum a_k$ be a series whose sequence of partial sums is bounded. If $\{b_k\}$ is a nonincreasing null sequence of nonnegative numbers, then the series $\sum a_k b_k$ converges.

In the proof of this theorem we rely on an estimate of Abel:

13.7 / Abel's Lemma Let the series $\sum a_k$ have partial sums s_n such that $\alpha \leq s_n \leq \beta$. If the sequence $\{b_k\}$ is as above, then

$$\alpha b_1 \leq \sum_{k=1}^{n} a_k b_k \leq \beta b_1.$$

Proof of Lemma A simple manipulation shows that, for each $n \in \mathfrak{N}$,

$$\sum_{k=1}^{n} a_k b_k = \sum_{k=1}^{n} s_k(b_k - b_{k+1}) + s_n b_{n+1}.$$

By assumption, $b_k - b_{k+1} \geq 0$ and $s_k \leq \beta$ for each value of k. Hence,

$$\begin{aligned}\sum_{k=1}^{n} a_k b_k &\leq \beta \sum_{k=1}^{n} (b_k - b_{k+1}) + \beta b_{n+1} \\ &= \beta[(b_1 - b_2) + (b_2 - b_3) + \cdots + (b_n - b_{n+1}) + b_{n+1}] \\ &= \beta b_1.\end{aligned}$$

The other inequality in the lemma is established in a like manner.

Proof of 13.6 We shall simply show that the sequence of partial sums of the series $\sum a_k b_k$ is a Cauchy sequence. Now, according to the statement of the theorem, there is a constant β such that $|s_n| \leq \beta$ for all $n \in \mathfrak{N}$. It follows that for arbitrary integers $1 \leq m \leq n$,

$$\left|\sum_{k=m}^{n} a_k\right| = |s_n - s_{m-1}| \leq |s_n| + |s_{m-1}| \leq 2\beta.$$

Abel's Lemma, when applied to the series $\sum_{k=m}^{\infty} a_k$ and sequence $\{b_m, b_{m+1}, \ldots\}$, shows that

$$\left|\sum_{k=m}^{n} a_k b_k\right| \leq 2\beta b_m.$$

Finally, we use the fact that the sequence $\{b_n\}$ is a null sequence. Namely, if $\epsilon > 0$ is an arbitrary number, then there is an integer n_ϵ such that $b_k < \epsilon/2\beta$ whenever $k > n_\epsilon$. Hence, for integers $m > n_\epsilon$,

$$\left|\sum_{k=m}^{n} a_k b_k\right| < \epsilon,$$

and this completes the proof.

For the comparison test to be useful we should have at our disposal a representative collection of *majorant* or *dominant* series with which others can be compared. To obtain such sequences we need only the somewhat specialized but nevertheless extremely important class of series with non-negative terms: these are analyzed next.

EXERCISES

13-1 The following is meaningful in view of Theorem 13.3: Suppose $s = \sum a_i$, $s_n = \sum_{i=0}^{n} a_i$. The number

$$r_n = s - s_n = \sum_{i=n+1}^{\infty} a_i$$

is said to be the nth *remainder* of the series. Prove that the sequence of remainders of a convergent series is a null sequence.

13-2 Establish the following counterpart of Exercise 11-1(3): If $s = \sum a_i$ and $t = \sum b_i$, then

$$\lim_{n\to\infty} \sum_{i=0}^{n} (a_i + b_i) = s + t.$$

13-3 Show by means of an example that $\sum a_i = s$ and $\sum b_i = t$ does *not* imply $\sum a_i \cdot b_i = s \cdot t$.

13-4 Prove: If $\sum a_i = s$, then for each $c \in \mathcal{R}$

$$\lim_{n\to\infty} \sum_{i=0}^{n} c \cdot a_i = c \cdot s.$$

13-5 Prove or disprove: The convergence of $\sum a_i$ and $\sum b_i$ implies the convergence of $\sum a_i \cdot b_i$.

13-6 Let $\{k_n\}$ be an arbitrary subsequence of $\{k\}$. Then $\sum a_k$ converges if and only if

$$\{\sum_{i=1}^{k_n} a_{n+i}\}$$

is a null sequence. Prove this fact.

13-7 Let $\{\alpha_n\}$ be a nonincreasing null sequence of positive numbers, and consider the series

$$\sum_{k=1}^{\infty} \alpha_k \sin kx.$$

If, for a given number x, we could exhibit a constant, $\beta(x)$, such that

$$| \sum_{k=1}^{n} \sin kx \mid \leq \beta(x),$$

then it would follow from the Dirichlet Test 13.6 that the above series converges for this value of x. Show that such a constant exists for each x which is not an integral multiple of 2π.

14 / SERIES WITH NONNEGATIVE TERMS

From Theorem 11.6 we derive the fundamental theorem:

14.1 / Boundedness Test A series of nonnegative terms converges when and only when the sequence of its partial sums is bounded.

Let us illustrate the utility of this and the preceding tests.

14.2 / *Examples*

(**1**) Consider the series $\sum x^k$ $(x \geq 0)$. When $x \neq 1$, then for each n

$$s_n = \sum_{k=0}^{n} x^k = \frac{1 - x^{n+1}}{1 - x},$$

as is easily verified by carrying out the division in the right side; when $x = 1$, then

$$s_n = \sum_{k=0}^{n} 1 = n + 1.$$

The test in 14.1 now tells us that the series diverges when $x \geq 1$, and converges for $0 \leq x < 1$ (since then $\{x^k\}$ is known to be a null sequence). In particular,

$$\sum_{k=0}^{\infty} x^k = \frac{1}{1-x} \qquad (0 \leq x < 1).$$

(2) Let us prove once and for all that the *harmonic series* $\sum \frac{1}{n}$ diverges to ∞. Obviously the partial sums s_n of this series form an increasing sequence:

$$s_1 < s_2 < s_3 < s_4 < \cdots.$$

Hence, the assertion will be proved once we show that a subsequence of $\{s_n\}$ is unbounded. For this we observe that each block

$$\frac{1}{2^{k-1}+1} + \cdots + \frac{1}{2^k}$$

contains 2^{k-1} members: Replacing each member by $\frac{1}{2^k}$ (which is the smallest in the block), we arrive at the estimate

$$\frac{1}{2^{k-1}+1} + \cdots + \frac{1}{2^k} > 2^{k-1}\frac{1}{2^k} = \frac{1}{2}.$$

Thus, for $n = 2^k$ we have

$$\begin{aligned} s_n &= 1 + \frac{1}{2} + \frac{1}{3} + \cdots + \frac{1}{2^k} \\ &= 1 + \frac{1}{2} + \left(\frac{1}{3} + \frac{1}{4}\right) + \left(\frac{1}{5} + \frac{1}{6} + \frac{1}{7} + \frac{1}{8}\right) + \cdots \\ &\quad + \left(\frac{1}{2^{k-1}+1} + \cdots + \frac{1}{2^k}\right) \\ &> 1 + \frac{1}{2} + 2\cdot\frac{1}{4} + 4\cdot\frac{1}{8} + \cdots + 2^{k-1}\cdot\frac{1}{2^k} \\ &= 1 + \frac{1}{2} + \frac{1}{2} + \cdots + \frac{1}{2} > \frac{k}{2} \end{aligned}$$

and the result follows from an application of Theorem 14.1.

(**3**) The sequence

$$\sum_{k=1}^{\infty} \frac{1}{k^r}$$

will be shown to diverge when $r \leq 1$, converge when $r > 1$.

The case $r \leq 1$ is disposed of when we note that $\frac{1}{k^r} \geq \frac{1}{k}$ and appeal to Exercise 13-6. Otherwise we wish to emulate the argument in (2) above, this time showing that each s_n is small. The student should have no difficulties in seeing that for $n < 2^k$

$$s_n = \sum_{k=1}^{n} \frac{1}{k^r} < 1 + \left(\frac{1}{2^r} + \frac{1}{3^r}\right) + \left(\frac{1}{4^r} + \cdots + \frac{1}{7^r}\right) + \cdots$$

$$+ \left(\frac{1}{2^{kr}} + \cdots + \frac{1}{(2^{k+1} - 1)^r}\right)$$

$$< 1 + 2 \cdot \frac{1}{2^r} + 4 \cdot \frac{1}{4^r} + \cdots + 2^k \cdot \frac{1}{2^{kr}}$$

$$= 1 + \frac{1}{2^{r-1}} + \frac{1}{(2^{r-1})^2} + \cdots + \frac{1}{(2^{r-1})^k} = t_k.$$

Owing to (1) above, $t_k \leq \dfrac{2^{r-1}}{2^{r-1} - 1}$, thereby showing the sequence $\{s_n\}$ to be bounded.

(**4**) The sequence $\sum \dfrac{1}{\sqrt{n^2 + 1}}$ diverges since, for each n,

$$\frac{1}{\sqrt{n^2 + 1}} \geq \frac{1}{\sqrt{n^2 + n^2}} = \frac{1}{\sqrt{2} \cdot n}.$$

On the other hand, the sequence $\sum \dfrac{1}{\sqrt{n^{2+\delta} + 1}}$ converges for each positive number δ, since

$$\frac{1}{\sqrt{n^{2+\delta} + 1}} < \frac{1}{n^{1+\delta/2}}$$

(see (3) above).

In line with Theorem 14.1 we have a rather fascinating theorem due to Cauchy:

14.3 / Cauchy Condensation Test Let $\{a_k\}$ be a nonincreasing sequence of nonnegative terms: $a_1 \geq a_2 \geq \cdots \geq 0$. Then the series $\sum_{k=1}^{\infty} a_k$ converges if and only if the series

$$\sum_{i=0}^{\infty} 2^i a_{2^i} = a_1 + 2a_2 + 4a_4 + 8a_8 + \cdots$$

converges.

This theorem demonstrates that a very "meager" subsequence already determines the convergence behavior of the sequence. Owing to the monotone character of the sequence $\{a_i\}$ and Theorem 13.2 we have in this theorem an estimate on the rate in which the a_i must tend to zero.

Proof In the proof we use the arguments employed in Examples 14.2(2) and 14.2(3). Namely, for $n < 2^k$,

$$\begin{aligned} s_n = \sum_{i=1}^{n} a_i \leq a_1 + (a_2 + a_3) + (a_4 + \cdots + a_7) + \cdots \\ + (a_{2^k} + \cdots + a_{2^{k+1}-1}) \\ \leq a_1 + 2a_2 + 4a_4 + \cdots + 2^k a_{2^k}, \end{aligned}$$

the inequalities following due to the monotonicity conditions $a_n \geq a_{n+1}$. Setting

$$t_n = \sum_{i=1}^{n} 2^i a_{2^i}$$

we arrive at the inequality

$$s_n \leq t_n. \tag{i}$$

At the same time we find that, for $n > 2^k$,

$$\begin{aligned} s_n \geq a_1 + a_2 + (a_3 + a_4) + \cdots + (a_{2^{k-1}+1} + \cdots + a_{2^k}) \\ \geq a_1 + a_2 + 2a_4 + \cdots + 2^{k-1} a_{2^k} \geq \tfrac{1}{2} t_n, \end{aligned}$$

i.e.,

$$t_n \leq 2s_n. \tag{ii}$$

In addition to the inequalities (i) and (ii), we know that the sequences $\{s_n\}$ and $\{t_n\}$ are both nondecreasing. Hence, both are either bounded or unbounded and the theorem follows.

14.4 / *Examples*

(1) Consider the series

$$\sum_{n=2}^{\infty} \frac{1}{n \cdot \ln n}$$

(here $n \geq 2$ since $\ln 1 = 0$). Applying the condensation test we obtain the series

$$\frac{1}{\ln 2} + \frac{1}{\ln 4} + \frac{1}{\ln 8} + \cdots + \frac{1}{\ln 2^n} + \cdots$$

which must converge if the original series is to converge. For each admitted value of n, however,

$$\frac{1}{\ln 2^n} = \frac{1}{n \cdot \ln 2},$$

so that the series diverges (to ∞).

(2) Next, consider the series

$$\sum_{n=2}^{\infty} \frac{1}{n(\ln n)^r}.$$

We can state without further ado that this series diverges for $r \leq 1$. When $r > 1$ we again call on the test in 14.3: Since the series

$$\sum_{n=1}^{\infty} 2^n \frac{1}{2^n(\ln 2^n)^r} = \sum_{n=1}^{\infty} \frac{1}{(n \ln 2)^r} = \frac{1}{(\ln 2)^r} \sum_{n=1}^{\infty} \frac{1}{n^r}$$

converges for $r > 1$, so does the original series.

Another extremely useful convergence criterion, again due to Cauchy, is the so-called integral test. Consider a nonincreasing sequence $\{a_k\}$ $(a_k \geq 0)$ and a nonincreasing continuous function $f : \mathbb{R} \to \mathbb{R}$ such that for each k,

$$f(k) = a_k.$$

We recall that f is nonincreasing when $x_1 < x_2$ implies $f(x_1) \geq f(x_2)$. From calculus we know that

$$a_k \geq \int_k^{k+1} f(x)\, dx \geq a_{k+1} \qquad (k = 0, 1, 2, 3, \ldots)$$

(see Figure 13). Consequently, we find upon summation that

$$s_k = \sum_{i=0}^{k} a_i \geq \int_0^{k+1} f(x)\, dx \geq \sum_{i=1}^{k+1} a_i = s_{k+1} - a_0. \tag{iii}$$

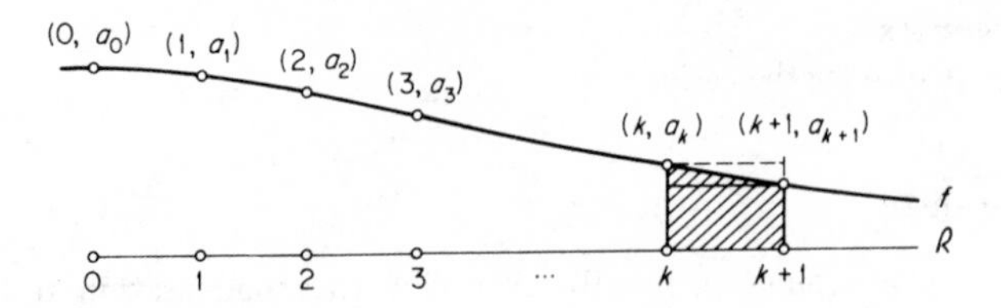

Figure 13

These inequalities show that the sequence $\{s_k\}$ of partial sums is bounded if and only if the sequence of integrals

$$I_k = \int_0^{k+1} f(x)\,dx$$

is bounded. Setting

$$\int_0^\infty f(x)\,dx = \lim_{k\to\infty} I_k$$

we formulate the ensuing theorem:

14.5 / Integral Test If $f : \mathcal{R} \to \mathcal{R}$ is a nonincreasing function satisfying the condition $f(k) = a_k$ ($a_k \geq 0$) for $k = 0, 1, 2, 3, \ldots$, then the series

$$\sum_{k=0}^{\infty} a_k$$

and the integral

$$\int_0^\infty f(x)\,dx$$

have the same convergence behavior.

The inequalities in (iii) can be written in the form

$$0 \leq s_k - I_{k+1} \leq a_0 - a_{k+1}.$$

Since $\{a_k\}$ is a null sequence when the series converges, we find that

$$\int_0^\infty f(x)\,dx \leq \sum_{k=0}^{\infty} a_k \leq \int_0^\infty f(x)\,dx + a_0. \tag{iv}$$

14.6 / *Examples*

(**1**) $\displaystyle\int_1^k \frac{dx}{x} = \ln k$ and since $\ln k \to \infty$ it follows that $\displaystyle\sum \frac{1}{k} \to \infty$.

The student should compare this demonstration with the more cumbersome one previously given.

(2) When $r \neq 1$, then

$$\int_1^k \frac{dx}{x^r} = \frac{1}{r-1}\left(1 - \frac{1}{k^{r-1}}\right).$$

The results in Example 14.2(3) are immediate consequences of this simple calculation.

(3) Consider the series

$$\sum_{k=1}^{\infty} ke^{-k}.$$

Then

$$\int_1^k xe^{-x}\,dx = \frac{2}{e} - (k+1)\frac{1}{e^k} \to \frac{2}{e} \qquad \text{as} \qquad k \to \infty$$

and it follows that the series in question converges. Furthermore, the estimates in (iv) tell us that

$$\frac{2}{e} \leq \sum_{k=1}^{\infty} ke^{-k} \leq \frac{3}{e}.$$

EXERCISES

14-1 Prove: If $\sum a_k$ and $\sum b_k$ are such that there is an integer n such that $a_k \geq b_k \geq 0$ for all $k > n$ and if $\sum b_k$ diverges, then so does $\sum a_k$.

14-2 Prove: If $\sum a_k$ $(a_1 \geq a_2 \geq \cdots > 0)$ converges, then $\{ka_k\}$ is a null sequence, but not conversely.

14-3 Derive the following test from the Cauchy integral test: Let g be a differentiable function defined for all $x \geq 0$ such that $g(x) \to \infty$ as $x \to \infty$. If for some number r, $g'(x)f[g(x)]$ is nonincreasing for $x > r$, then the series

$$\sum f(k)$$

and

$$\sum g'(k)f[g(k)]$$

have the same convergence behavior.

Remark: This is actually a very versatile test since it offers us a great variety of tests by using different functions g. The result is attributed to N. V. Bugaev.

14-4 Determine the behavior of each of the following series:

(a) $\sum \frac{1}{k(k+1)}$

(b) $\sum \frac{\ln k}{k^2}$

(c) $\sum \frac{\ln \ln k}{k(\ln k)^2}$

(d) $\sum \frac{1}{(\ln k)^k}$

(e) $\sum \frac{1}{k^{1+(1/k)}}$

(f) $\sum \frac{1}{k \cdot \ln k \cdot \ln \ln k}$

(g) $\sum \frac{1}{k \ln k (\ln \ln k)^r}$.

14-5 Let $\{a_k\}$ be a given sequence with members $0 \leq a_k \leq 1$. Show that the series $\sum a_k x^k$ converges for all values $0 \leq x < 1$.

14-6 Prove or disprove: The series $\sum a_k$ ($a_k \geq 0$) converges if and only if the series $\sum a_k/(1 + a_k)$ converges.

14-7 Referring to Exercise 13-1, let $\sum a_k$ be a convergent series of positive terms, r_k its kth remainder. Show that

(1) $\sum \frac{a_k}{\sqrt{r_k}}$ converges;

(2) $\sum \frac{a_k}{r_k}$ diverges.

14-8 Prove or disprove: Referring to Theorem 12.13, let $\{a_k\}$ be a given sequence, $\{t_k\}$ its sequence of arithmetic means. If $\sum a_k$ converges, then so does $\sum t_k$.

14-9 Prove or disprove: If $\sum a_k$ ($a_k \geq 0$) converges, then an integer $r > 0$ can be found such that also $\sum k^r a_k$ converges.

14-10 Let $\sum b_k$ converge, $\sum c_k$ diverge ($b_k > 0$, $c_k > 0$); let $\sum a_k$ be a given series of positive terms. Deduce from the Comparison Test that:

(a) $\sum a_k$ converges if there is an integer n such that

$$\frac{a_{k+1}}{a_k} \leq \frac{b_{k+1}}{b_k} \qquad (k > n);$$

(b) $\sum a_k$ diverges if, for some integer n,

$$\frac{a_{k+1}}{a_k} \geq \frac{c_{k+1}}{c_k} \qquad (k > n).$$

(*Hint*: Write down the respective quotients for $k = n + 1$, $n + 2, \ldots$ and multiply them together.)

14-11 Let $\sum a_k$ be a series of positive terms. Prove that the series converges if there is an integer n such that

$$\frac{a_{k+1}}{a_k} \leq c < 1 \qquad (k > n)$$

for some constant c, and diverges if

$$\frac{a_{k+1}}{a_k} \geq c \geq 1 \qquad (k > n).$$

15 / ALTERNATING SERIES

15.1 / *Definition* The series $\sum a_k$ is *alternating* if every two consecutive members have opposite signs:

$$\operatorname{sgn} a_k + \operatorname{sgn} a_{k+1} = 0 \qquad (k = 0, 1, 2, 3, \ldots).$$

Without loss of generality we only consider series of the form

$$\sum(-1)^k a_k \qquad (a_k \geq 0).$$

For such series we have the following:

15.2 / Leibnitz Test If $\{a_k\}$ $(a_k \geq a_{k+1} \geq 0)$ is a null sequence, then $\sum(-1)^k a_k$ converges.

Proof It suffices to show that the subsequences $\{s_{2k}\}$ and $\{s_{2k+1}\}$ of the sequence $\{s_k\}$ of partial sums converge to the same limit. Consider the first subsequence. The monotone character of $\{a_k\}$ assures the validity of the inequalities

$$a_{2k} - a_{2k+1} \geq 0 \tag{i}$$

and

$$-(a_{2k+1} - a_{2k+2}) \leq 0 \tag{ii}$$

for each admitted value of k. Therefrom we deduce that

$$a_0 - a_1 \leq (a_0 - a_1) + (a_2 - a_3) + \cdots + (a_{2k-2} - a_{2k-1}) + a_{2k} = s_{2k}$$

and

$$s_{2k+2} = s_{2k} - (a_{2k+1} - a_{2k+2}) \leq s_{2k}.$$

Hence, the sequence $\{s_{2k}\}$ is seen to be nonincreasing and bounded, and its convergence is guaranteed by Theorem 14.1.

Regarding the other sequence, we find that

$$s_{2k+1} = a_0 - (a_1 - a_2) - (a_3 - a_4) - \cdots$$
$$- (a_{2k-1} - a_{2k}) - a_{2k+1} \leq a_0$$

and

$$s_{2k+1} = s_{2k-1} + (a_{2k} - a_{2k+1}) \geq s_{2k-1},$$

being nondecreasing and bounded, also the sequence $\{s_{2k+1}\}$ converges.

To verify that the subsequences have a common limit, we note that

$$s_{2k} - s_{2k+1} = a_{2k+1}.$$

Since $\{a_{2k+1}\}$ is a null sequence, it follows that $s_{2k} - s_{2k+1} \to 0$ thereby completing the proof.

Setting $s = \sum(-1)^k a_k$ we have at once the following corollary:

15.3 / Corollary For each value of k

$$s_{2k+1} \leq s \leq s_{2k}.$$

Proof This is evident from the fact that

$$s_{2k+1} \leq s_{2k+1} + (a_{2k+2} - a_{2k+3}) + \cdots \qquad \text{by (i)}$$

and

$$s_{2k} \geq s_{2k} - (a_{2k+1} - a_{2k+2}) - \cdots \qquad \text{by (ii)}.$$

15.4 / *Examples*

(**1**) The series in (0.4) is seen to converge for each $p > 0$; in particular, the alternating harmonic series converges. From the above inequalities we deduce that

$$-1 + \frac{1}{2^p} - \frac{1}{3^p} \leq \sum \frac{(-1)^k}{k^p} \leq -1 + \frac{1}{2^p},$$

for now we have

$$a_2 - a_3 \leq s + a_1 \leq a_2.$$

(**2**) The series

$$\sum_{k=0}^{\infty} \frac{(-1)^k}{(k^p + 1)^q}$$

converges for $p > 0$ and $q > 0$ [In contrast, see Example 14.2(4)], and diverges when either $p < 0$ or $q < 0$.

(3) The series

$$\sum \frac{(-1)^k}{(\ln k)^r}$$

converges for each $r > 0$.

Corollary 15.3 also offers a very effective estimate of the remainder (See Exercise 13-1). Namely, setting $r_k = s - s_k$, $s = \sum(-1)^k a_k$, then the corollary yields the estimates

$$0 \leq r_{2k+1} \leq s_{2k} - s_{2k+1} = a_{2k+1}$$

and

$$-a_{2k+1} = s_{2k+1} - s_{2k} \leq r_{2k} \leq 0,$$

from which we easily deduce that

$$|r_k| \leq a_k.$$

To illustrate the usefulness of such estimates consider the series $s = \sum \dfrac{(-1)^k}{k!}$. Here

$$|s - s_4| \leq \frac{1}{4!} = \frac{1}{24}$$

$$|s - s_5| \leq \frac{1}{5!} = \frac{1}{120}$$

$$|s - s_6| \leq \frac{1}{6!} = \frac{1}{720};$$

and we see, in particular, that this series converges very rapidly.

16 / ABSOLUTE CONVERGENCE

Since infinite series formally resemble finite sums it is natural to inquire to what extent the basic arithmetic rules apply to them. Consider, for example, associativity. While the indiscriminate insertion or removal of parentheses in arbitrary series is barred it is always permitted in convergent series.

16.1 / Theorem Consider the series $\sum a_i$, a subsequence $\{i_k\}$ of the integers ($i_0 = 0$), and the series $\sum b_i$ composed of members

$$b_i = a_{i_k} + a_{i_k+1} + \cdots + a_{i_{k+1}-1}.$$

If $\sum a_i$ converges, then so does $\sum b_i$. Moreover, the two series converge to the same limit.†

Rearrangements, on the other hand, are not admissible even in the class of convergent series. To be precise we formulate the following:

16.2 / *Definition* Let $\sum a_i$ be a given series. If the sequence $\{i_k\}$ is a one-to-one function of $\mathfrak{N}$ *onto* $\mathfrak{N}$, then $\sum a_{i_k}$ is a *rearrangement* of the original series.

A rearrangement, in other words, effects only the relative order of the a_i but not their occurence nor the possible recurrence of their numerical values.

The dependence of arbitrary convergent series on the order of their entries is illustrated in the example:

16.3 / *Example* Consider the alternating harmonic series $\sum(-1)^n/n$ (which is known to converge). We assert that for each $\xi \in \mathfrak{R}^E$ there is a rearrangement of the series converging to it.

To begin with we observe that each of the series

$$\sum \frac{1}{2n+1} \qquad \text{and} \qquad \sum \frac{1}{2n}$$

diverges to ∞ (This statement should be verified at this point). Next, let ξ be given. For the sake of argument suppose $1 \leq \xi < \infty$. Then there is a smallest odd integer n_1 for which

$$1 + \frac{1}{3} + \cdots + \frac{1}{n_1} > \xi$$

and hence

$$1 + \frac{1}{3} + \cdots + \frac{1}{n_1} - \frac{1}{2} \leq \xi.$$

Again, there is a smallest odd integer n_2 such that

$$1 + \frac{1}{3} + \cdots + \frac{1}{n_1} - \frac{1}{2} + \frac{1}{n_1+2} + \cdots + \frac{1}{n_2} > \xi$$

and accordingly

$$1 + \frac{1}{3} + \cdots + \frac{1}{n_1} - \frac{1}{2} + \frac{1}{n_1+1} + \cdots + \frac{1}{n_2} - \frac{1}{4} \leq \xi.$$

† We know, of course, that the convergence of $\sum b_i$ warrants no conclusion concerning the convergence of $\sum a_i$ (see, however, Exercise 16-1).

The student should experience no difficulties in showing that the rearrangement resulting from this procedure is, indeed, as claimed (See also Exercise 16-2).

The question that we are thus led to ask is this: Can one characterize a class of series which are invariant under rearrangements of their members, in the sense that all rearrangements of a given series of the class share a common limit? Since rearrangements give rise to different sequences of partial sums we ask, in effect, to describe that class of series in which the sequences of partial sums corresponding to the rearrangements of each series are equivalent. To answer this question we introduce:

16.4 / ***Definition*** Let $\sum a_i$ be a given convergent series. Then $\sum a_i$ converges *unconditionally* when every one of its rearrangements converges, *conditionally* otherwise.

An examination of the features of the series in Example 16.3 suggests that we also consider the following:

16.5 / ***Definition*** The series $\sum a_i$ converges *absolutely* when $\sum | a_i |$ converges, *nonabsolutely* otherwise.

In particular, every convergent series whose terms have constant sign converges absolutely. From Theorem 13.5 we have at once:

16.6 / **Corollary** An absolutely convergent series converges.

The characterization we are after can now be given:

16.7 / **Theorem** The series $\sum a_i$ converges unconditionally if and only if it converges absolutely. Furthermore, all rearrangements have a common limit in this case.

In the proof we call on the following observation of Riemann:

16.8 / **Riemann's Theorem** If $\sum a_i$ converges nonabsolutely, then for every value $\xi \in \mathcal{R}^E$ there is a rearrangement of $\sum a_i$ converging to it.

Proof Let us begin the proof of Riemann's theorem by showing that $\sum a_i$ contains infinitely many positive and infinitely many negative terms. With this aim in mind we introduce the elements

$$b_i = \begin{cases} a_i & \text{when} \quad a_i \geq 0 \\ 0 & \text{when} \quad a_i < 0 \end{cases}$$

and

$$c_i = \begin{cases} -a_i & \text{when} \quad a_i \leq 0 \\ 0 & \text{when} \quad a_i > 0. \end{cases}$$

Then $b_i \geq 0$ and $c_i \geq 0$ for all $i \in \mathfrak{N}$, and we claim that both series $\sum b_i$ and $\sum c_i$ diverge.

If the contrary were true, then the series

$$\sum(b_i + c_i) = \sum | a_i |$$

would converge, thereby contradicting the hypothesis that $\sum a_i$ converges nonabsolutely. On the other hand, if, say, $\sum b_i$ were convergent and $\sum c_i$ were divergent, then the equation

$$\sum_{i=1}^{k} a_i = \sum_{i=1}^{k} (b_i - c_i) = \sum_{i=1}^{k} b_i - \sum_{i=1}^{k} c_i,$$

which holds for each $k \in \mathfrak{N}$, would tell us that $\sum a_i$ diverges. Hence, we can conclude that our claim was justified. In particular, we have supported the claim made at the beginning of the proof.

According to Theorem 13.2 $\{a_i\}$ is a null sequence. From this point on we can, therefore, fashion the proof along the lines of Example 16.3.

Proof of 16.7 To prove Theorem 16.7 we note that Theorem 16.8 says, in effect, that nonabsolute convergence implies conditional convergence. But this is equivalent to the statement that unconditional convergence implies absolute convergence. Thus, we have only to prove the converse statement. To do this, let the series $\sum a_i$ be absolutely convergent. Then for each $\epsilon > 0$ there is an integer k such that

$$\sum_{i=m}^{n} | a_i | < \epsilon \qquad (n > m \geq k).$$

Let $\sum a_{p_j}$ be a rearrangement of $\sum a_i$. Then for each $i \in \mathfrak{N}$ there is a unique $j(i) \in \mathfrak{N}$ such that $i = p_{j(i)}$. If

$$q = \max(j(0), j(1), \ldots, j(k)),$$

then evidently

$$\{0, 1, 2, \ldots, k\} \subset \{p_0, p_1, p_2, \ldots, p_q\}.$$

Now consider the elements $a_{p_0}, \ldots, a_{p_q}$. We can select an integer r

sufficiently large so as to guarantee the inclusion

$$\{a_{p_0}, a_{p_1}, \ldots, a_{p_q}\} \subset \{a_0, a_1, \ldots, a_r\}.$$

If $r \geq n \geq k$, where k is the integer fixed above, then

$$\left| \sum_{i=0}^{n} a_i - \sum_{j=0}^{n} a_{p_j} \right| \leq \sum_{i=k+1}^{r} |a_i| < \epsilon,$$

and it follows that the series in question have the same limit. In particular, the rearrangement converges and the proof is therefore complete.

A specific rearrangement problem is brought about when we attempt to associate an infinite series with a *double sequence* $\{a_{ij}\}$ $(i, j \in \mathfrak{N})$. Looking at the series

$$\sum_{i=0}^{\infty} \sum_{j=0}^{\infty} a_{ij}$$

and

$$\sum_{j=0}^{\infty} \sum_{i=0}^{\infty} a_{ij}$$

one wishes to know when they converge to the same limit (see Figure 14). An incentive for considering such a situation is provided, for instance, by Taylor's Theorem in Section 31.

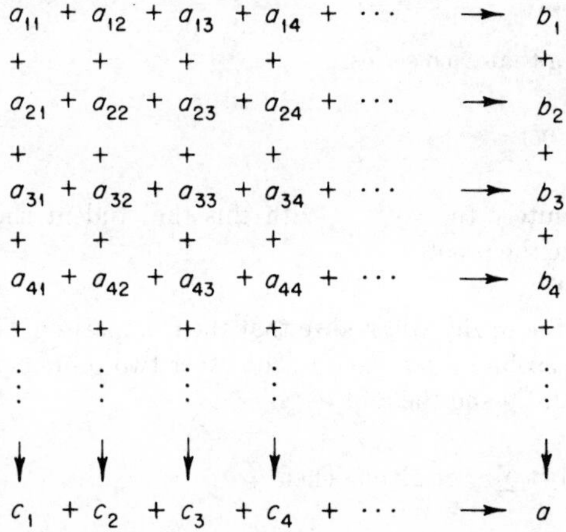

$$\begin{array}{ccccccccccc}
a_{11} & + & a_{12} & + & a_{13} & + & a_{14} & + & \cdots & \rightarrow & b_1 \\
+ & & + & & + & & + & & & & + \\
a_{21} & + & a_{22} & + & a_{23} & + & a_{24} & + & \cdots & \rightarrow & b_2 \\
+ & & + & & + & & + & & & & + \\
a_{31} & + & a_{32} & + & a_{33} & + & a_{34} & + & \cdots & \rightarrow & b_3 \\
+ & & + & & + & & + & & & & + \\
a_{41} & + & a_{42} & + & a_{43} & + & a_{44} & + & \cdots & \rightarrow & b_4 \\
+ & & + & & + & & + & & & & + \\
\vdots & & \vdots & & \vdots & & \vdots & & & & \vdots \\
\downarrow & & \downarrow & & \downarrow & & \downarrow & & & & \downarrow \\
c_1 & + & c_2 & + & c_3 & + & c_4 & + & \cdots & \rightarrow & a
\end{array}$$

Figure 14

16.9 / Theorem Let the double sequence $\{a_{ij}\}$ $(i, j \in \mathfrak{N})$ be given and suppose

(a) $\sum_{j=1}^{\infty} | a_{ij} | = a_i \qquad (i \in \mathfrak{N}),$

and

(b) $\sum_{i=1}^{\infty} a_i$ converges.

Then

$$\sum_{j=1}^{\infty} \sum_{i=1}^{\infty} a_{ij} = \sum_{i=1}^{\infty} \sum_{j=1}^{\infty} a_{ij}.$$

Proof The statement in (a) implies that, for each fixed value of i,

$$\sum_{j=1}^{\infty} a_{ij} \qquad (i \in \mathfrak{N})$$

converges absolutely to, say, b_i. According to (b) this series converges absolutely to some number a. Next let it be shown that each of the "column series" in Figure 14 converges absolutely. Namely, let n be a fixed but otherwise arbitrary positive integer. Then for each $m \in \mathfrak{N}$,

$$\sum_{i=1}^{n} | a_{ij} | \leq \sum_{j=1}^{n} \sum_{i=1}^{m} | a_{ij} | = \sum_{i=1}^{m} \sum_{j=1}^{n} | a_{ij} | \leq \sum_{i=1}^{m} a_i,$$

thereby showing that each series

$$\sum_{i=1}^{\infty} a_{ij} \qquad (j \in \mathfrak{N})$$

converges absolutely to, say, c_j. With this the student should be able to complete the proof.

Looking back, the reader will realize that the Comparison Test 13.5 is actually a test for absolute convergence. The other two prominent tests discussed here are the ratio and the root tests.

16.10 / Ratio Test Let $\sum a_k$ be given. Then

(**1**) $\sum a_k$ converges when $\limsup_{k \to \infty} \left| \dfrac{a_{k+1}}{a_k} \right| < 1;$

(**2**) $\sum a_k$ diverges if there is an integer n such that

$$\left|\frac{a_{k+1}}{a_k}\right| \geq 1 \qquad \text{for all values} \quad k \geq n;$$

(**3**) the test is inconclusive when

$$\liminf_{k\to\infty} \left|\frac{a_{k+1}}{a_k}\right| \leq 1 \leq \limsup_{k\to\infty} \left|\frac{a_{k+1}}{a_k}\right|.$$

Proof Items (1) and (2) require no further justification (see Exercise 14.11). The assertion made in (3) is readily disposed of with a simple example. Namely, for each of the series $\sum 1/n$ and $\sum 1/n^2$ we find that

$$\lim_{n\to\infty} \left|\frac{a_{n+1}}{a_n}\right| = 1,$$

yet they have a different convergence behavior.

16.11 / **Root Test** Consider the series $\sum a_k$ and let $a^* = \limsup_{k\to\infty} \sqrt[k]{|a_k|}$.

(**1**) the series converges when $a^* < 1$;
(**2**) the series diverges when $a^* > 1$;
(**3**) the test fails when $a^* = 1$.

Proof Consider the assertion made in (1). Owing to Theorem 12.10 there is an integer n for each $\epsilon > 0$ such that

$$|a_k|^{1/k} < a^* + \epsilon \qquad (k > n).$$

Since ϵ can be so chosen that $b = a^* + \epsilon < 1$, it follows that $|a_k| < b^k$ for all values $k > n$. The series $\sum b^k$ converges by virtue of Example 14.2(1), and an application of Theorem 13.5 completes the proof of (1).

The second case is argued as follows: There is a subsequence $\{|a_{n_k}|^{1/n_k}\}$ of $\{|a_k|^{1/k}\}$ converging to a^* thereby showing that $\{|a_k|\}$ is not a null sequence and the proof follows.

The third case is handled as (3) in the Ratio Test 16.10.

As another application of absolute convergence we consider briefly the question of convergence with regard to product series. From the several possible products one can form from a given pair of series we consider here only the so-called *Cauchy product*: When formally multiplying together term by term the power series $\sum a_k x^k$ and $\sum b_k x^k$ and collecting the coefficients of like powers of x, we arrive at the power series

$$a_0b_0 + (a_0b_1 + a_1b_0)x + (a_0b_2 + a_1b_1 + a_2b_0)x^2 + \cdots + (a_0b_k + a_1b_{k-1} + \cdots + a_kb_0)x^k + \cdots.$$

Particular importance is, therefore, attached to the coefficient series (which simply corresponds to the case $x = 1$).

16.12 / *Definition* Let $\sum a_k$ and $\sum b_k$ be given; for each k set

$$c_k = \sum_{j=0}^{k} a_j b_{k-j}.$$

The series $\sum c_k$ is the *product* of $\sum a_k$ and $\sum b_k$.

Let us first show that the product of convergent series need not converge.

16.13 / *Example* Consider the series

$$\sum \frac{(-1)^k}{\sqrt{k+1}}.$$

Let $\sum c_k$ in Definition 16.12 be formed by multiplying this series by itself in the indicated manner. Then for each k one gets

$$c_k = (-1)^k \sum_{j=0}^{k} \frac{1}{\sqrt{(k-j+1)(j+1)}}.$$

For each admitted value of j, however,

$$(k-j+1)(j+1) = kj - j^2 + k + 1 \leq k^2 + k + 1 \leq (k+1)^2$$

and consequently

$$|c_k| \geq \frac{k+1}{\sqrt{(k+1)^2}} = 1.$$

Hence $\sum c_k$ diverges, since otherwise $\{c_k\}$ would have to be a null sequence.

It should be noted that the series in the above example converges nonabsolutely. The following, in fact, is true:

16.14 / Theorem Let the notation be that in Definition 16.12. If the series $\sum a_k$ and $\sum b_k$ converge absolutely to the respective limits a and b, then $\sum c_k$ converges absolutely to $c = a \cdot b$.

Proof Let us first show that the series converges. To do this, we note that, for each k_1

$$\begin{aligned} |c_0| + |c_1| + \cdots + |c_k| &\leq |a_0b_0| + (|a_0b_1| + |a_1b_0|) \\ &\quad + \cdots + (|a_0b_{k_1}| + \cdots + |a_{k_1}b_0|) \\ &\leq (|a_0| + \cdots + |a_{k_1}|)(|b_0| + \cdots + |b_{k_1}|) \\ &\leq (\sum |a_k|)(\sum |b_k|) \\ &= c', \end{aligned}$$

c' being some constant. Theorem 14.1 now tells us that the series $\sum |c_k|$ converges, and owing to Corollary 16.6 $\sum c_k$ converges. Thus, it remains only to show that the series converges to the appropriate limit. Here we use Theorem 16.7 which permits us to use a particular rearrangement of the members in $\sum c_k$. Listing the terms a_ib_i in a matrix,

$$\begin{matrix} a_0b_0 & a_0b_1 & a_0b_2 & \cdots \\ a_1b_0 & a_1b_1 & a_1b_2 & \cdots \\ a_2b_0 & a_2b_1 & a_2b_2 & \cdots \\ \vdots & \vdots & \vdots & \end{matrix}$$

we use that rearrangement for which the kth partial sum s_k consists of all entries in the $k \times k$ matrix standing in the upper left corner. Thus,

$$s_k = (\sum_{j=0}^{k} a_j)(\sum_{j=0}^{k} b_j)$$

and it follows at once that $\lim_{k\to\infty} s_k = a \cdot b$.

Mertens has shown that the conclusion of Theorem 16.14 remains valid when only one of the series converges absolutely (see Exercise 16-8).

EXERCISES

16-1 Referring to Theorem 16.1, prove the following: If $\sum b_n$ converges and if the sequence of members

$$c_n = |a_{k_n}| + |a_{k_n+1}| + \cdots + |a_{k_{n+1}-1}|$$

is a null sequence, then $\sum a_n$ converges, and in fact, $\sum a_n = \sum b_n$.

(*Hint*: For each integer k there is one and only one integer n such that $k_{n-1} < k \leq k_n - 1$ $(n \geq 1)$; let each k have the appropriate n associated with it.)

16-2 Construct a rearrangement of the alternating harmonic series which diverges to ∞.

16-3 Prove the following generalization of Theorem 16.7: If $\sum a_i$ converges nonabsolutely, then for every pair of numbers $x \leq y$ in $\mathcal{R}^E$ there is a rearrangement $\sum a_{n_i}$ for the series, with partial sums s_i, such that

$$\liminf_{i\to\infty} s_i = x$$

and

$$\limsup_{i\to\infty} s_i = y.$$

16-4 Can you suggest a product for series other than the one considered?

16-5 In the following power series determine the values of x for which they converge:

(a) $\sum \frac{x^k}{k}$; (c) $\sum \frac{1}{1+x^k}$;

(b) $\sum \frac{(-x)^k}{k}$; (d) $\sum k^x \cdot x^k$.

16-6 Prove that if $\sum a_k x^k$ converges absolutely for some value $x = x_0$, then it converges absolutely for all values x such that $|x| < |x_0|$.

16-7 Show that the Root Test 16.11 is of wider applicability than the Ratio Test 16.10: For any sequence $\{a_k\}$,

(a) $$\liminf_{k\to\infty} \left|\frac{a_{k+1}}{a_k}\right| \leq \liminf_{k\to\infty} \sqrt[k]{|a_k|}$$

(b) $$\limsup_{k\to\infty} \sqrt[k]{|a_k|} \leq \limsup_{k\to\infty} \left|\frac{a_{k+1}}{a_k}\right|.$$

16-8 Establish the following theorem of Mertens: If $\sum a_k$ converges absolutely to a, and $\sum b_k$ converges to b, then $\sum c_k$ converges to ab, the notation being that of 16.12.

(*Hint*: Designating the partial sums of the series involved by α_k, β_k and γ_k, show that $\lim(\gamma_{2k} - \alpha_k\beta_k) = 0$ and $\lim(\gamma_{2k+1} - \alpha_k\beta_k) = 0$.)

16-9 Prove: If $\sum a_k^2$ and $\sum b_k^2$ converge, then $\sum a_k b_k$ converges absolutely.

16-10 Show that the convergence of $\sum a_k^2$ and $\sum b_k^2$ guarantees the inequality

$$\sum a_k b_k \leq (\sum a_k^2)^{1/2}(\sum b_k^2)^{1/2}.$$

The following exercises pertain to the remainder r_k of $\sum a_k$.

16-11 If $| a_{k+1}/a_k | \leq \alpha < 1$ for all values $k > n$, n being some fixed integer, then

$$| r_k | \leq \frac{\alpha}{1 - \alpha} | a_k | \qquad (k > n).$$

16-12 If $\sqrt[k]{| a_k |} \leq \alpha < 1$ for $k > n$ (n fixed), then

$$| r_k | \leq \frac{\alpha^{k+1}}{1 - \alpha} \qquad (k > n).$$

16-13 Consider the series

$$\sum_{k=2}^{\infty} \frac{\sin kx}{\ln k}.$$

Does this series converge for each real number x?

FIVE

THE STRUCTURE OF POINT SETS

Generally speaking, this chapter is concerned with classifications of point sets in $\mathcal{R}$. Although we discuss only subsets of the real line most of our arguments and results apply with little or no modification to point sets in arbitrary metric spaces. The considerations to follow will bring into play the intimate structural features of point sets. Systems of neighborhoods (see Definition 11.2) will constitute a very important tool in this chapter. The role which the neighborhood plays in the theory of convergence was already recognized in Theorem 11.3. Its use is advantageous in preparing for eventual generalizations and abstractions, for the concept of a neighborhood can be completely divorced from that of a metric. The reader should attempt to discover, in fact, how this can be done.

17 / BASIC NOTIONS

Let us begin our investigation with the following definition:

17.1 / ***Definition*** Given a set A and a point x, then x is a *limit point* of A if every neighborhood of x contains a point $a \in A$, $a \neq x$.

It is noted that x may or may not belong to A; this notwithstanding, A must contain at least one member other than x. One has at once the following theorem:

17.2 / Theorem If x is a limit point of A then every neighborhood of x contains infinitely many elements of A.

Proof Suppose some neighborhood of x contains only finitely many points of A. Then it has a member $a \neq x$ such that $d(x, a) \leq d(x, a')$ for all $a' \in A$, $a' \neq x$. In this case, any neighborhood of x of radius $\leq d(x, a)$ contains no point of A different from x and the proof follows.

A given set may contain points which do not satisfy Definition 17.1.

17.3 / *Definition* The point $a \in A$ is an *isolated point* of A if it is not a limit point of A.

Isolated points may be found among the members of both finite and infinite sets, but only the latter have limit points. An interval, for example, consists exclusively of limit points; a finite set contains only isolated points; the set $\{1/n \mid n \in \mathfrak{N}\}$ also has nothing but isolated points yet it has a limit point. Isolated points have a very simple characterization:

17.4 / Theorem The point $a \in A$ is an isolated point of A if and only if there is a neighborhood $N(a; r)$ such that

$$N(a; r) \cap (A \setminus \{a\}) = \varnothing.$$

Infinite sets are not characterized by the existence of limit points as seen from the set $\mathfrak{N}$ which has no limit points in $\mathfrak{R}$.† Limit points are always present with infinite bounded sets. This is the subject of the following well-known theorem:

17.5 / Bolzano–Weierstrass Theorem Every infinite bounded subset of $\mathfrak{R}$ has at least one limit point.

Proof Let A be an infinite bounded set. There are finitely many closed intervals of length 1 whose union contains A. At least one of these, say I_1, contains infinitely many members of A: this follows from the fact that the finite union of finite sets is a finite set. Likewise, there is a closed interval $I_2 \subset I_1$, of length $\frac{1}{2}$, which has infinitely many members of A. Inductively we can select nested intervals

$$I_1 \supset I_2 \supset I_3 \supset \cdots \supset I_n \supset \cdots$$

† Regarded as a subset of $\mathfrak{R}^E$, however, the set $\mathfrak{N}$ has a limit point but we shall not elaborate on this point here.

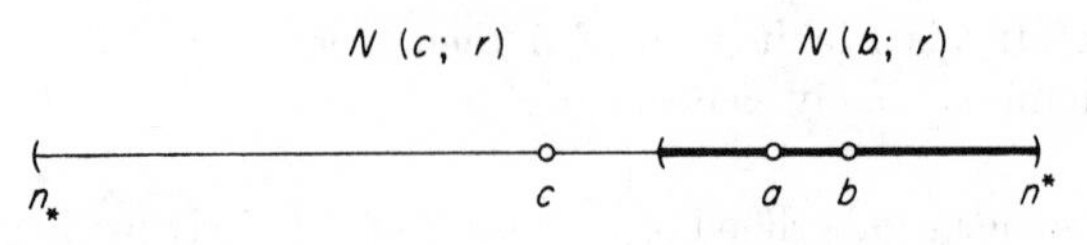

Figure 15

such that $m(I_n) = 1/n$ and each I_n contains infinitely many members of A. Owing to Lemma 9.14 the infinite intersection of the I_n consists of a single point. By Definition 17.1 this point is a limit point of the set A.

The set $\mathcal{R}^*$ was already distinguished by the property that each member of $\mathcal{R}$ can be approximated with its members. This is expressed by saying that $\mathcal{R}^*$ is dense in $\mathcal{R}$:

17.6 / *Definition* Given a set $X \subset \mathcal{R}$, the set $A \subset X$ is *dense in* X when every point of X is a limit point of A or a point of A. A set which is dense in $\mathcal{R}$ is referred to as *everywhere dense.*

The transitive nature of denseness is established in the following:

17.7 / Theorem Let the sets $A \subset B \subset C$ be given. If A is dense in B and B is dense in C, then A is dense in C.

Proof Let c be an isolated point of C. Then c has a neighborhood which contains no other points of C. Since $B \subset C$ and B is dense in C, it follows that $c \in B$. A similar argument shows that $c \in A$. Now let c be a limit point in C; let $N(c; r)$ be any neighborhood of c (see Figure 15). Then this neighborhood must contain some point $b \in B$, $b \neq c$. Being an open interval, $N(c; r)$ contains neither its glb n_* nor its lub n^*, so that this point b cannot coincide with either one of these points. Setting $r' = \min\{d(b, n_*), d(b, n^*), d(b, c)\}$, let $N(b; r')$ be the neighborhood of b of radius r. Again by hypothesis, $N(b; r')$ contains a point $a \in A$, $a \neq b$ and $a \neq c$: since $N(b; r') \subset N(c; r)$ we deduce that $a \in N(c; r)$ and the proof follows with an application of Theorem 17.2.

Contrasting Definition 17.6 is:

17.8 / *Definition* The set $A \subset X$ is *nowhere dense in* X if the following holds:

For every open interval I intersecting X there is an open interval $J \subset I$ such that

$$J \cap X \neq \emptyset$$

whereas

$$J \cap A = \varnothing.$$

The set is simply said to be *nowhere dense* when $X = \mathcal{R}$.

Although not explicitly stated, the case $X = A$ is clearly excepted in the above definition; it should also be pointed out that not all sets necessarily satisfy the demands in Definition 17.6 or Definition 17.8 when the underlying set X is given. The interval $(0, \frac{1}{2})$, for example, is neither dense nor nowhere dense in $(0, 1)$. A finite set is nowhere dense in any interval containing it; the Cantor set to be constructed in a later section will be shown to be uncountable yet nowhere dense. Without difficulty we prove the following:

17.9 / Theorem The set A is nowhere dense if and only if for each nondegenerate closed interval I there is a nondegenerate closed interval $J \subset I$ such that $A \cap J = \varnothing$.

The real line is endowed with two properties which emerge out of the above considerations. The first is that $\mathcal{R}$ contains a countable dense set: $\mathcal{R}$ is said to be *separable*, and separability carries with it the following structure theorem:

17.10 / Theorem There is a countable collection $\mathfrak{U}$ of open intervals in $\mathcal{R}$ with the property that any open interval in $\mathcal{R}$ is expressible as the union of members of $\mathfrak{U}$.

The set $\mathfrak{U}$ is called a *countable basis* for the open sets. A more sophisticated version of this theorem is considered later.

Proof Let A be a given open interval in $\mathcal{R}$. Suppose there is a countable collection of open intervals B_n, having the property that for each $x \in A$ there is an index $n = n(x)$ such that

$$x \in B_{n(x)}$$

and

$$B_{n(x)} \subset A.$$

Then clearly

$$\bigcup_{x \in A} B_{n(x)} \subset A.$$

On the other hand A is the union of its members, implying that

$$A = \bigcup_{x \in A} \{x\} \subset \bigcup_{x \in A} B_{n(x)},$$

and the equality

$$A = \bigcup_{x \in A} B_{n(x)}$$

follows. Thus, the theorem will be proved once we exhibit a collection of open intervals which admits the above for each open interval B. It is here that the separability of $\mathcal{R}$ comes into play.

Let $\{t_m\}$ be any countable dense subset of $\mathcal{R}$. With each t_m we associate the countable collection of neighborhoods $N(t_m; 1/n)$ with radius $1/n$. Owing to Theorem 4.13 the totality $\mathfrak{U}$ of all such neighborhoods is itself countable and we assert that it is endowed with the desired properties.

As in the proof of Theorem 17.7 we deduce that for each $x \in A$ there is a number $r > 0$ such that $N(x; r) \subset A$. There is an integer n such that $1/n < r/2$, and an integer m such that $t_m \in N(x; 1/n)$. Since under these circumstances $d(t_m, x) < 1/n$, one sees that $x \in N(t_m; 1/n)$. But $N(t_m; 1/n) \subset N(x, r) \subset A$ (see Figure 16), and this completes the proof.

To describe the other property we introduce the Baire concept of categories:

17.11 / ***Definition*** The set A is of the *first category* if it is expressible as the countable union of nowhere dense sets. Otherwise A is of the *second category.*

For example, the case $A = \varnothing$ is clearly admitted in Definition 17.11. Hence $\varnothing$ is of the first category. Also of this category is any countable set, since such can be expressed as the countable union of singletons. We notice, in particular, that a union of nowhere dense sets may itself be everywhere dense.

The following is a weak version of the *Baire Category Theorem*:

17.12 / **Theorem** The set $\mathcal{R}$ is of the second category.

Proof The proof consists of showing that no set of the first category contains an interval. In the process we utilize the nested interval

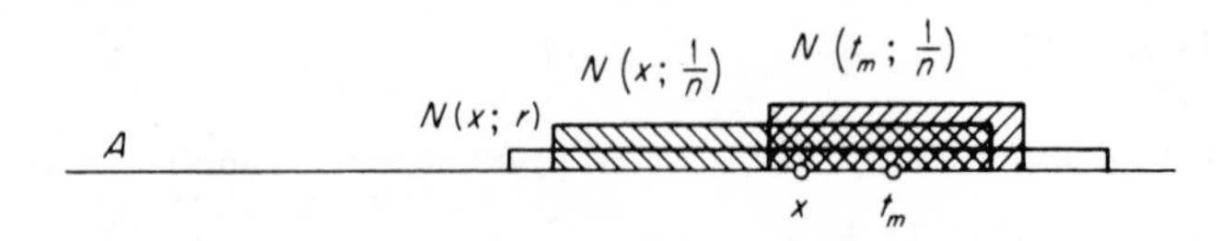

Figure 16

property of the real numbers and Theorem 17.9. All closed intervals to be used in the proof are assumed to be nondegenerate.

Thus, let A be of the first category. Then there are nowhere dense sets $A_n \subset A$ such that

$$A = \bigcup_{n \in \mathfrak{N}} A_n.$$

Suppose A contains an open interval I: A_1 being nowhere dense guarantees the existence of a closed interval $I_1 \subset I$ such that

$$I_1 \cap A_1 = \emptyset.$$

Without loss of generality we may assume that $m(I_1) \leq 1$. Similarly, there is the closed interval $I_2 \subset I_1$, $m(I_2) \leq \frac{1}{2}$, such that

$$I_2 \cap A_2 = \emptyset.$$

By induction, we select a nested sequence of closed intervals

$$I_1 \supset I_2 \supset I_3 \supset \cdots \supset I_n \supset \cdots$$

such that

$$m(I_n) \leq 2^{-n+1}$$

and

$$I_n \cap A_n = \emptyset.$$

Owing to Lemma 9.14 the intersection $\bigcap I_n$ consists of a single real number a: This number, while in I, is excluded from each A_n since $a \in I_n$ for all $n \in \mathfrak{N}$. Hence $a \notin A$, as was to be shown.
We further have:

17.13 / Theorem If I is any nondegenerate interval and A a set of the first category, then the complement I/A is of the second category.

Indeed, writing $I = A \cup (I/A)$, we see that it suffices to show that the union of sets of the first category is a first category set. This is done in the following theorem:

17.14 / Theorem The countable union of sets of the first category is a set of the first category.

Proof If $\{A_n\}$ $(n \in \mathfrak{N})$ is a collection of sets of the first category, then each is expressible as a countable union of nowhere dense sets, say,

$$A_n = \bigcup_{k \in \mathfrak{N}} A_{nk}.$$

In this case, however,

$$A = \bigcup_{n \in \mathfrak{N}} A_n = \bigcup_{n \in \mathfrak{N}} \bigcup_{k \in \mathfrak{N}} A_{nk}$$

so that the set A is a countable union of nowhere dense sets.

17.15 / ***Remark*** We conclude this section with an interesting observation which will play an important role later on. It is this: Let the rational numbers be listed in some sequence, $a_1, a_2, a_3, \ldots$; this is possible owing to their countability. Selecting now an arbitrarily small number $\epsilon > 0$ we associate with the point a_k the neighborhood $N(a_k; \epsilon/2^{k+1})$: this neighborhood has length $\epsilon \cdot 2^{-k}$. For the total length of all of the neighborhoods we have the infinite series

$$\epsilon \cdot \sum_{k=1}^{\infty} \frac{1}{2^k} = \epsilon.$$

We are thus led to the remarkable conclusion that the totality of all rational numbers occupies no space on the line to which a nonzero length can be attached. The set of rational points on the line is said to have *measure zero*. It will be shown in Section 20 that even sets of cardinality $\aleph$ may have this property.

The following conclusion is immediate: By the above argument we can show that the set of all algebraic numbers in the interval $[0, 1]$ has measure 0. The interval itself being of length 1, it follows that this is the "measure" of the set of transcendental (nonalgebraic) numbers in this interval. These ideas will be developed in Chapter Nine.

EXERCISES

17-1 Show that if A is dense in X and $x \in X$, then every neighborhood of x intersects A.

17-2 We say that A is *dense with respect to* X if every point of X is a limit point of A. Prove that if A is dense with respect to B, B is dense with respect to C, then A is dense with respect to C.

17-3 Show that every set in $\mathfrak{R}$ of isolated points is at most countable.

17-4 Prove the following converse of Theorem 17.10: If $\mathfrak{R}$ has a countable basis for its open sets, then $\mathfrak{R}$ is separable.

17-5 Show that if A is of the first category, then so is every one of its subsets.
(*Hint*: Express A as $A = \bigcup_{n \in \mathfrak{N}} A_n$ where each A_n is nowhere dense.

17-6 If B is any subset of A, put $B_n = B \cap A_n$ for each n. Show:

(a) Every subset of a nowhere dense set is nowhere dense;
(b) $B = \bigcup_{n \in \mathfrak{N}} B_n$.)

17-7 Given disjoint closed intervals I_1 and I_2, then there are open intervals G_1 and G_2 such that $G_1 \supset I_1$, $G_2 \supset I_2$ yet $G_1 \cap G_2 = \varnothing$. Prove this statement.

17-8 Show that an arbitrary collection of pairwise disjoint nondegenerate intervals is at most countable.

18 / CLOSED SETS

Closed sets were first encountered in the characterization of sets of subsequential limits (Section 12). Definition 12.5 is now restated as:

18.1 / *Definition* The set A is *closed* when every limit point of A is a member of A.

Note, however, that the present setting is $\mathcal{R}$ rather than $\mathcal{R}^E$. To handle closed sets effectively we associate with each such set the following two sets:

18.2 / *Definitions*

(1) The *derived set* of A is the set A' consisting of its limit points.
(2) The *closure* of A is the set $\bar{A} = A \cup A'$.

18.3 / *Examples*

(1) If A is empty or finite, then $A' = \varnothing$ and $\bar{A} = A$; the same relations hold when $A = \mathfrak{N}$.

(2) In case $A = \mathcal{R}^*$ one sees at once that $A' = \mathcal{R} = \bar{A}$; for any interval I, $I' = \bar{I}$.

(3) When $A = \left\{\frac{1}{n} \,\middle|\, n \in \mathfrak{N}\right\}$, $A' = \{0\}$ and $\bar{A} = \left\{\frac{1}{n} \,\middle|\, n \in \mathfrak{N}\right\} \cup \{0\}$.

In the light of the above definition we can simply state that A is closed when $A' \subset A$. That the sets A' and $\bar{A}$ attached to A are, in fact, closed is the subject of the next two propositions.

18.4 / Theorem For any set A, A' is closed.

Proof The theorem needs no justification when A is empty. Suppose, therefore, that $A \neq \varnothing$. Knowing from the definition that A' contains all limit points of A, it remains only to verify that it contains its own limit points. This, however, was already proved in Lemma 12.4.

18.5 / Theorem The set A is closed if and only if $A = \bar{A}$.

Proof Indeed, when A is closed, then $A' \subset A$ so that also $\bar{A} = A \cup A' \subset A$; on the other hand $A \subset A \cup A' = \bar{A}$ and the desired equality follows. Conversely, suppose $A = \bar{A}$: We shall simply show $\bar{A}$ to be closed. To accomplish this, consider any convergent sequence $\{a_k\}$ with range in $\bar{A}$. This sequence necessarily has a subsequence $\{a_{n_k}\}$ with range in A or in A'. Whichever the case, we are assured that the limit of the subsequence is in A', and Exercise 8-7 then tells us that $\{a_n\}$ converges in $\bar{A}$. Thus $\bar{A}$ is closed.

The set $\bar{A}$ has the following characterization: If $x \in \bar{A}$ and $N(x; r)$ is any neighborhood thereof, then either $x \in A$ or else $N(x; r)$ contains a member of A: The reader should have no difficulty in verifying this fact. Owing to the last theorem, $\bar{\bar{A}} = (\overline{\bar{A}}) = \bar{A}$ holds for an arbitrary set A.

It is now time to investigate the behavior of closed sets under the basic set operations. Evidently neither $A \setminus B$ nor $A \triangle B$ need be closed when A and B are closed. Regarding the operations of taking unions and intersections we have the following results:

18.6 / Theorem The union of finitely many closed sets is closed.

Proof The required argument is essentially that employed in the second part of the proof of Theorem 18.5. Namely, if $A_1, \ldots, A_r$ is a given collection of closed sets, then any sequence $\{a_k\}$ with range in their union contains a subsequence $\{a_{n_k}\}$ with range in but one of the sets, say, A_s. If $\{a_k\}$ converges, then $\lim_{k \to \infty} a_k = \lim_{k \to \infty} a_{n_k} = a$ (Exercise 8-7) where $a \in A_s$, thereby completing the proof.

That this theorem is best possible is illustrated with the following example:

18.7 / *Example* Consider the countable collection of closed sets A_k, where for each $k \in \mathfrak{N}$,

$$A_k = \left[\frac{1}{k+1}, \frac{1}{k}\right].$$

Then $\bigcup_{k \in \mathfrak{N}} A_k = (0, 1]$, a set which is not closed. On the other hand the union of closed sets $B_k = [k, k+1]$ is closed.

18.8 / Theorem The arbitrary intersection of closed sets is always closed.

Proof Let $\{A_t\}$ be a given collection of closed sets with t varying over some index set T; let $A = \bigcap_{t \in T} A_t$. When A is empty or finite there is nothing to prove. Otherwise consider an arbitrary sequence $\{a_n\}$ with range in A: this sequence has its range in each A_t. When the sequence converges, its limit is therefore in each A_t, and consequently also in A. Thus A is closed.

The bounded closed subsets of $\mathfrak{R}$ are distinguished from all others with the Heine–Borel property of which we present here an elementary version. It is described in terms of the following nomenclature:

18.9 / *Definition* Given an arbitrary set $A \subset \mathfrak{R}$, the system $\mathfrak{Q}$ of open intervals is an *open covering* of A if A is contained in their union. The set A is *compact* when every open covering contains a finite subcovering.

By an open subcovering is meant, of course, a subsystem of the original system which is itself a covering. The statement that A has the *Heine–Borel* property will be regarded as synonymous with saying that A is compact. It is clear that in dealing with compact sets we need consider only coverings with intervals of finite length.

18.10 / *Examples*

(1) Consider the open interval $(0, 1)$ and the family $\mathfrak{Q}$ of intervals

$$Q_n = \left(\frac{1}{2^{n+1}}, \frac{1}{2^{n-1}}\right) \qquad (n \in \mathfrak{N}).$$

This family is easily seen to be an open covering of $(0, 1)$ (see Figure 17). It is, however, endowed with the property that, for each value of n, $1/2^n \in Q_m$ if and only if $m = n$. Hence the open interval $(0, 1)$ is not compact: It is readily seen that this conclusion applies to every open interval and half open interval.

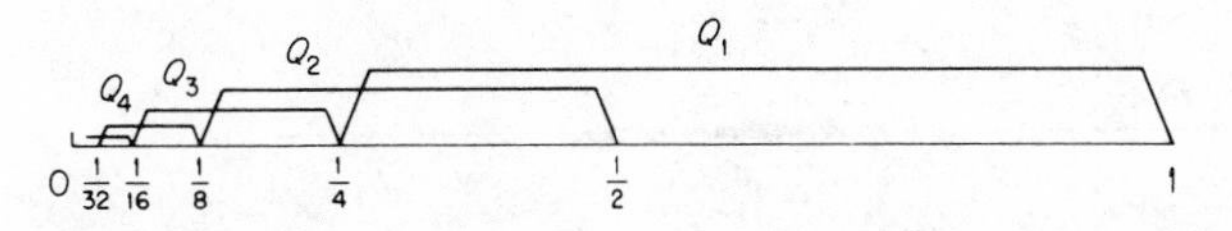

Figure 17

(2) Let us show that also closed but unbounded sets fail to qualify for compactness. Specifically, consider the interval $[1, \infty)$: containing all its limit points, the set is clearly closed. This time let the family $\mathfrak{Q}$ consist of the members $Q_n = (n-1, n+1)$ $(n \in \mathfrak{N})$. As before, the set in question is contained in the union of the Q_n, yet each integer n is the member of one and only one open interval of the family, namely $n \in Q_n$. Our claim is thus justified.

The two examples just examined preclude from further consideration all but one candidate—the closed bounded set. Without further ado we state:

18.11 / Theorem A subset $A \subset \mathfrak{R}$ is compact if and only if it is closed and bounded.

Proof Since the theorem is trivially true for finite sets, let A be infinite. Suppose A is compact: Being contained in a finite number of intervals it is bounded, and it remains only to show that the set is closed. For this it suffices to demonstrate that every point $x \notin A$ has, in fact, a neighborhood which does not intersect A, for no such x can be a limit point of the set.

Let x be a fixed but otherwise arbitrary point in $\mathfrak{R} \setminus A$. Since A is bounded it is contained in the closed interval $[a_*, a^*]$, where a_* is the glb of A, a^* its lub. Let $B = (-b, b)$ be any open interval containing $[a_*, a^*]$ as well as the points $x - 1$ and $x + 1$. With the point x we now associate the open intervals

$$Q_{2n-1} = \left(-b, x - \frac{1}{n}\right)$$

$$Q_{2n} = \left(x + \frac{1}{n}, b\right)$$

where n ranges over $\mathfrak{N}$ (See Figure 18). Clearly

$$\bigcup_{n \in \mathfrak{N}} Q_n = B \setminus \{x\}$$

and since $x \notin A$ it follows that the system $\{Q_n\}$ is an open covering of A. But A is compact, so that there is a finite subcovering of $\{Q_n\}$.

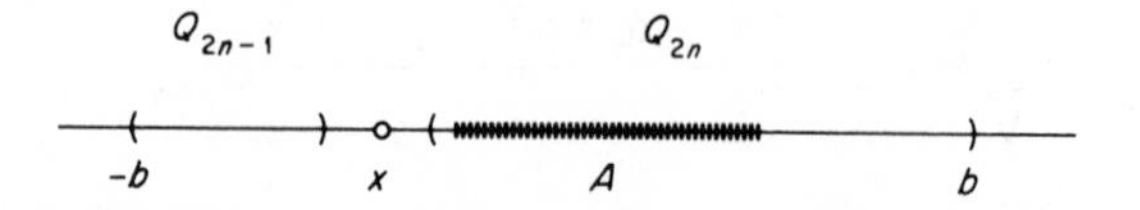

Figure 18

Specifically, there is an integer k such that the family $\{Q_1, \ldots, Q_k\}$ is an open covering of A. This, however, implies that the neighborhood

$$N(x) = \left(x - \frac{1}{k}, x + \frac{1}{k}\right)$$

of x does not intersect A, as was to be shown.

Conversely, assume A to be closed and bounded. The compactness of A will be shown by exhibiting a contradiction which results each time we concede that an open covering $\mathfrak{Q}$ has no finite subcovering. The argument runs like this: The boundedness of A guarantees that it can be covered with a finite collection of closed intervals I_k of lengths not exceeding 1. In view of the above assumption there must be at least one interval, say, I_{n_1}, such that

$$A_1 = I_{n_1} \cap A$$

has no finite subcovering in $\mathfrak{Q}$; in the absence of such an I_{n_1} the entire set A would surely have a finite subcovering in $\mathfrak{Q}$. Focusing our attention on A_1 above, we cover I_{n_1} (and hence A_1) with a finite family of closed intervals I'_k of lengths not exceeding $\frac{1}{2}$. When repeated, the preceding argument produces an interval $I_{n_2} \subset \{I'_k\}$ such that the set

$$A_2 = I_{n_2} \cap A$$

has no finite subcovering in $\mathfrak{Q}$. An iteration now yields a nested sequence of closed intervals I_{n_k},

$$I_{n_1} \supset I_{n_2} \supset \cdots \supset I_{n_k} \supset \cdots,$$

where $m(I_{n_k}) \leq 2^{-k}$, and a corresponding sequence of nested sets $A_k = I_{n_k} \cap A$. These sets have the property that not one of them is covered by finitely many members of $\mathfrak{Q}$.

Now, Lemma 9.14 tells us that the intervals intersect in a single point x,

$$x = \bigcap_{k \in \mathfrak{N}} I_{n_k}.$$

Since clearly $x \in A$, it lies in some member Q_p of $\mathfrak{Q}$. Let $Q_p = (q_1, q_2)$. The number $r = \min\{x - q_1, q_2 - x\}$ is positive and so

$$A_k \subset I_{n_k} \subset Q_p$$

for all values $1/2^k < r$. This means, however, that this Q_p contains all but finitely many of the A_k: Since the construction of the sets A_k bars this possibility we arrived at a contradiction. This contradiction resulted from the assumption that A is not compact. Hence A is compact. —

Open coverings of compact sets cannot be entirely arbitrary; as already discovered by Lebesgue, they are related to the sets in question. This is described in the following theorem:

18.12 / Lebesgue Covering Theorem Let A be a given compact subset of $\mathcal{R}$; let $\mathfrak{Q} = \{Q_\alpha\}$ be an arbitrary open covering thereof. Associated with A and $\mathfrak{Q}$ is a number $\lambda = \lambda(A; \mathfrak{Q})$ with the following property: For every pair of points x_1 and x_2 of A for which $d(x_1, x_2) < \lambda$ there is a member $Q_\alpha \in \mathfrak{Q}$ containing them both.

Thus λ is a *Lebesgue number* of the covering. Clearly, λ is not unique.

Proof For each $x \in A$ there is a Q_α containing it. Holding x momentarily fixed, let $\alpha = \alpha(x)$ be the index relating Q_α to x. Let g_* be the glb of $Q_{\alpha(x)}$, g^* be its lub; set $r(x) = \frac{1}{2}\min\{x - g_*, g^* - x\}$ (see Figure 19). Then this number $r = r(x)$ has the property that if $d(x, y) < 2r(x)$, then $y \in Q_{\alpha(x)}$. Setting

$$N(x; r(x)) = \{y \in A \mid d(x, y) < r(x)\};$$

then obviously the neighborhood lies in $Q_{\alpha(x)}$.

Now consider the collection $\mathfrak{M}$ of such neighborhoods of members of A:

$$\mathfrak{M} = \{N(x; r(x)) \mid x \in A\}.$$

This family is clearly an open covering of A and consequently has among its members a finite subcovering, say,

$$N(x_1; r(x_1)), \ldots, N(x_k; r(x_k)),$$

and we let

$$\lambda = \min\{r(x_1), \ldots, r(x_k)\}.$$

Consider arbitrary points x and y of A such that $d(x, y) < \lambda$. For some integer m, $1 \leq m \leq k$, we know that $x \in N(x_m; r(x_m))$. We also know that $d(x, x_m) < r(x_m)$, and hence

$$d(y, x_m) \leq d(x, y) + d(x, x_m) < \lambda + r(x_m) \leq 2r(x_m).$$

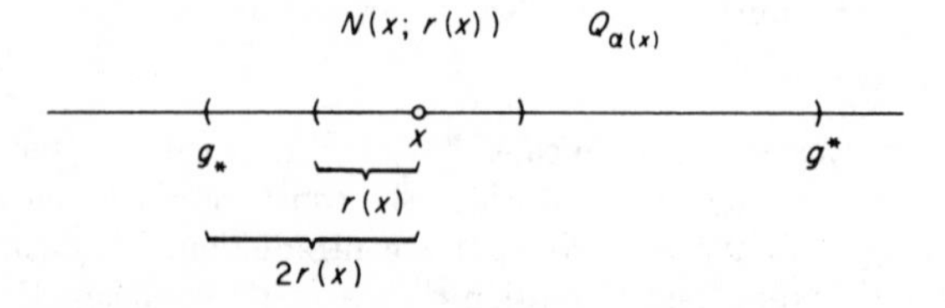

Figure 19

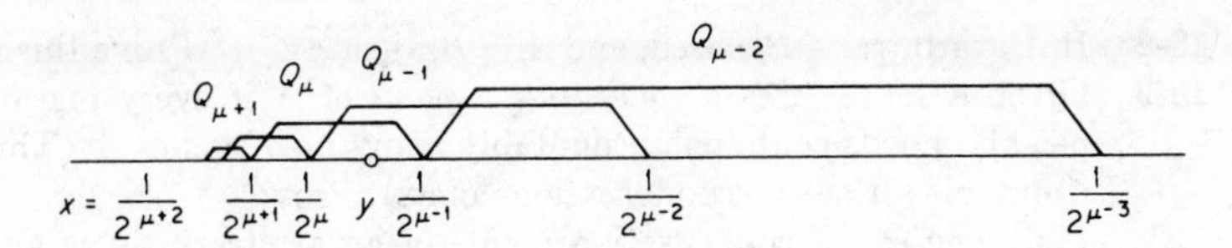

Figure 20

From the definition of $r(x_m)$ it follows, however, that $y \in Q_{\alpha(x_m)}$, thereby completing the demonstration.

18.13 / ***Example*** Consider the situation in Example 18.10(1). We assert that no Lebesgue number λ exists for the covering there. Indeed, let λ be an arbitrary positive number. Select an integer μ such that $3 \cdot 2^{-\mu} < \lambda$ and let $x = 2^{-\mu-2}$, $y = x + 2^{-\mu}$. Then $x \in Q_{\mu+2}$ but to no other member Q_ν, while it is readily seen that $y \in Q_\mu \cap Q_{\mu-1}$ and $Q_{\mu+2} \cap Q_\mu = Q_{\mu+2} \cap Q_{\mu-1} = \varnothing$ (see Figure 20).

Together with Exercise 18-13 this example shows that the Lebesgue Covering Theorem characterizes, in fact, compact sets. The student should explicitly formulate the very important theorem stating the equivalence between the concepts of being closed and bounded, compact and having a Lebesgue number.

EXERCISES

18-1 Show that, for any set A, $\bar{A}$ is the *smallest* closed set containing it: If B is any closed set such that $B \subset \bar{A}$ and $B \neq \bar{A}$, then $A \not\subset B$.

18-2 Show that $A \subset B$ implies that $A' \subset B'$. Demonstrate by means of an example that the converse statement is false.

18-3 Verify the equality $(A \cup B)' = A' \cup B'$, A and B being arbitrary subsets of $\mathfrak{R}$.

18-4 Let $\{A_n\}$ be a countable collection of sets $A_n \subset \mathfrak{R}$: show that

$$\Big(\bigcup_{n \in \mathfrak{N}} A_n\Big)' \supset \bigcup_{n \in \mathfrak{N}} A_n'.$$

Produce an example showing the converse to be false.

18-5 Verify that the set of irrational numbers is not closed.

18-6 Let the sets $A \subset B$ be given. Show that if $B \setminus A$ is finite, then $A' = B'$.

18-7 Construct sets A and B such that

(a) $A \subset B$;
(b) $B \setminus A$ is countable;
(c) $A' = B'$.

18-8 If A is arbitrary, B closed, and $A \subset B$, then $\bar{A} \subset B$. Prove this fact.

18-9 Given a set A, x is a *condensation point* of A if every one of its neighborhoods contains uncountably many members of A (Thus, a countable set has no condensation points). Prove:

(**1**) The set of condensation points of an arbitrary set is always closed.

(**2**) If A_c stands for the set of condensation points of A, then $A_c \subset A'$.

18-10 Use Theorem 18.12 to prove the following generalization of Theorem 9.13(2): If $\{A_k\}$ is a collection of nonempty compact nested subsets of $\mathcal{R}$, then $\bigcap_{k=1}^{\infty} A_k$ is not empty. This result is due to Cantor.

18-11 Is the set

$$A = \bigcup_{n \in \mathfrak{N}} \left[\frac{1}{2^{n+1}}, \frac{1}{2^{n-1}}\right]$$

compact?

18-12 Construct an open covering for $[0, 1]$ having the Lebesgue number $\lambda = \frac{1}{10}$.

18-13 Construct an open covering for $[1, \infty)$ having no Lebesgue number. (*Hint*: Consider a countable covering $\{Q_n\}$ no three members of which intersect and such that the lengths of the pairwise intersections tend to zero as n increases.)

18-14 A family of sets has the *finite intersection property* if every finite subfamily has a nonempty intersection. Show that the set $A \subset \mathcal{R}$ is compact if and only if every family of closed subsets of A with the finite intersection property has a nonempty intersection.

(*Hint*: Let $\{Q_\alpha\}$ be an open covering of A and consider the sets $Q_\alpha{}^c$.)

18-15 Let A be a given set. Its *boundary* $B(A)$ is the set

$$B(A) = \bar{A} \cap \overline{A^c}.$$

Show that the boundary of a closed set is nowhere dense.

19 / OPEN SETS

Open intervals, in general, and neighborhoods, in particular, certainly played a very important role in much of the preceding developments. When examining the applications of these sets in the various proofs and examples, the student will discover that their being intervals was purely incidental whereas their innate structure as sets of interior points was fundamental:

19.1 / *Definition* Given a set A, then $x \in A$ is an *interior point* of A if x has a neighborhood in A.

We see that open intervals (and neighborhoods) are merely examples of open sets:

19.2 / *Definition* The set A is *open* when every point of A is an interior point of A.

It is true that the results in which open intervals have been used could have been proved with arbitrary open sets. The student is strongly urged to closely examine at least some of the instances in this chapter to convince himself that this is, indeed, so. The extensiveness of the class of open sets and their structure are explored below. We begin with:

19.3 / Theorem The arbitrary union of open sets is an open set.

Proof Imagine a family $\{A_\sigma\}$ $(\sigma \in S)$ of open sets and put

$$A = \bigcup_{\sigma \in S} A_\sigma.$$

If $a \in A$, then this point belongs to some member of the family, say, A_τ. Since A_τ is open there is a neighborhood $N(a; r)$ of a such that $N(a; r) \subset A_\tau$ and hence $N(a; r) \subset A$. Thus, a is an interior point of A.

19.4 / *Example* We already know that the property of being open is not preserved in taking arbitrary intersections of open sets. For instance, if $A_n = \left(-\frac{1}{n}, \frac{1}{n}\right)$, then $\bigcap_{n \in \mathfrak{N}} A_n = \{0\}$, but the singleton is not open; if $B_n = \left(-\frac{n+1}{n}, \frac{n+1}{n}\right)$, then $\bigcap_{n \in \mathfrak{N}} B_n = [-1, 1]$. Finite intersections of open sets, however, cause no difficulty:

19.5 / Theorem The intersection of finitely many open sets is an open set.

Proof Suppose

$$A = \bigcap_{k=1}^{n} A_k,$$

where each A_k is open. The theorem is trivial when A is empty, since the empty set is open. Otherwise let $a \in A$. Then $a \in A_k$ for each admitted value of k and hence there are neighborhoods

$$N(a; r_k) = (a_k, b_k)$$

of x such that

$$N(a; r_k) \subset A_k$$

for $k = 1, 2, \ldots, n$. If we set

$$\alpha = \max\{a_1, a_2, \ldots, a_n\}$$

and

$$\beta = \min\{b_1, b_2, \ldots, b_n\},$$

then quite clearly the inclusions

$$(\alpha, \beta) \subset N(a; r_k) \subset A_k$$

ensue, so that

$$a \in (\alpha, \beta) \subset A,$$

i.e., a is an interior point of A.

Recalling now the definition of a complement we prove:

19.6 / Theorem The complement of an open set is closed; the complement of a closed set is open.

Proof Consider the first assertion. Let A be a given open set, let A^c be its complement. We use an argument already exploited in the proof of Theorem 18.11. Namely, if $a \in A$, then the point has a neighborhood $N(a; r) \subset A$ which obviously does *not* intersect A^c. Hence a cannot be a limit point of A^c and the proof follows.

Regarding the second claim, let A be closed. If $a \in A^c$ then a is excluded from A' since $a \notin A$ and so, in particular, $a \notin \bar{A}$. This means, however, that a has a neighborhood $N(a; r)$ such that $N(a; r) \cap A = \varnothing$; this, in turn, implies that $N(a; r) \subset A^c$, thereby proving A^c to be open.

Like open intervals, arbitrary open sets have a simple characterization in terms of a basis. Specifically, Theorem 17.10 can be replaced with the following theorem:

19.7 / Theorem There is a countable family of open intervals such that every open set in $\mathcal{R}$ can be represented as the union of its members.

In view of Theorem 17.10 one need only establish:

19.8 / Lemma Every open set in $\mathcal{R}$ can be realized as the finite or countable union of disjoint open intervals.

Proof Let $A \subset \mathcal{R}$ be open. With each fixed point $a \in A$ we associate the open interval I_a as follows: Since A is an open set, a is an interior point of A and hence there is at least one open interval J containing a. The union

$$I_a = \bigcup_{\substack{J_\rho \subset A \\ a \in J_\rho}} J_\rho \tag{i}$$

represents an open interval (Prove this statement!). In fact, if

$$J_\rho = (\alpha_\rho, \beta_\rho),$$

then

$$I_a = (\operatorname{glb}_\rho \alpha_\rho, \operatorname{lub}_\rho \beta_\rho)$$

(Prove also this statement).

Now, if a_1 and a_2 are arbitrary points of A, then either $I_{a_1} = I_{a_2}$ or else $I_{a_1} \cap I_{a_2} = \varnothing$. For if $x \in I_{a_1} \cap I_{a_2}$, then $I_{a_1} \cup I_{a_2}$ is an open interval containing both a_1 and a_2: Owing to (i), this implies that $I_{a_1} \cup I_{a_2} \subset I_{a_1}$ and $I_{a_1} \cup I_{a_2} \subset I_{a_2}$. Hence $I_{a_1} = I_{a_2}$.

Finally, in each interval I_a we can fix a rational number. Hence there are at most countably many such intervals. It goes without saying that each $a \in A$ is contained in one of the intervals.

Note: The intervals (i) are said to be the *components* of the set in question.

19.9 / Corollary Every closed set is the complement of at most countably many disjoint open intervals.

EXERCISES

19-1 Associated with each point set A is its *interior* A° which is comprised of all interior points of A. Prove the following:

(**1**) For any set A, A° is the largest open set contained in it, in the sense that if $A^\circ \subset B$ and B is open, then $B \not\subset A$.

(**2**) If $A \subset B$, then $A^\circ \subset B^\circ$. Produce an example to show that the converse is false.

(**3**) Characterize the sets A for which $\overline{A^\circ} = A$.

(4) Prove or disprove: $(A \cup B)^\circ = A^\circ \cup B^\circ$ and $(A \cap B)^\circ = A^\circ \cap B^\circ$.

(5) Let B be an arbitrary set. If a given open set A is such that $A \subset B$, then also $A \subset B^\circ$.

(6) With the concept of interior of a set we can describe nowhere dense sets as those sets whose closure has no interior: Prove this statement.

19-2 Prove that every closed set can be realized as the intersection of countably many open sets.

19-2a Can the set $\{1/n \mid n \in \mathfrak{N}\}$ be expressed as the intersection of countably many open sets?

19-2b Can an arbitrary set be expressed as the intersection of countably many open sets?

19-3 We extend the concept of neighborhood as follows: Given an arbitrary point $x \in \mathfrak{R}$, then any open set containing it is a neighborhood thereof. Satisfy yourself that all theorems in which the restricted neighborhoods were used remain true in the new setting.

19-4 Let D represent the aggregate of all open subsets of $\mathfrak{R}$. What is the cardinality of D?

19-5 Show that an arbitrary set with interior is of the second category.

19-6 Verify that the open sets of $\mathfrak{R}$ have a countable basis. Specifically, show that any basis for the open intervals of $\mathfrak{R}$ is at the same time a basis for its open sets.

19-7 The following question is intriguing: Can an open interval be expressed as the union of pairwise disjoint nondegenerate closed intervals? According to Exercise 17-7 there are at most countably many intervals in such a union.

20 / PERFECT SETS

The elements of a closed set fall into the two subsets A' and $A \setminus A'$, the first consisting of its limit points, the second consisting of its isolated points. That this classification is truly exhaustive is substantiated by the elementary decomposition $A = A' \cup (A \setminus A')$. Important as a source of examples and counterexamples is the class of closed sets having no isolated points. We formally state:

20.1 / *Definition* The set A is *perfect* if and only if $A = A'$.

Every closed interval, for example, is perfect; so are the sets $\varnothing$ and $\mathfrak{R}$. We deem it desirable to study these sets through a concrete and nontrivial construction of a perfect set which displays their highly interesting features. The example is due to Cantor.

20.2 / The Cantor Set The set $\mathcal{P}$ to be constructed will be shown to possess these qualities:

(**1**) $\mathcal{P}$ is nowhere dense;
(**2**) $\mathcal{P}$ is perfect;
(**3**) $\mathcal{P}$ has measure zero;
(**4**) $\mathcal{P}$ has cardinality $\aleph$.

The properties have been listed so that the student can keep them in mind while he goes through the construction which follows.

Starting with the closed unit interval $E = [0, 1]$, let P_1 be the set remaining after the removal of the open interval $(\frac{1}{3}, \frac{2}{3})$:

$$P_1 = [0, \tfrac{1}{3}] \cup [\tfrac{2}{3}, 1].$$

Next, discard the open middle third of each of the components of P_1, thus forming the set P_2 whose components are

$$[0, \tfrac{1}{9}], \quad [\tfrac{2}{9}, \tfrac{3}{9}], \quad [\tfrac{6}{9}, \tfrac{7}{9}], \quad [\tfrac{8}{9}, 1]$$

(see Figure 21). When iterated, these steps lead to a nested sequence of closed sets P_n,

$$P_1 \supset P_2 \supset P_3 \supset \cdots,$$

each P_n consisting of 2^n closed intervals of length 3^{-n}. The set

$$\mathcal{P} = \bigcap_{n \in \mathfrak{N}} P_n$$

is said to be a *Cantor Set*.

Regarding the properties of the set, we first note that $\mathcal{P}$ is not vacuous: this is guaranteed by Theorem 9.13(2). To show that $\mathcal{P}$ is nowhere dense in E we observe that no interval of the form

$$\left(\frac{2n-1}{3^m}, \frac{2n}{3^m}\right) \qquad (m, n \in \mathfrak{N})$$

intersects $\mathcal{P}$. Also, given an arbitrary interval $(a, b) \subset E$, then the inequalities

$$a < \frac{2n-1}{3^m} < \frac{2n}{3^m} < b$$

are satisfied when $3^{-m} < b - a$ and $3^m a + 1 < 2n < 3^m b$: Selecting a sufficiently large value of m to assure that the interval $(3^m a + 1, 3^m b)$ contains an even integer, we find that every open interval in E contains an interval which does not intersect $\mathcal{P}$. This shows $\mathcal{P}$ to be as claimed in (1).

Moving on to verify (2), we begin with the observation that $\mathcal{P}$ is closed: this follows from Theorem 18.8. The set will be shown to be perfect once we demonstrate that it has no isolated points. For this, let p be an

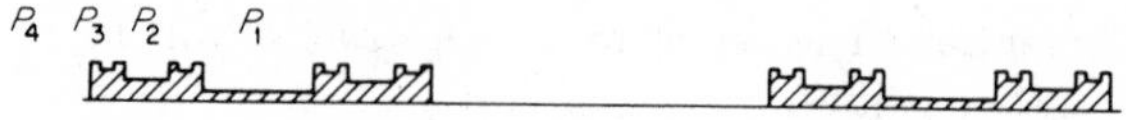

Figure 21

arbitrary point of $\mathcal{P}$; let $N(p; r)$ be any neighborhood of p. Each set P_n contains a component I_n such that

$$p = \bigcap_{n \in \mathfrak{N}} I_n.$$

Because $m(I_n) = 3^{-n}$, it follows that $I_n \subset N(p; r)$ whenever $3^{-n} < r$. But each I_n contains at least two members of $\mathcal{P}$ (why?) and (2) is therefore established.

The assertion in (3) is to be interpreted as follows: The set $E \setminus \mathcal{P}$ is the countable union of disjoint open intervals. Let the intervals be I_n. Then the measure of $E \setminus \mathcal{P}$ is defined to be

$$\mu = \sum_{n=1}^{\infty} m(I_n)$$

and the measure of $\mathcal{P}$ is taken to be $1 - \mu$ (see Remark 17.15).

With this in mind we recall that in obtaining P_1 we discarded an interval of length $\frac{1}{3}$; in the next step we discarded two intervals, each of length 3^{-2}. In general, 2^{n-1} intervals of length 3^{-n} each have been discarded in the nth step. Thus, the total length of the intervals eliminated in this step is

$$3^{-1} + 2 \cdot 3^{-2} + \cdots + 2^{n-1} \cdot 3^{-n} = 1 - (\tfrac{2}{3})^n,$$

and (3) follows.

To dispose of the remaining assertion we employ a normalized representation of the numbers $x \in [0, 1]$ to the base 3:

$$x = \sum_{n=1}^{\infty} a_n 3^{-n} \qquad (a_n = 0, 1, 2).$$

It is quite apparent that if $x \in E \setminus P_1$, then necessarily $a_1 = 1$. Setting $\frac{1}{3} = 0.0222\cdots$ and $\frac{2}{3} = 0.2000\cdots$, it follows that for *all* points x in P_1, $a_1 = 0, 2$. In a like fashion we realize that if $x \in E \setminus P_2$, then $a_2 = 1$, and with a suitable normalization we find that $a_2 = 0, 2$ for all points $x \in P_2$. Continuing in this manner, we can draw the following conclusion: $x \in \mathcal{P}$ if and only if the entries a_n in its expansion to the base 3 assume the values 0 and 2; that is, each such $x \in \mathcal{P}$ has a representation $x = 0.a_1a_2a_3\cdots$, where $a_n = 0, 2$.

Using the base 2, each $x \in E$ has a representation $x = 0.b_1b_2b_3\cdots$ in which $b_n = 0, 1$. The association $0 \leftrightarrow 0$, $2 \leftrightarrow 1$ establishes a one-to-one correspondence between $\mathcal{P}$ and E, and thus $\mathcal{P}$ is uncountable.

That (4) above is a characteristic of nontrivial perfect sets is the subject of the next theorem:

20.3 / Theorem A perfect set is either empty or uncountable.

Proof The impossibility of having a finite perfect set is self-evident. Thus, suppose P is countable; let its members be $p_1, p_2, p_3, \ldots$. Assume momentarily that there is a nested sequence of compact intervals I_k,

$$I_1 \supset I_2 \supset I_3 \supset \cdots$$

such that for each value of k

$$p_k \notin I_k,$$

but

$$J_k = I_k \cap P \neq \varnothing.$$

Then no point of P is included in every set J_k and hence

$$\bigcap_{k \in \mathfrak{N}} J_k = \varnothing.$$

On the other hand, the sets J_k are themselves nested and compact and, as such, have a nonempty intersection, as follows from Exercise 18-10. Having thus arrived at a contradiction, we are compelled to agree that P has cardinality $\aleph$ provided we can produce the intervals in question. This is done inductively as follows:

Let N be an arbitrary neighborhood of p_1. The fact that this point is a limit point of P assures the existence of an open interval $N_1 \subset N$ such that $\bar{N}_1 \subset N$, $p_1 \notin \bar{N}_1$, yet $N_1 \cap P \neq \varnothing$ (see Figure 22); set $\bar{N}_1 = I_1$. Now, N_1 contains a member of P, that is, a limit point of P. This permits us to emulate the above argument: We now have an open interval $N_2 \subset N_1$ such that $\bar{N}_2 \subset N_1$, $p_2 \notin \bar{N}_2$, but $N_2 \cap P \neq \varnothing$. Using an induction hypothesis, suppose the open set N_k $(k \geq 2)$ is already fixed. Then it contains a limit point of P and an interval N_{k+1} with the qualifications $\bar{N}_{k+1} \subset N_k$, $p_{k+1} \notin \bar{N}_{k+1} \cap P \neq \varnothing$. We set $\bar{N}_{k+1} = I_{k+1}$. Clearly, the set of intervals $I_1, I_2, I_3, \ldots$ is as demanded.

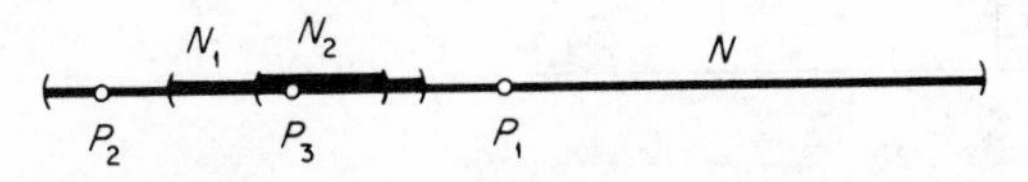

Figure 22

Another very interesting property possessed by the Cantor set $\mathcal{P}$ is this: For any number $\alpha \in [0, 1]$ there are points x and y in $\mathcal{P}$ such that $d(x, y) = \alpha$. More generally, with every nonempty set A we associate the *difference set* $D(A)$:

$$D(A) = \{x - y \mid x \in A, y \in A\}.$$

We have therefore:

20.4 / Theorem If $\mathcal{P}$ is as in 20.2, then $D(\mathcal{P}) = [-1, 1]$.

Specifically, this theorem demonstrates that the difference set of a nowhere dense set may generate a continuum. The verification of this theorem is merely sketched here; it is carried out in the following manner: for each value of $\alpha \in [0, 1]$ consider the function $x - y = \alpha$ in the product set $E \times E$. It clearly suffices to show that every such line intersects the set $\mathcal{P} \times \mathcal{P}$ (see Figure 23). Clearly, for each point $(x, y) \in \mathcal{P} \times \mathcal{P}$ there are components $I_k \subset P_k$ and $J_k \subset P_k$ such that $(x, y) = \bigcap_{k \in \mathfrak{N}} I_k \times J_k$. From the geometry of the construction it is clear that:

(a) every line must intersect at least one corner square;
(b) if a line intersects a given corner square $I_k \times J_k$, then it will intersect a corner square $I_{k+1} \times J_{k+1}$ contained therein.

From this we conclude that every line intersects a member of $\mathcal{P} \times \mathcal{P}$.

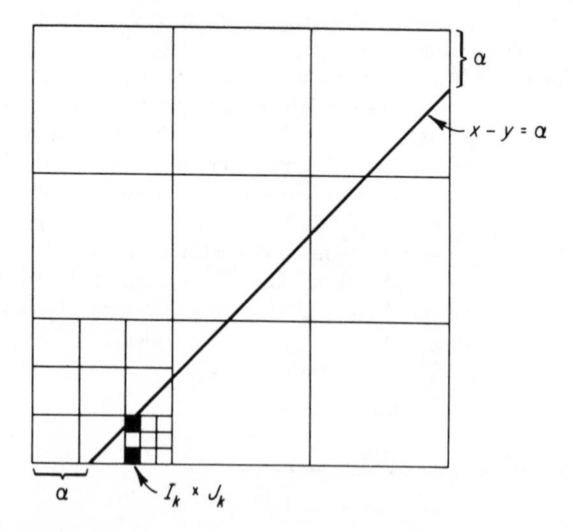

Figure 23

EXERCISES

20-1 Using the argument in 20.2, construct a Cantor Set in $[0, 1]$ having measure $\frac{1}{2}$ (that is, a set whose complement in $[0, 1]$ consists of intervals having total length $\frac{1}{2}$).

20-2 Deduce from the above exercise that for each $\epsilon \geq 0$ there is a Cantor set in $[0,1]$ having measure $1 - \epsilon$.

20-3 Consider the (normalized) representation of numbers $x \in [0, 1]$ to the base 5: $x = \sum a_k 5^{-k}$ $(0 \leq a_k \leq 4)$. Consider the set A of all expansions in which the a_k are restricted to the values 0, 2, 4: is A a Cantor set?

20-4 Can you suggest the construction of a perfect set Q such that its difference set $D(Q)$ will contain *no* neighborhood of the origin?

20-5 Can you construct a first category set in $[0, 1]$ having measure 1?

20-6 Show that an arbitrary closed set is the union of a perfect set and a finite or countable set.

21 / DISTANCE BETWEEN POINT SETS

An often implied feature of $\mathcal{R}$ is its point separating quality, which is to say that any two distinct points possess mutually exclusive neighborhoods. It should be instructive for the student to hunt through Chapters Four and Five for those propositions in which the inference was indispensable. The corresponding separation of closed intervals was the subject of Exercise 17-6 and our aim in the present section is to show that a separating property actually extends to every pair of arbitrary closed sets. In the process we extend the notion of distance from singletons to arbitrary point sets, pursuing this line of thought: Given nonempty point sets A and B, we consider the distance function $d : A \times B \to \mathcal{R}$. The range of the function is bounded below by zero and hence has a greatest lower bound d_*. The function d is thus seen to associate with the pair (A, B) the unique number d_*. This motivates us to consider:

21.1 / *Definition* If A and B are given nonempty point sets, then the *distance* between them is the number

$$\delta(A, B) = \operatorname*{glb}_{\substack{a \in A \\ b \in B}} d(a, b).$$

Among the elementary properties of the function δ are these:

(a) Whenever the sets A and B intersect, then $\delta(A, B) = 0$; in general $\delta(A, B) \geq 0$.

(b) When A and B have a common limit point, then again $\delta(A, B) = 0$, this time regardless of whether or not the sets intersect.

21.2 / *Examples*

(**1**) Consider the sets $\mathfrak{N}$ and $M = \left\{k + \frac{1}{2k} \,\middle|\, k \in \mathfrak{N}\right\}$. Both sets are closed and disjoint, yet $\delta(M, \mathfrak{N}) = 0$.

(**2**) Let $A = (0, 1), B = (1, 2), C = (2, 3)$. Then

$$1 = \delta(A, C) \geq \delta(A, B) + \delta(B, C) = 0 + 0 = 0.$$

The situation depicted here shows that the triangle inequality associated with a metric is violated in this case; of course (a) above already told us that δ is not a metric.

The two examples show that there may exist no points for which the infimum is realized. We are now led to consider:

21.3 / Theorem Let A and B be nonempty point sets. If A is compact and B is closed, then there are points $a \in A$ and $b \in B$ such that

$$d(a, b) = \delta(A, B).$$

Proof The definition of the greatest lower bound (see, in particular, Exercise 9-4) tells us that for each $\nu \in \mathfrak{N}$ there are members $a_\nu \in A$ and $b_\nu \in B$ such that

$$\delta(A, B) \leq d(a_\nu, b_\nu) < \delta(A, B) + \frac{1}{\nu}.$$

From the compactness of A we deduce that the sequence $\{a_\nu\}$ is bounded. Owing to Exercise 9-12 this sequence has a convergent subsequence $\{a_{k_\nu}\}$; setting

$$\lim_{\nu \to \infty} a_{k_\nu} = a,$$

it is clear that $a \in A$.

Suppose $|a_\nu| \leq m$ for all $\nu \in \mathfrak{N}$. Consider the subsequence $\{b_{k_\nu}\}$ of $\{b_\nu\}$. Then for each fixed ν

$$|b_{k_\nu}| \leq |a_{k_\nu}| + d(a_{k_\nu}, b_{k_\nu}) \leq m + \delta(A, B) + \frac{1}{\nu}$$

$$\leq m + \delta(A, B) + 1,$$

thereby showing that the sequence $\{b_{k_\nu}\}$ is bounded. Again appealing to Exercise 9-12 we conclude that it has a convergent subsequence, say, $\{b_{nk_\nu}\}$. If we set

$$\lim_{\nu\to\infty} b_{nk_\nu} = b,$$

then clearly $b \in B$ (since the set is closed) and with the help of Exercise 8-7 we find that

$$d(a, b) = \lim_{\nu\to\infty} d(a_{nk_\nu}, b_{nk_\nu}) = \delta(A, B)$$

as was to be demonstrated.

The preceding examples show that the hypotheses of the theorem cannot be relaxed. With it we can prove:

21.4 / Separation Theorem Let A_1 and A_2 be arbitrary closed sets. If

$$A_1 \cap A_2 = \varnothing,$$

then there are open sets B_1 and B_2 such that

$$B_1 \supset A_1, \qquad B_2 \supset A_2$$

and

$$B_1 \cap B_2 = \varnothing.$$

Proof Consider the set A_1. Since a singleton is a compact set, exercise 21-1 will permit us to associate with each $x \in A_1$ a constant $\epsilon_1(x) > 0$ such that $\delta(x, A_2) = \epsilon_1(x)$. Attaching to each $x \in A_1$ the neighborhood $N(x; \frac{1}{3}\epsilon_1(x))$ we set

$$B_1 = \bigcup_{x\in A_1} N(x; \tfrac{1}{3}\epsilon_1(x))$$

owing to Theorem 19.3 this set is open, and evidently $B_1 \supset A_1$. Interchanging the indices 1 and 2 in the above argument we arrive at the set $B_2 \supset A_2$ and we assert that $B_1 \cap B_2 = \varnothing$.

Consulting Theorem 21.3, we conclude that for each point $x \in A_1$ there is a point $y \in A_2$ such that

$$\epsilon_1(x) = \delta(x, A_2) = d(x, y) \geq \delta(y, A_1) = \epsilon_2(y).$$

For each fixed $x \in A_1$, therefore,

$$\delta[N(x; \tfrac{1}{3}\epsilon_1(x)), B_2] = \delta[N(x; \tfrac{1}{3}\epsilon_1(x)), N(y; \tfrac{1}{3}\epsilon_2(y))]$$
$$= \epsilon_1(x) - \tfrac{1}{3}\epsilon_1(x) - \tfrac{1}{3}\epsilon_2(y) \geq \tfrac{1}{3}\epsilon_1(x) > 0,$$

with a similar calculation following for each $y \in A_2$. The theorem is therefore true.

EXERCISES

21-1 Let A be compact, B closed. If $\delta(A, B) = 0$, then $A \cap B \neq \emptyset$.

21-2 Given an arbitrary set A, and a number $\epsilon > 0$ let

$$B = \{x \mid \delta(x, A) < \epsilon\}$$

$$C = \bigcup_{a \in A} N(a; \epsilon).$$

Prove that $B = C$.

21-3 Show that the Separation Theorem 21.4 remains true for arbitrary sets A_1 and A_2 such that $\delta(A_1, A_2) > 0$.

21-4 Consider the following distance function: Given bounded sets A and B, define

$$\Delta(A, B) = \operatorname{lub}_{\substack{a \in A \\ b \in B}} d(a, b).$$

Does this function constitute a metric on the bounded subsets of $\mathcal{R}$?

21-5 The set $A \subset \mathcal{R}$ is *totally bounded* if, for each $\epsilon > 0$, there is a finite set $x_1, \ldots, x_n$, $x_j \in \mathcal{R}$, such that for each $a \in A$ there is an x_j for which $d(a, x_j) < \epsilon$: Show that a set $A \subset \mathcal{R}$ is bounded if and only if it is totally bounded.

22 / CONNECTED SETS

The closed sets and open sets of $\mathcal{R}$ were distinguished from each other through numerous properties. Yet a certain pathology is inherent here since a dual classification of sets was observed to occur; namely, $\emptyset$ and $\mathcal{R}$ were seen to be both open and closed. It is, therefore, of some importance to ascertain that the mentioned sets are, in fact, the only subsets of $\mathcal{R}$ having this double status. This is settled in the next theorem:

22.1 / Theorem The only subsets of $\mathcal{R}$ which are both open and closed are $\emptyset$ and $\mathcal{R}$ itself.

The characterization of $\mathcal{R}$ which is offered by this theorem is noteworthy since it is devoid of both algebraic and metric concepts. This theorem is replaced by an equivalent statement with the following argument:

Let $A \subset \mathcal{R}$ be both open and closed. Viewed as a closed set, A has an open complement (Theorem 19.6): The formula $\mathcal{R} = A \cup (\mathcal{R} \setminus A)$ reveals that $\mathcal{R}$ can be expressed as the union of two disjoint open sets in this case. Conversely, if $\mathcal{R} = A \cup B$, where $A \cap B = \emptyset$ and the sets are open, then they are complementary to each other, and as such must be both open and closed. This leads us to state:

22.2 / Theorem The set $\mathcal{R}$ cannot be represented as the union of two disjoint nonempty open sets.

Proof According to Lemma 19.8, if $\mathcal{R}$ is expressible as the union of two open sets, then it can be written as the union of at most countably many disjoint open intervals, say

$$\mathcal{R} = \bigcup I_k. \tag{i}$$

The completeness of $\mathcal{R}$ tells us that this union cannot consist of two nonempty intervals: This fact is most vividly brought out in Dedekind's completion of the real numbers (Sections 5 and 9). Hence, if no I_k equals $\mathcal{R}$, there must be at least three nonempty members in the right side of (i), one of which must be bounded. Let it be I_1. Taking into account that the endpoints of I_1 are limit points of I_1 and interior points of members I_k, it follows that either some of the intervals have nonempty intersections or else (i) is false. In any case, the theorem follows.

22.3 / *Definition* The set $X \subset \mathcal{R}^E$ is *disconnected* if there are disjoint open sets A and B such that

(a) $A \cap X \neq \varnothing$,
(b) $B \cap X \neq \varnothing$,
(c) $A \cup B \supset X$.

Otherwise X is said to be *connected.*

Note, in particular, that the empty set and a singleton are connected: Other examples are given below. The connected subsets of $\mathcal{R}$ have a simple property:

22.4 / Lemma The set $A \subset \mathcal{R}$ is connected when and only when the following is true: If $a_1 < a_2$ belong to A, then so does every point x such that $a_1 < x < a_2$.

Indeed, if the last statement were false for some admitted point $a_1 < b < a_2$, then the sets $B_1 = \{x \in \mathcal{R} \mid x < b\}$ and $B_2 = \{x \in \mathcal{R} \mid x > b\}$ would satisfy Definition 22.3. Conversely, when A is disconnected it is covered with at most countably many disjoint open intervals I_k. Among these we can find two nonempty members which intersect A, say I_1 and I_2, $b =$ lub of $I_1 <$ glb of I_2. But if we select points $a_1 \in I_1 \cap A$ and $a_2 \in I_2 \cap A$, then evidently $b \notin A$ yet $a_1 < b < a_2$.

From this lemma we deduce at once:

22.5 / Corollary The only connected subsets of $\mathcal{R}$ are (bounded and unbounded) intervals.

One is easily convinced that connected subsets in the n-fold product space $\mathbb{R}^n$ are not simply described when $n > 1$.

22.6 / *Example* Consider subsets of $\mathbb{R}$. While the union of intervals need not be an interval, their (possibly empty) intersection always is. The complement of a bounded interval is disconnected, that of an unbounded interval is connected.

22.7 / Theorem If $\{A_\sigma\}$ is an arbitrary family of connected subsets of $\mathbb{R}$ having a nonempty intersection, then their union is connected.

Proof Setting $A = \bigcup A_\sigma$, suppose B_1 and B_2 to be disjoint open sets such that $B_1 \cap A \neq \emptyset$, $B_2 \cap A \neq \emptyset$ and $A \subset B_1 \cup B_2$. Consider an arbitrary point $a \in \bigcap A_\sigma$. Then this point belongs to one of the covering sets of A, say, B_1. At the same time there is a value σ such that $A_\sigma \cap B_2 \neq \emptyset$ and since, in particular, $a \in A_\sigma$, it follows that also $A_\sigma \cap B_1 \neq \emptyset$. All this leads us to the realization that

$$A_\sigma = (A_\sigma \cap B_1) \cup (A_\sigma \cap B_2) = A_\sigma \cap (B_1 \cup B_2)$$

which implies, in turn, that $A_\sigma \subset B_1 \cup B_2$. The disconnectedness of A_σ is hereby contradicted and the theorem is therefore true.

EXERCISES

22-1 Let $A \subset \mathbb{R}$ be a nonempty set and let $a \in A$ be fixed. Let $C(a)$ stand for the union of all connected subsets of A which contain a. The set $C(a)$ is the *connected component* of a in A. Prove:
(1) $C(a)$ is the largest connected subset of A containing a: If $B \subset A$ is connected and $a \in B$, then $B \subset C(a)$.
(2) If $x \in C(a)$, then $C(x) = C(a)$.
(3) If $x \notin C(a)$, then $C(x) \cap C(a) = \emptyset$.
(4) $C(a)$ is closed in A: If $x \in A$ is a limit point of $C(a)$, then $x \in C(a)$.

22-2 Show that the decomposition of an open subset of $\mathbb{R}$ into the countable union of pairwise disjoint open intervals is unique.

22-3 Consider a finite family of connected sets $A_i \subset \mathbb{R}$, $1 \leq i \leq n$. If

$$A_i \cap A_{i+1} \neq \emptyset$$

for $i = 1, 2, \ldots, n - 1$, then

$$\bigcup_{i=1}^{n} A_i$$

is connected. Can the finiteness restriction be lifted?

PART THREE

FUNCTIONS OF A REAL VARIABLE

SIX

CONTINUITY

The notion of continuity is known to the reader from calculus. Continuity, it will be recalled, is a local property, which is to say that a given function may be continuous at some points but not at others. Indeed, continuity implies the existence of a certain limit *at a point* (The derivative is another local property which again entails the existence at a point of a particular limit: it is studied in Chapter Seven).

The reader may wish to review the nomenclature and examples in Section 3.

23 / LIMITS OF FUNCTIONS

Given a set $A \subset \mathcal{R}$ and a function $f: A \to \mathcal{R}$, then an arbitrary sequence $\{x_n\}$ with range in A uniquely specifies the sequence $\{f(x_n)\}$ with range in $\mathcal{R}$; particular significance being attached to the case in which all equivalent convergent sequences $\{x_n\}$ produce equivalent convergent sequences $\{f(x_n)\}$. Suppose $x_n \to a$. From the various alternatives which present themselves here our current interest concerns the case in which $f(a)$ need not equal $\lim f(x_n)$ and, in fact, may not even exist. Against this background we introduce the following sets: Let $a \in \mathcal{R}$ be fixed and let $\delta > 0$

be a given number. Then

$$N^-(a;\delta) = \{x \mid a-\delta < x < a\},$$

$$N^+(a;\delta) = \{x \mid a < x < a+\delta\},$$

$$N^*(a;\delta) = N^-(a;\delta) \cup N^+(a;\delta).$$

The last set is said to be a *deleted neighborhood* of a (see Figure 24).

23.1 / ***Definitions*** Let $A \subset \mathcal{R}$ be a given set, $f: A \to \mathcal{R}$ a given function and a a limit point of A.

(**1**) f has a *left limit* at a, written

$$\lim_{x \to a-} f(x) = f(a-),$$

if for each $\epsilon > 0$ there is a $\delta > 0$ such that

$$f(x) \in N(f(a-);\epsilon)$$

whenever

$$x \in N^-(a;\delta) \cap A;$$

(**2**) f has a *right limit* at a, written

$$\lim_{x \to a+} f(x) = f(a+),$$

if for each $\epsilon > 0$ there is a $\delta > 0$ such that

$$f(x) \in N(f(a+);\epsilon)$$

whenever

$$x \in N^+(a;\delta) \cap A;$$

(**3**) f has a *limit* b at a, written

$$\lim_{x \to a} f(x) = b,$$

if for each $\epsilon > 0$ there is a $\delta > 0$ such that

$$f(x) \in N(b;\epsilon)$$

when

$$x \in N^*(a;\delta) \cap A.$$

These definitions are illustrated in Figure 24. It must be made clear that b need not belong to the range of f. The student should prove:

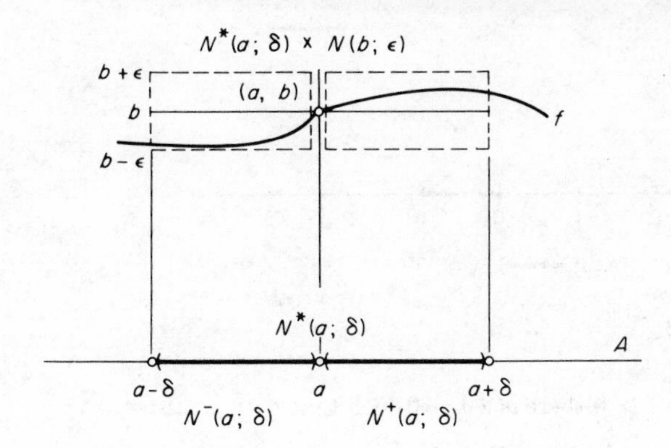

Figure 24

23.2 / Theorem If $f: A \to \mathcal{R}$ is given and if a is a limit point of A, then f has a limit b at a if and only if

$$f(a-) = f(a+) = b.$$

23.3 / Theorem The limit of a function at a point is unique whenever it exists.

23.4 / *Examples*

(**1**) Consider the *step function*

$$f(x) = [x] \qquad (x \in \mathcal{R}),$$

$[x]$ standing for the integral part of x (See Figure 25). For each integer n,

$$f(n-) = n - 1, \qquad f(n+) = n, \qquad f(n) = n.$$

Thus, the left and right limits exist at each point but disagree at all integral points. At the latter points, therefore, f has no limit.

(**2**) Let $g(x) = 1/x$ $(x \neq 0)$. Then

$$g(0-) = -\infty \qquad \text{and} \qquad g(0+) = \infty;$$

that is, neither the left nor the right limits exist at the point $x = 0$. The existence of a limit can be put in a metric setting:

23.5 / Theorem Let $f: A \to \mathcal{R}$ be a given function and a a limit point of A. For f to have a limit b at $x = a$ it is necessary and sufficient that

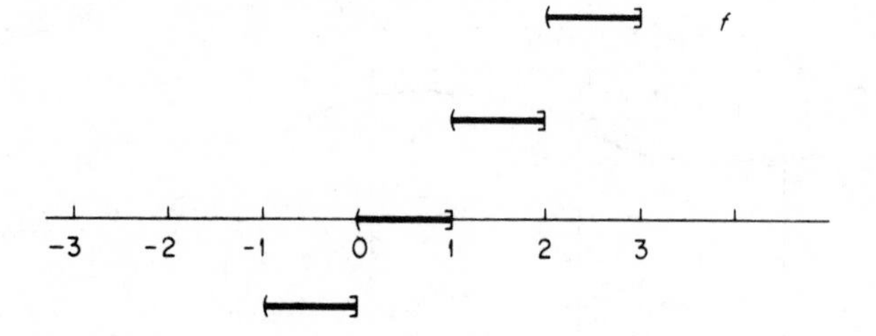

Figure 25

for each $\epsilon > 0$ there is a $\delta > 0$ such that

$$0 < d(x, a) < \delta \qquad (x \in A)$$

implies

$$d(f(x), b) < \epsilon.$$

To make available to us the machinery developed in Chapters Three and Four for handling sequences of numbers we express Definition 23.1 equivalently in terms of limits of sequences:

23.6 / Corollary In the setting of the above theorem, f has a limit b at $x = a$ if and only if

$$\lim_{n\to\infty} f(x_n) = b.$$

for *every* sequence $\{x_n\}$ with range in $A \setminus \{a\}$ and limit a.

The last corollary makes the following additional examples readily accessible to us:

23.7 / *Examples*

(**1**) An arbitrary polynomial $p(x)$ has a limit at every point of $\mathcal{R}$; moreover, $\lim_{x\to a} p(x) = p(a)$ for each $a \in \mathcal{R}$.

(**2**) The function

$$f_1(x) = \begin{cases} 1 & \text{when } x \text{ is rational} \\ -1 & \text{when } x \text{ is irrational} \end{cases}$$

has a limit nowhere, since every x can be approximated with rational numbers r_n, for which $f(r_n) \to 1$, or with irrational numbers s_n, for which $f(s_n) \to -1$.

(**3**) The function

$$f_2(x) = \begin{cases} x & \text{when } x \text{ is rational} \\ -x & \text{when } x \text{ is irrational} \end{cases}$$

has a limit at the point $x = 0$ but at no other point; in fact, $\lim_{x \to 0} f_2(x) = f_2(0)$.

(**4**) The function h is said to be *nowhere constant* if, given an arbitrary point x_0, then every neighborhood $N(x_0; r)$ thereof contains a point x such that $f(x) \neq f(x_0)$. Constructed now is a nowhere constant function h having the property that $\lim_{x \to a} h(x) = 0$ at every point $a \in \mathcal{R}$. To carry this out it is agreed to consider only rational numbers $\frac{m}{n}$ for which m and n are relatively prime and $n > 0$. With this understanding we define the function h for each x as follows:

$$h(x) = \begin{cases} \frac{1}{n} & \text{when } x = \frac{m}{n} \\ 1 & \text{at } x = 0 \\ 0 & \text{elsewhere.} \end{cases}$$

Since the function vanishes at every irrational point but differs from zero at each rational point, it is clear that it is nowhere constant.

To substantiate the other claim made we prove:

Lemma Let x be an arbitrary irrational number; let $\{m_\nu/n_\nu\}$ $(m_\nu, n_\nu \in \mathfrak{N})$ be a sequence converging to it. Then $|n_\nu| \to \infty$ as $\nu \to \infty$.

Proof If the sequence $\{n_\nu\}$ were bounded, then so would be the sequence $\{m_\nu\}$. The sequence $\{m_\nu/n_\nu\}$, as a consequence, could then contain only finitely many distinct members. This implies that its limit, should it exist, must be a member of the sequence, i.e., a rational number.

With this lemma the student can easily show that $\lim_{x \to a} h(x) = 0$ at each $a \in \mathcal{R}$. We are aware of the fact that $\lim_{x \to a} h(x) = h(a)$ when a is irrational, $\lim_{x \to a} h(x) \neq h(a)$ otherwise.

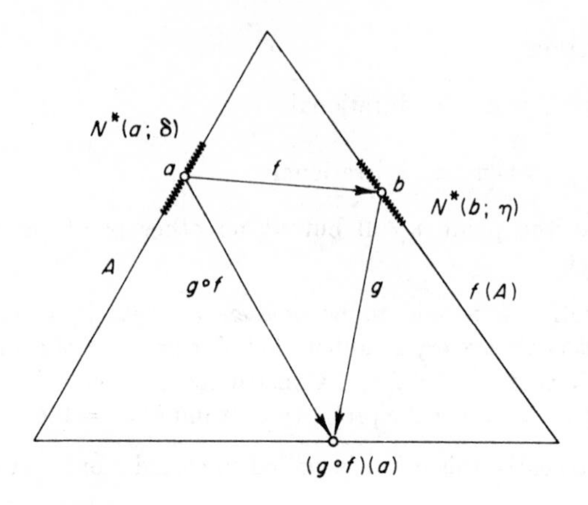

Figure 26

The preservation of limits under algebraic compositions of functions is, of course, expected (See Exercise 23-1). For other compositions we have:

23.8 / Theorem Consider the composite function $g \circ f$, where f has domain A and g is defined on $f(A)$. Let a be a limit point of A. Let

(a) $\lim_{x\to a} f(x) = b,$

(b) $\lim_{y\to b} g(y) = c,$

(c) either $g(b) = c$ or else $f(x) \neq b$ throughout some neighborhood of a.

Then

$$\lim_{x\to a} (g \circ f)(x) = c.$$

The first condition in (c) means, of course, that g is continuous at the point $y = b$ (see Section 24). Figure 26 should aid in understanding this theorem (see also Exercise 3.6).

Proof From (b) we know that for each $\epsilon > 0$ there is an $\eta > 0$ such that

$$d(g(y), c) < \epsilon$$

whenever

$$y \in N^*(b; \eta) \cap f(A); \qquad \text{(i)}$$

owing to (a) there is a $\delta > 0$ for each $\eta > 0$ such that

$$d(f(x), b) < \eta \tag{ii}$$

whenever

$$x \in N^*(a; \delta) \cap A.$$

Now, the relations (i) and (ii) are both true or both false when either $f(x) \neq b$ throughout a neighborhood of a, or when $g(b) = c$. In any event, these inequalities may be combined to read "for each $\epsilon > 0$ there is a $\delta > 0$ such that $d(g \circ f(x), c) < \epsilon$ for all points $x \in N^*(a; \delta) \cap A$."

The necessary presence of condition (c) in the theorem is verified with a simple example. Namely, take the functions

$$f(x) = \begin{cases} 0 & \text{for } x \neq 0 \\ 1 & \text{for } x = 0, \end{cases}$$

$g = f$. Then

$$g \circ f(x) = \begin{cases} 1 & \text{for } x \neq 0 \\ 0 & \text{for } x = 0, \end{cases}$$

and evidently $\lim_{x \to 0} g \circ f(x) = 1$. At the same time we see that $\lim_{x \to 0} f(x) = 0$, $\lim_{y \to 0} g(y) = 0$, $g(0) \neq 0$ and $f(0) \neq 0$.

EXERCISES

23-1 Let A be a given set; let f and g be functions with domain A. Prove: If x_0 is a limit point of A,

$$\lim_{x \to x_0} f(x) = a, \qquad \lim_{x \to x_0} g(x) = b,$$

then

(1) $\lim_{x \to x_0} (f + g)(x) = a + b;$

(2) $\lim_{x \to x_0} (f \cdot g)(x) = a \cdot b;$

(3) $\lim_{x \to x_0} \left(\frac{f}{g}\right)(x) = \frac{a}{b} \qquad (b \neq 0, \quad g(x) \neq 0).$

23-2 Consider the Cantor set $\mathcal{P}$ (Section 20) and its characteristic function $\chi_{\mathcal{P}}$ (see Example 3.7). Show that $\chi_{\mathcal{P}}$ has a limit at every point in $\mathcal{R}\backslash\mathcal{P}$, no limit at points of $\mathcal{P}$.

23-3 Given an arbitrary closed set $A \subset \mathcal{R}$, is there a function having a limit at each point of $\mathcal{R} \setminus A$, no limit at the points of A?

24 / CONTINUOUS FUNCTIONS

It was amply illustrated in the last section that the limit of a function at a point is not necessarily related to the functional value at the point (should such exist). When the two do agree the function is said to be continuous.

24.1 / ***Definition*** Let f be defined on A and let $a \in A$. Then f is *continuous at* a if for each $\epsilon > 0$ there is a $\delta > 0$ such that

$$f(x) \in N(f(a);\epsilon)$$

whenever

$$x \in N(a;\delta) \cap A.$$

Otherwise f is *discontinuous* at a.

Without further ado we formulate:

24.2 / Theorem Let $f: A \to \mathcal{R}$ and $a \in A$ be given. Then the following statements are equivalent:

(**1**) f is continuous at a;

(**2**) for each $\epsilon > 0$ there is a $\delta > 0$ such that

$$d(f(x), f(a)) < \epsilon$$

whenever

$$d(x, a) < \delta \qquad (x \in A);$$

(**3**) if a is a limit point of A, then

$$\lim_{x \to a} f(x) = f(a).$$

The above statements should be compared with the analogous ones in the previous section, then proved. Particular attention should be paid to the effect which the removal of the restriction "$x \neq a$" has on the involved arguments. A useful paraphrasing of Definition 24.1 is obtained as follows:

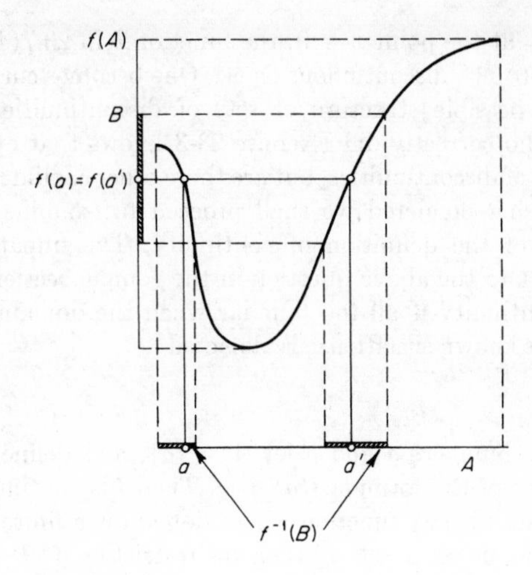

Figure 27

Given a function $f: A \to \mathcal{R}$, we introduce the *inverse image* $f^{-1}(B)$ of a set $B \subset f(A)$:

$$f^{-1}(B) = \{x \in A \mid f(x) \in B\}$$

(see Figure 27).

24.3 / Corollary The function $f: A \to \mathcal{R}$ is continuous at the point $a \in A$ if and only if for every neighborhood V of $f(a)$ there is a neighborhood U of a such that

$$U \cap A \subset f^{-1}(V).$$

Proof The proof should be clear from an inspection of Figure 27.

The continuity at a point $x = a$ of algebraic compositions of functions which are continuous there, is a direct consequence of the definition and Exercises 23-1; the continuity of compositions $g \circ f$ was already proved in Theorem 23.8.

24.4 / *Definition* The function f is *continuous on* A if f is continuous at each point of A; f is *discontinuous on* A if A has a point of discontinuity of f; the function is *nowhere continuous* on A (*everywhere discontinuous* on A) if no point of A is a point of continuity of f.

Thus, the function f_1 in Example 23.7(2) is nowhere continuous (on $\mathcal{R}$), the function $|f_1|$ is everywhere continuous; the function f_2 in

23.7(3) is continuous only at the point $x = 0$; the function h in 23.7(4) is seen to be continuous on $\mathcal{R} \setminus \mathcal{R}^*$, discontinuous on $\mathcal{R}^*$. One becomes curious at this point about the possible structure of sets of discontinuities of functions. The student who correctly did Exercise 23-3 knows that every closed set qualifies as a set of discontinuities, but are there other candidates? While this question is being pondered we shall proceed to examine the immediate consequences of the definition of continuity. The impatient reader will find the answer to the above question in the coming section.

The concept of continuity is all too familiar when the domains in question are intervals; less known situations are depicted:

24.5 / ***Examples***

(**1**) For arbitrary numbers a and b let $A = \{a\}$, and define the function f by means of the formula $f(a) = b$. Then f is continuous on A. Also continuous is any function whose domain is a finite set.

(**2**) If f is continuous on a set A, then its restriction $f \mid B$ (see Section 3) to an arbitrary subset $B \subset A$ is also continuous; on the other hand, f may be discontinuous on A yet have a continuous restriction to B. Thus, for instance, the function $f: \mathcal{R}^* \to \mathcal{R}$ defined as $f(x) = x$ for each $x \in \mathcal{R}^*$ is continuous. On the other hand, the restriction of the function h in Example 23.7(4) to $\mathcal{R}^*$ is everywhere discontinuous.

Consider a set A and a function $f: A \to \mathcal{R}^*$. For each real number α the function f associates with A the subsets

$$A[f < \alpha] = \{x \in A \mid f(x) < \alpha\}$$

and

$$A[f > \alpha] = \{x \in A \mid f(x) > \alpha\}$$

(see Figure 28). A very useful characterization of continuity is offered by the next theorem:

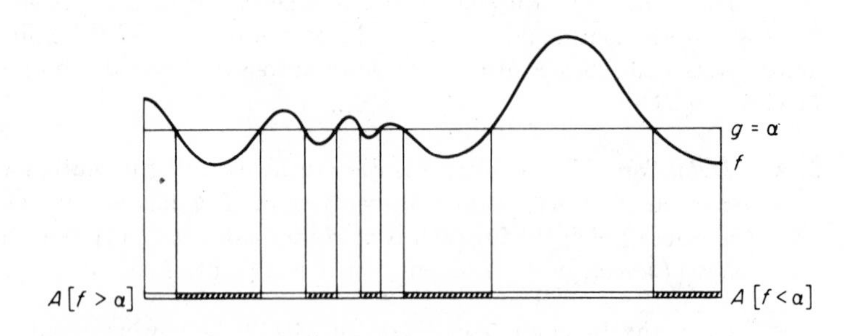

Figure 28

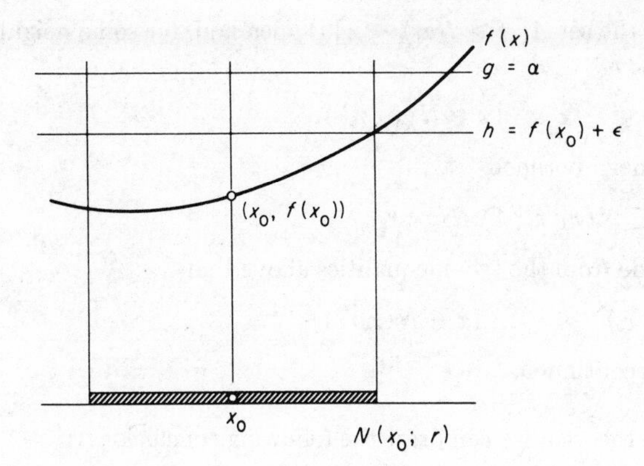

Figure 29

24.6 / Theorem Let A be a given open interval. The function $f: A \to \mathcal{R}$ is continuous if and only if the sets $A[f < \alpha]$ and $A[f > \alpha]$ are open for every number α.

Proof Assuming first f to be continuous, consider an arbitrary constant α. Then both sets in question are empty when and only when $f(x) = \alpha$ for all $x \in A$, but in this case the sets are open and f is continuous. Thus, suppose the set $A[f < \alpha] \neq \varnothing$. Then there is a point $x_0 \in A$ at which $f(x_0) < \alpha$: For the positive constant $\epsilon = \frac{1}{2}(\alpha - f(x_0))$ also $f(x_0) + \epsilon < \alpha$. The number ϵ being so specified, the continuity of f at x_0 guarantees that a $\delta > 0$ can be found such that

$$d(f(x), f(x_0)) < \epsilon \tag{i}$$

for all values x satisfying the relation $d(x, x_0) < \delta$ (See Figure 29). Finally, since A is open, the point x_0 has a neighborhood $N(x_0; r) \subset A$ $(r \leq \delta)$. From (i) we now deduce that

$$f(x) < f(x_0) + \epsilon < \alpha \qquad (x \in N(x_0; r)).$$

Hence, $N(x_0; r) \subset A[f < \alpha]$, showing that $A[f < \alpha]$ is open. The demonstration that $A[f > \alpha]$ is open is entirely analogous to the above. Conversely, regard the sets $A[f < \alpha]$ and $A[f > \alpha]$ to be open for each α; fix an arbitrary point $x_0 \in A$ and an $\epsilon > 0$. Then, in particular, $A[f < f(x_0) + \epsilon]$ is an open set, implying that there is a neighborhood $N(x_0; r_1) \subset A$ of x_0 such that

$$f(x) < f(x_0) + \epsilon \qquad (x \in N(x_0; r_1)).$$

Similarly, the set $A[f > f(x_0) - \epsilon]$ is open and, for some neighborhood $N(x_0; r_2) \subset A$,

$$f(x) > f(x_0) - \epsilon \qquad (x \in N(x_0; r_2)).$$

Taking a neighborhood

$$N(x_0; r) \subset N(x_0; r_1) \cap N(x_0; r_2),$$

we conclude from the two inequalities above that

$$d(f(x), f(x_0)) < \epsilon \qquad (x \in N(x_0; r)).$$

Hence f is continuous.

From this theorem we can draw the following conclusion:

24.7 / Corollary Let A be a given set: let $B \subset A$ be an arbitrary open set. A necessary condition for a function $f: A \to \mathbb{R}$ to be continuous is that the sets $B[f < \alpha]$ and $B[f > \alpha]$ be open for every α.

The corollary will be proved once we show that f is continuous on B when and only when the sets $B[f < \alpha]$ and $B[f > \alpha]$ are open. This can be done by the student with the help of Lemma 19.8.

Extending Corollary 24.3 we prove:

24.8 / Theorem Given an arbitrary set $A \subset \mathbb{R}$, then a function $f: A \to \mathbb{R}$ is continuous if and only if for each open set V there is an open set U such that $U \cap A = f^{-1}(V)$

Proof First suppose f is continuous. Let V be an open set intersecting $f(A)$ and let a be an arbitrary point in $f^{-1}(V)$. Then, for some $\epsilon > 0$, $f(a)$ has a neighborhood $V_\epsilon \subset V$ of radius ϵ. The continuity of f at a tells us that to this V_ϵ there corresponds a neighborhood $U_{\delta(a)}$ of a of radius $\delta(a) > 0$ such that $f(x) \in V_\epsilon$ when $x \in U_{\delta(a)} \cap A$. Associating such a $U_{\delta(a)}$ with each $a \in f^{-1}(V)$ and setting

$$U = \bigcup_{a \in f^{-1}(V)} U_{\delta(a)},$$

then U is open and clearly $U \cap A = f^{-1}(V)$.

Conversely, assume each open set V to determine an open set U as required. If $a \in A$ is arbitrary and V is any neighborhood of $f(a)$, then there is an open set U such that $U \cap A = f^{-1}(V)$. Specifically, a is an interior point of U and for any $x \in U \cap A$, $f(x) \in V$: according to Corollary 24.3 f is continuous at a, and the theorem follows.

Intuitively, a continuous function f with a connected domain is thought to be a certain connected subset of a product space, which intimates that the range is also connected. At least for functions in $\mathcal{R} \times \mathcal{R}$ this seems to be straightforward, and is proved in Darboux's Theorem:

24.9 / Intermediate Value Theorem Let f be continuous on an interval A. If $y_1 < y_2$ are in the range of f, then so is every intermediate value $y_1 < y < y_2$.

Proof Let $f(x_1) = y_1$, $f(x_2) = y_2$. The inequality $x_1 < x_2$ can be admitted without loss of generality and we let $B = (x_1, x_2)$. Let y be as asserted and consider the intervals $B[f < y]$ and $B[f > y]$: These sets are known to be nonempty, open, and disjoint (Theorem 24.6). If there is no $x \in B$ such that $f(x) = y$, then evidently $B = B[f < y] \cup B[f > y]$, but this statement contradicts the connectedness of B.

That this theorem has no true converse is illustrated in the next example:

24.10 / *Example* Consider the function $f:[0, 1] \to [0, 1]$ defined pointwise as follows:

$$f(x) = \begin{cases} x & \text{when } x \text{ is rational} \\ 1 - x & \text{when } x \text{ is irrational.} \end{cases}$$

This function is seen to be a one-to-one mapping between the given intervals, and hence satisfies the Intermediate Value Theorem. On the other hand, the function is clearly discontinuous, except at $x = \frac{1}{2}$.

In general, the continuous image of a connected subset of $\mathcal{R}^n$ is connected but we do not dwell on this point here.

We readily verify:

24.11 / Theorem If $f:A \to \mathcal{R}$ and $g:A \to \mathcal{R}$ are continuous on A, then so are the functions $f + g$, $f \cdot g$, and f/g, the latter function being restricted to those points of A at which $g \neq 0$.

The proof of this proposition is deduced from Exercises 23-1. The student should observe that at the isolated points of A, if any, there is nothing to prove. Next we establish the continuity of the distance function δ introduced in Definition 21.1:

24.12 / Theorem Let A be an arbitrary subset of $\mathcal{R}$. Then the function $\delta(x, A)$ is continuous everywhere.

Proof Let x be a fixed but otherwise arbitrary point. Given any point y and $a \in A$ we call on the triangle inequality $d(x, a) \leq d(y, a) + d(y, x)$ to provide the relations

$$\delta(x, A) = \operatorname*{glb}_{a \in A} d(x, a) \leq \operatorname*{glb}_{a \in A} [d(y, a) + d(y, x)]$$

$$= \operatorname*{glb}_{a \in A} d(y, a) + d(y, x) = \delta(y, A) + d(y, x).$$

Similarly it is shown that

$$\delta(y, A) \leq \delta(x, A) + d(y, x).$$

When combined, the inequalities just derived show that

$$\delta(x, A) - d(y, x) \leq \delta(y, A) \leq \delta(x, A) + d(y, x).$$

Hence

$$\lim_{y \to x} \delta(y, A) = \delta(x, A)$$

as was to be demonstrated.

An important problem concerns the extension of a function f, defined on a domain A, to a function g, which is defined on a certain domain $B \supset A$ and which coincides with f on A, while at the same time preserving continuity or other properties originally attributed to f. The Tietze Extension Theorem takes the first step in this direction. The version given here remains valid in its formulation when A is taken to be an arbitrary metric space rather than a subset of $\mathcal{R}$, but the proof becomes considerably more complicated.

24.13 / Tietze Extension Theorem Consider a closed set $A \subset \mathcal{R}$ and a continuous bounded function $f: A \to \mathcal{R}$. There is a continuous function $g: \mathcal{R} \to \mathcal{R}$ with the following properties:

(1) $g \mid A = f$;

(2) $\operatorname*{glb}_{x \in \mathcal{R}} g(x) = \operatorname*{glb}_{x \in A} f(x)$;

(3) $\operatorname*{lub}_{x \in \mathcal{R}} g(x) = \operatorname*{lub}_{x \in A} f(x)$.

The function g is the *continuous extension* of f to $\mathcal{R}$.

Proof Using the fact that $\mathcal{R} \setminus A$ is the union of at most countably many disjoint open intervals $I_k = (a_k, b_k)$ whose endpoints are in A, we merely set

$$g(x) = \begin{cases} f(x) & \text{for } x \in A \\ \dfrac{(b_k - x)f(a_k) + (x - a_k)f(b_k)}{b_k - a_k} & (x \in I_k). \end{cases} \tag{ii}$$

The function is clearly as demanded in the theorem (Verify this assertion).

In concluding this section we present a simple proof of the Separation Theorem 21.4, using the theorem just proved. Thus, supposing A_1 and A_2 to be disjoint nonempty closed sets, let $f: A_1 \cup A_2 \to \mathcal{R}$ be defined as follows:

$$f(x) = \begin{cases} 1 & \text{for } x \in A_1 \\ 0 & \text{for } x \in A_2. \end{cases}$$

Then $f(x)$ is continuous on its domain and hence has an extension $g(x)$ as in (ii). Owing to the Intermediate Value Property 24.9 this function g assumes the value $\frac{1}{2}$: The sets $A[g < \frac{1}{2}]$ and $A[g > \frac{1}{2}]$ are open, disjoint, and evidently $A[g < \frac{1}{2}] \supset A_2$, $A[g > \frac{1}{2}] \supset A_1$.

The student is asked, however, to answer this question: Is the Separation Theorem 21.4 implicitly present in this argument?

EXERCISES

24-1 Let I be a given interval. The function $f: I \to \mathcal{R}$ is said to be *Lipschitz continuous* if there is a constant $\alpha > 0$ such that

$$d(f(x_1), f(x_2)) \leq \alpha d(x_1, x_2)$$

for all points $x, y \in I$ (See Figure 30). Show that a Lipschitz continuous function is continuous.

24-2 Let $a \in A$ be a limit point of A. Prove that $f: A \to \mathcal{R}$ is continuous at a if and only if

$$f(a-) = f(a+) = \lim_{x \to a} f(x).$$

24-3 Referring to Theorem 24.6, let A be a closed interval. Prove that $f: A \to \mathcal{R}$ is continuous if and only if the sets $A[f \leq \alpha]$ and $A[f \geq \alpha]$ are closed.

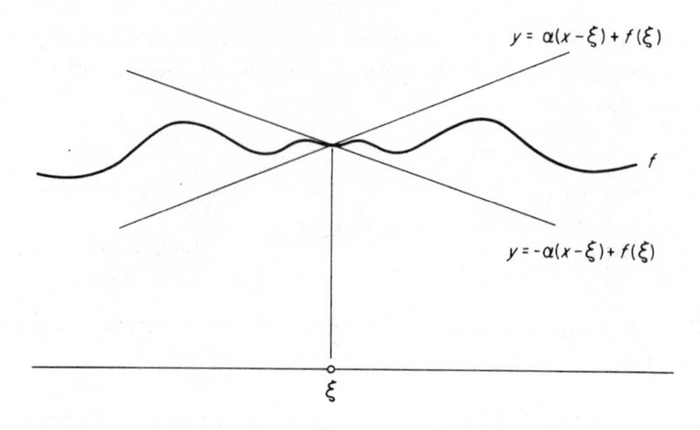

Figure 30

24-4 Let A be a closed interval; let f be continuous on A. For a given constant α, let $A_\alpha = \{x \in A \mid f(x) = \alpha\}$. Then prove that A_α is closed.

24-5 Given an arbitrary closed set $A \subset \mathcal{R}$, can you construct a continuous function $f: \mathcal{R} \to \mathcal{R}$ such that $f(x) = 1$ for $x \in A$, $f(x) \neq 1$ on $\mathcal{R} \setminus A$?

24-6 Prove Theorem 24.11.

24-7 When the set A in Theorem 24.8 is open this theorem reads: f is continuous if and only if for every open set $V \subset f(A)$, $f^{-1}(V)$ is open. Prove it.

24-8 Verify the continuity of the given functions on the prescribed intervals:

(**1**) $f_1(x) = \dfrac{1}{x}, \qquad x \in (0, 1);$

(**2**) $f_2(x) = e^x, \qquad x \in \mathcal{R};$

(**3**) $f_3(x) = \sin \dfrac{1}{x}, \qquad x \in (0, 1);$

(**4**) $$f_4(x) = \begin{cases} x \sin \dfrac{1}{x} & (x \neq 0), \\ 0 & \text{at } x = 0, \end{cases} \qquad x \in [0, 1].$$

24-9 Given a closed interval I and a continuous function $f: I \to I$, then $x \in I$ is a *fixed point* of f if $f(x) = x$. Show that f has a fixed point.

24-10 Show that a continuous function on $[0, 1]$ cannot assume each value in its range exactly twice.

24-11 Give an example of a continuous nowhere constant function on $(0, 1)$ which takes on every value in its range infinitely often.

24-12 Prove: If f is continuous on a compact set, then f is bounded.

24-13 Does f have a continuous extension to $[a, b]$ if f is continuous and bounded on the open interval (a, b)?

24-14 Let A be the union of two disjoint open intervals. Let $f: A \to \mathcal{R}$ be continuous and bounded. Then prove that f need not have a continuous extension to $\bar{A}$.

24-15 Prove: If f and g are continuous functions on $[0, 1]$ and if $f = g$ on a dense subset of $[0, 1]$, then $f = g$ everywhere.

25 / THE NATURE OF DISCONTINUITIES

In addition to the situations already encountered in Examples 23.4, 24.5, and 24.10 consider also:

25.1 / ***Example*** The function $h(x) = \sin \frac{1}{x}$ $(x > 0)$ is bounded yet has no right limit at $x = 0$ (see Figure 31). To see this, consider for example the sequences

$$x_n = \frac{2}{(1 + 4n)\pi}$$

and

$$y_n = \frac{2}{(-1 + 4n)\pi}.$$

Then

$$\lim_{n\to\infty} \sin \frac{1}{x_n} = 1$$

whereas

$$\lim_{n\to\infty} \sin \frac{1}{y_n} = -1.$$

In fact, for any number $a \in [-1, 1]$ there is a sequence $\{z_n\}$ such that $\frac{1}{z_n} \in (0, 1)$ and $\lim_{n\to\infty} \sin \frac{1}{z_n} = a$.

Discontinuities are classified according to the following definition:

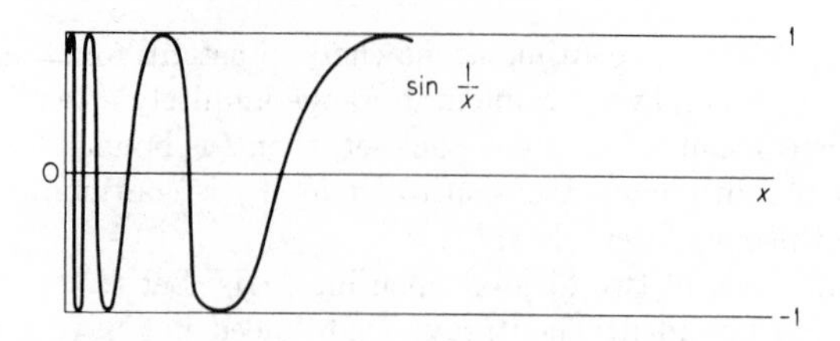

Figure 31

25.2 / ***Definition*** Assume f to be defined on the interval (a, b). Then f has a *discontinuity of the first kind* at the point $x \in (a, b)$ if:
(**1**) f is discontinuous at x;
(**2**) both limits $f(x-)$ and $f(x+)$ exist.

When (1) holds without (2), then f is said to have a *discontinuity of the second kind* at x.

A discontinuity of the first kind is also termed a *simple discontinuity*. Of the examples in 23.4 only the function in (1) has a simple discontinuity. We note that such discontinuities can occur under two sets of circumstances:

when $f(x-) = f(x+) \neq f(x)$, or

when $f(x-) \neq f(x+)$.

Clearly, the value $f(x)$ has no effect on the second case.

Now that we have classified the possible continuities that a function may have, let us investigate the structure of the sets on which they can occur. Given an arbitrary closed set A, then the function

$$f(x) = \begin{cases} 1 & \text{when} \quad x \in A \cap \mathcal{R}^* \\ -1 & \text{when} \quad x \in A \setminus \mathcal{R}^* \\ 0 & \text{when} \quad x \in \mathcal{R} \setminus A \end{cases}$$

is discontinuous on A, continuous on $\mathcal{R} \setminus A$. Thus, an arbitrary closed set can be the set of discontinuities of a function: so can any finite union of closed sets, and it is natural to inquire if this is true, in fact, for countable unions of closed sets.

Consider an arbitrary countable collection of closed sets A_k. The sets

$$B_n = \bigcup_{k=1}^{n} A_k$$

are closed, nested:

$$B_1 \subset B_2 \subset B_3 \subset \cdots,$$

and with the stipulation that $B_0 = \varnothing$ it is possible to write

$$A = \bigcup_{k \in \mathfrak{N}} A_k = \bigcup_{k \in \mathfrak{N}} (B_k \setminus B_{k-1}),$$

where the sets $(B_k \setminus B_{k-1})$ are pairwise disjoint. Put

$$B_k \setminus B_{k-1} = R_k \cup Q_k,$$

where

$$R_k = (B_k \setminus B_{k-1}) \cap \mathfrak{R}^*$$

$$Q_k = (B_k \setminus B_{k-1}) \setminus \mathfrak{R}^*.$$

The function

$$f(x) = \begin{cases} \dfrac{1}{k} & \text{for} \quad x \in R_k \\ -\dfrac{1}{k} & \text{for} \quad x \in Q_k \\ 0 & \text{for} \quad x \in \mathfrak{R} \setminus A \end{cases}$$

is asserted to be discontinuous on A, continuous on $\mathfrak{R} \setminus A$. The latter claim is easily verified, for consider an arbitrary point $x_0 \in \mathfrak{R} \setminus A$. Given an $\epsilon > 0$, select an integer n so large that $\dfrac{1}{n} < \epsilon$; next, take a neighborhood $N(x_0; \delta)$ which does not intersect the set B_n. Since the sets B_k are nested, it follows that $B_k \cap N(x_0; \delta) = \varnothing$ for integers $1 \leq k \leq n$. According to the definition of f, therefore, $-\dfrac{1}{k} < f(x) < \dfrac{1}{k}$ when $x \in N(x_0; \delta)$. In particular,

$$d(f(x_0), f(x)) = |f(x)| < \frac{1}{n} < \epsilon \qquad \text{when} \quad x \in N(x_0; \delta),$$

so that f is continuous at the point $x = x_0$: Since this point was arbitrary, it follows that f is continuous on $\mathfrak{R} \setminus A$.

The discontinuity of f on A is verified in two steps: If x_0 is an arbitrary point of A, then there is a value of k such that $x_0 \in B_k \setminus B_{k-1}$.

If x_0 is not an interior point of A, then every neighborhood $N(x_0; \delta)$ intersects $\mathfrak{R} \setminus A$. Hence, every neighborhood of x_0 contains points x at which $f(x) = 0$. Since $|f(x_0)| = \dfrac{1}{k}$, it follows that f cannot be continuous at $x = x_0$.

If x_0 is an interior point of A, suppose first $x_0 \in R_k$. Then an arbitrary neighborhood $N(x_0; \delta)$ contains points x where either $f(x) = -\frac{1}{k}$, $f(x) < \frac{1}{k}$ or $f(x) > \frac{1}{k}$. Be this as it may, $f(x_0) = \frac{1}{k}$ and consequently

$$d(f(x_0), f(x)) \geq \frac{1}{k(k+1)}$$

for some $x \in N(x_0; \delta)$, and since δ was arbitrary, f cannot be continuous at $x = x_0$. The final case $x_0 \in Q_k$ is disposed of in a like manner.

Now let f be defined on $\mathcal{R}$ and let D be its set of discontinuities. If $a \in D$ then there is an $\epsilon > 0$ for which every $N(a; \delta)$ has a point x such that

$$d(f(a), f(x)) > \epsilon.$$

Selecting an integer k such that $\epsilon > \frac{1}{k}$, it follows that each neighborhood $N(a; \delta)$ contains an x for which

$$d(f(a), f(x)) > \frac{1}{k}.$$

This suggests that a set A_k be associated with each $k \in \mathfrak{N}$ as follows: $a \in A_k$ if and only if each $N(a; \delta)$ contains points x_1 and x_2 such that

$$d(f(x_1), f(x_2)) > \frac{1}{k}.$$

Let $A = \bigcup_{k \in \mathfrak{N}} A_k$; we have just shown that $D \subset A$.

Conversely, let us show that $A \subset D$. To do this, let k be arbitrary but fixed, let $a \in A_k$ and let the neighborhood $N(a; \delta)$ be given. If, for each $x \in N(a; \delta)$, we have $d(f(a), f(x)) < \frac{1}{2k}$, then for arbitrary points x_1 and x_2 in the neighborhood

$$d(f(x_1), f(x_2)) \leq d(f(a), f(x_1)) + d(f(a), f(x_2))$$
$$\leq \frac{1}{2k} + \frac{1}{2k} = \frac{1}{k}.$$

The definition of A_k dictates, therefore, that for some $x \in N(a; \delta)$, $d(f(a), f(x)) \geq \frac{1}{2k}$. Hence $a \in D$.

To summarize, the preceding discussion has shown that the set of discontinuities of an arbitrary function is expressible as the union of certain sets A_k. In concluding this argument, each A_k will be demonstrated to be closed. Namely, fixing an arbitrary value of k, let $\{a_m\}$ be a convergent sequence with range in A_k; let its limit be a. If $N(a; \delta)$ is any neighborhood thereof, then, for some value of m, $a_m \in N(a; \delta)$, so that $N(a; \delta)$ is also a neighborhood of this a_m. But $a_m \in A_k$ and so there are points x_1 and x_2 in $N(a; \delta)$ such that $d(f(x_1), f(x_2)) > \frac{1}{k}$; that is, $a \in A_k$. Hence A_k is closed.

This leads us to introduce:

25.3 / ***Definition*** The set A is of *type* F_σ if it is the countable union of closed sets.

The preceding results can now be stated as follows:

25.4 / Theorem The set D is the set of discontinuities of a function $f: \mathcal{R} \to \mathcal{R}$ if and only if D is of type F_σ.

Not every set is of type F_σ. To be sure, we have:

25.5 / Theorem The set $\mathcal{R} \setminus \mathcal{R}^*$ is not of type F_σ.

Proof Assume, on the contrary, that there is a countable collection of closed sets A_k for which

$$\mathcal{R} \setminus \mathcal{R}^* = \bigcup_{k \in \mathfrak{N}} A_k.$$

Since $\mathcal{R}^*$ is dense in $\mathcal{R}$, the set $\mathcal{R} \setminus \mathcal{R}^*$ has no interior points (see Definition 19.1), implying that the closed subsets of $\mathcal{R} \setminus \mathcal{R}^*$ are nowhere dense (why?). This means, in turn, that $\mathcal{R} \setminus \mathcal{R}^*$ and consequently $\mathcal{R}$ itself are of the first category (Theorem 17.14), a conclusion contradicting Theorem 17.12.

Thus, no function can be continuous on $\mathcal{R} \setminus \mathcal{R}^*$, discontinuous on $\mathcal{R}^*$.

EXERCISES

25-1 Show that every interval is of type F_σ.

25-2 Prove: If A is a set of type F_σ, then either A is of the first category or else it contains an interval.

25-3 The function f is *nondecreasing* if $f(x) \leq f(y)$ whenever $x \leq y$. Given an arbitrary countable subset $A \subset [0, 1]$, construct a nondecreasing function $f: [0, 1] \to \mathcal{R}$ having A as its set of discontinuities.

25-4 Referring to the construction leading to Theorem 25.4, discuss the following:

(a) When was the closure of the sets B_k essential?
(b) What was the relevance of the nested property of the sets B_k in the argument?

25-5 Given a function $f: A \to \mathcal{R}$, show that $\lim_{x \to a-} f(x) = f(a-)$ if and only if $\lim_{k \to \infty} f(x_k) = f(a-)$ for every nondecreasing sequence $\{x_k\}$ with range in A and a as its limit.

25-6 Let $f: I \to \mathcal{R}$ be given. For each set $A \subset I$ we associate with f the number

$$\omega(f; A) = \operatorname*{lub}_{x \in A} f(x) - \operatorname*{glb}_{x \in A} f(x),$$

called the *oscillation of f on A*. Evidently

$$\omega(f; A) \geq 0.$$

For each $\xi \in I$ we introduce the number

$$\omega(f; \xi) = \operatorname*{glb}_{\delta > 0} \omega(f; N(\xi; \delta) \cap I),$$

called the *oscillation of f at ξ*.
(**1**) Show that f is continuous at ξ if and only if $\omega(f; \xi) = 0$.
(**2**) Prove or disprove: $\omega(f; \xi) = \omega(g; \xi)$ for each $\xi \in I$ if and only if $f - g =$ constant.
(**3**) Prove or disprove: If $A = B \cup C$ and $B \cap C = \emptyset$, then

$$\omega(f; A) \leq \omega(f; B) + \omega(f; C).$$

(**4**) Produce functions f and g for which

$$\omega(f + g; A) \neq \omega(f; A) + \omega(g; A).$$

(**5**) Produce functions h and k such that

$$\omega(h + k; A) = \omega(h; A) + \omega(k; A)$$

for each $A \subset I$.
(**6**) If

$$A_n = \left\{x \in I \mid \omega(f; x) \geq \frac{1}{n}\right\},$$

then show that A_n is closed.

26 / MONOTONIC FUNCTIONS

The discontinuities of functions which are never decreasing or never increasing are particularly easy to describe: they are always of the first kind. We state the following definition:

26.1 / *Definition* Let there be given an interval I and a function $f: I \to \mathbb{R}$. Then f is:

(1) *nondecreasing* on I when $x_1 < x_2$ implies that

$$f(x_1) \leq f(x_2);$$

(2) *nonincreasing* on I when $x_1 < x_2$ implies that

$$f(x_1) \geq f(x_2).$$

The function is *increasing* when strict inequalities hold in (1), *decreasing* when such is the case in (2).

The terminology *strictly increasing* and *strictly decreasing* is also used in the literature instead of increasing and decreasing, respectively. The function is said to be *monotonic* when it falls into either of the categories (1) or (2), *strictly monotonic* when strict inequalities hold: clearly, a strictly monotonic function is monotonic. Examples of monotonic functions abound. In fact, the student may be surprised to discover that all continuous functions he can think of are at least *piecewise monotonic*, which is to say that their domain can be written as the union of at most countably many intervals on each of which they are monotonic. In contrast, the function in Example 23.7(4) is nowhere monotonic and the following intriguing question poses itself: Is there a *continuous* function $f: \mathbb{R} \to \mathbb{R}$ which is nowhere monotonic? While the answer is by no means trivial, the student should tackle this problem anyway, for the effort will be rewarded in the very least with a better understanding of the more subtle properties of continuous functions. Another possible source for the answer is Gelbaum–Omstead's "Counter Examples in Analysis."

We prove the basic theorem:

26.2 / Theorem Let f be nondecreasing on the open interval I. Then, for each $x \in I$, the limits $f(x-)$ and $f(x+)$ exist and stand in the relation

$$f(x-) \leq f(x) \leq f(x+)$$

to $f(x)$.

Let $x \in I$ be fixed; let

$$I^-(x) = \{y \in I \mid y < x\},$$

$$I^+(x) = \{y \in I \mid y > x\}.$$

Then neither of the sets is empty and the monotone character of f implies that

$$\operatorname*{lub}_{y \in I^-(x)} f(y) \leq f(x) \leq \operatorname*{glb}_{y \in I^+(x)} f(y).$$

In view of these relations, it suffices to establish:

26.3 / Lemma For each $x \in I$,

$$\operatorname*{lub}_{y \in I^-(x)} f(y) = f(x-)$$

and

$$\operatorname*{glb}_{y \in I^+(x)} f(y) = f(x+).$$

Proof Let us verify the first of these equalities. If x is fixed and α stands for the least upper bound in question, then $\alpha \leq f(x)$. If $\epsilon > 0$ is given, then the definition of least upper bound guarantees the existence of a point $x_0 \in I^-(x)$ such that

$$\alpha - \epsilon < f(x_0) \leq \alpha.$$

The monotonicity of f provides the relations

$$f(x_0) \leq f(y) \leq \alpha$$

whenever $x_0 < y < x$. Hence, we conclude that

$$\alpha - \epsilon < f(y) \leq \alpha$$

whenever $x_0 < y < x$, but this says that

$$f(x-) = \lim_{y \to x-} f(y) = \alpha.$$

The second part of the lemma is proved in a like manner.

A comparison of the theorem with Definition 25.2 yields:

26.4 / Corollary The only discontinuities of a monotonic function are of the first kind.

Another basic fact concerning monotonic functions is formulated in the next theorem:

26.5 / Theorem Let the notation be that of 26.2. If $x < y$ are in I, then $f(x+) \leq f(y-)$.

Proof Let $J = I^+(x) \cap I^-(y)$; then $J \neq \emptyset$. From the proof of 26.3 we conclude that

$$\operatorname*{lub}_{z \in J} f(z) = \operatorname*{lub}_{z \in I^-(y)} f(z) = f(y-)$$

and similarly

$$f(x+) = \operatorname*{glb}_{z \in I^+(x)} f(z) = \operatorname*{glb}_{z \in J} f(z).$$

Since $\operatorname*{glb}_{z \in J} f(z) \leq \operatorname*{lub}_{z \in J} f(z)$, the result follows.

With the help of this theorem we prove:

26.6 / Corollary The set of discontinuities of a monotonic function $f: I \to \mathcal{R}$ is at most countable.

Proof For the sake of argument suppose f is nondecreasing. If x is a point of discontinuity of f, then $f(x+) - f(x-) > 0$. Let $r(x)$ be an arbitrary rational number such that $f(x-) < r(x) < f(x+)$. Following this prescription we associate a rational number with each point where f is discontinuous, and Theorem 26.5 tells us that if $x < y$, then $r(x) \neq r(y)$. Having thus established a one-to-one correspondence between the set of discontinuities of f and a subset of $\mathcal{R}^*$, we proved the corollary.

In the language of measure theory we can say that a monotonic function is continuous everywhere in its domain save possibly for a set of measure zero, or continuous *almost everywhere* (see Remark 17.15).

This corollary has a converse, namely, for every countable set A there is a monotonic function $f: \mathcal{R} \to \mathcal{R}$ which is discontinuous on A, continuous on $\mathcal{R} \setminus A$.

26.7 / *Example* Let A be an arbitrary countable set with members a_n ($n \in \mathfrak{N}$), and consider an arbitrary convergent series,

$$\sum_{n \in \mathfrak{N}} \epsilon_n = \alpha,$$

of positive terms. Define for each value of n the step function

$$f_n(x) = \begin{cases} 0 & \text{for} \quad x < a_n \\ \epsilon_n & \text{for} \quad x \geq a_n \end{cases}$$

and pointwise define the function f to be

$$f(x) = \sum_{n \in \mathfrak{N}} f_n(x).$$

Regarding this function it is asserted that

(a) f is nondecreasing on $\mathcal{R}$;
(b) $f(a_k+) - f(a_k-) = \epsilon_k$ for each k; that is, f is discontinuous on A;
(c) f is continuous on $\mathcal{R} \setminus A$.

The student will easily convince himself that, for each x,

$$f(x) = \sum_{a_n \le x} \epsilon_n,$$

where the summation is extended over all values of n for which $a_n \le x$. It is also clear that $0 \le f \le \alpha$. Now, if $x < y$, then

$$f(x) = \sum_{a_n \le x} \epsilon_n = \sum_{a_n \le y} \epsilon_n - \sum_{x < a_n \le y} \epsilon_n = f(y) - \sum_{x < a_n \le y} \epsilon_n \le f(y)$$

since the ϵ_n are positive and hence (a) is established. Next, fix a point $a_k \in A$ and put in the preceding formula $y = a_k$. Then for $x < a_k$,

$$f(x) = \sum_{a_n \le a_k} \epsilon_n - \sum_{x < a_n \le a_k} \epsilon_n = f(a_k) - \sum_{x < a_n \le a_k} \epsilon_n.$$

It is clear that

$$\lim_{x \to a_k-} \sum_{x < a_n \le a_k} \epsilon_n = \epsilon_k,$$

and it follows that $f(a_k-) = f(a_k) - \epsilon_k$. On the other hand, for each $x \ge a_k$,

$$f(x) = \sum_{a_n \le a_k} \epsilon_n + \sum_{a_k < a_n \le x} \epsilon_n = f(a_k) + \sum_{a_k < a_n \le x} \epsilon_n.$$

Since

$$\lim_{x \to a_k+} \sum_{a_k < a_n \le x} \epsilon_n = 0,$$

it is seen that $f(a_k+) = f(a_k)$. With the preceding result we find that

$$f(a_k+) = f(a_k) = f(a_k-) + \epsilon_k,$$

and (b) is thus verified. The verification of (c), which follows along similar lines, is left as an exercise.

EXERCISES

26-1 Given an arbitrary countable set A, construct a *strictly* monotonic function $f: \mathcal{R} \to \mathcal{R}$ which is discontinuous precisely on A.

26-2 Show that if $f:[0, 1] \to \mathcal{R}$ is strictly monotonic and continuous, then f maps its domain one-to-one onto its range.

26-3 Prove that the inverse of a continuous strictly monotonic function on an interval exists and is continuous.

26-4 Construct a continuous increasing function $f:[0, 1] \to [0, 1]$ which maps a set of measure $\frac{1}{2}$ onto a set of measure 0.
(*Hint*: Consult Section 20 and the Tietze Extension Theorem.)

27 / UNIFORM CONTINUITY

The local character of continuity is a matter of definition: the terms for continuity were dictated for a fixed point. The resulting dependence of δ on both the given ϵ and the point is now illustrated with some known cases.

27.1 / *Examples*

(**1**) Let us begin with the function $f(x) = \dfrac{1}{x}$ defined on $E = (0, 1)$:

Although unbounded, this function is continuous throughout its domain. The dependence of δ on x is easily exhibited (See Figure 32). Take, for instance, $\epsilon = 1$; let $\delta > \frac{1}{4}$ be arbitrary but fixed, and we may safely assume that $\delta < \frac{1}{2}$. If $x = \delta$ and $y = x + \delta$, then both points belong to E, they satisfy the relation $d(x, y) = \delta$, yet

$$d(f(x), f(y)) = d\left(\frac{1}{\delta}, \frac{1}{2\delta}\right) = \frac{1}{2\delta} > 1.$$

In contrast, the points $x = \frac{1}{2}$ and $y = \frac{1}{2} + \delta$ are also in E, are such that $d(x, y) = \delta$, but this time

$$d(f(x), f(y)) = \frac{4\delta}{1 + 2\delta} < 1.$$

(**2**) We reexamine the function $f(x) = \sin \dfrac{1}{x}$ $(x \in E)$ (See Example 25.1). This function is bounded and continuous. Again let $\epsilon = 1$ and let $\delta > 0$ be arbitrary but fixed; select an odd positive integer k which meets the requirement $\dfrac{4}{3\pi k} < \delta$. If we take the points $x = \dfrac{2}{\pi k}$ and $y = \dfrac{2}{3\pi k}$, then $d(x, y) = \dfrac{4}{3\pi k} < \delta$, but $d\left(\sin \dfrac{1}{x}, \sin \dfrac{1}{y}\right) = 2 > \epsilon$.

From Theorem 23.8 we know, of course, that the function under discussion is continuous on E.

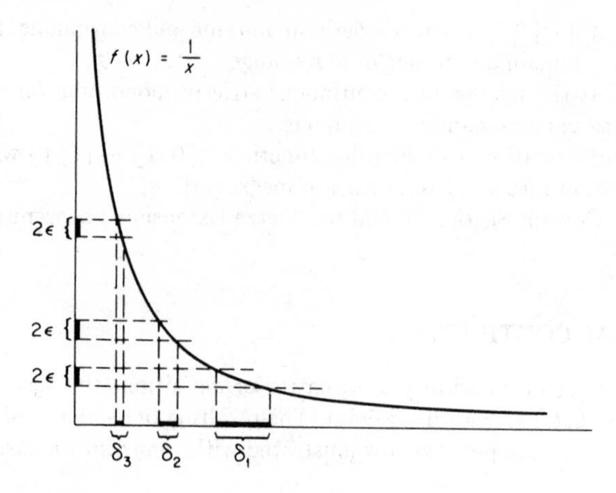

Figure 32

(3) A character different from that of the function above is possessed by the function $g(x) = x^2$ $(x \in E)$. Namely, if $\epsilon > 0$ is arbitrary and we let $\delta = \frac{1}{2}\epsilon$, then for all points $x, y \in E$ satisfying the inequality $d(x, y) < \delta$ we have

$$d(g(x), g(y)) = |x^2 - y^2| = |x + y||x - y|$$
$$\leq 2d(x, y) < 2\delta = \epsilon.$$

Continuity in this instance has another stature; it has a markedly global rather than a local character. As we already know, such a function is said to be uniformly continuous (see below). It should be pointed out that the function $g(x) = x^2$ with domain $\mathcal{R}$ no longer has this property: This should be proved by the student.

27.2 / *Definition* The function $f: A \to \mathcal{R}$ is *uniformly continuous* if for every $\epsilon > 0$ there is a $\delta > 0$ such that if $x, y \in A$ and $d(x, y) < \delta$, then $d(f(x), f(y)) < \epsilon$.

Clearly every uniformly continuous function is continuous, but it should be observed that uniform continuity at a point is a meaningless concept. Uniform continuity alone is not sufficient to guarantee boundedness as indicated by the function $f(x) = x$ with domain $\mathcal{R}$. When the domain is bounded, however, we have:

27.3 / Theorem If $f: A \to \mathcal{R}$ is uniformly continuous and A is bounded, then also f is bounded.

Proof Given an $\epsilon > 0$, let $\delta > 0$ be a number such as provided by Definition 27.2. Since A is bounded, it can be covered with finitely many open intervals I_k, say, $1 \leq k \leq n$, of radius $\frac{1}{2}\delta$. On each of the sets $A \cap I_k$, f is bounded since $d(f(x), f(y)) < \epsilon$ for arbitrary points $x, y \in A \cap I_k$. Hence f is bounded as asserted.

Again the simple function $f(x) = x$ $(0 < x < 1)$ shows that a uniformly continuous function on a bounded set need not realize its least upper bound or greatest lower bound. Thus, further restrictions have to be imposed on the domain: What is needed is compactness. Before proving this, however, let it be shown that the concepts of continuity and uniform continuity coincide for functions whose domain is compact.

27.4 / Theorem Let $f: A \to \mathcal{R}$ be continuous. If A is compact, then f is uniformly continuous.

Proof Let $\epsilon > 0$ be given. Owing to the continuity of f we can relate to each $x \in A$ a neighborhood $N(x; \delta(x))$ with the property that

$$d(f(x), f(y)) < \epsilon \qquad (y \in N(x; \delta(x)) \cap A).$$

Consider the neighborhood $M(x; \frac{1}{2}\delta(x))$. Then the family of sets $M(x; \frac{1}{2}\delta(x))$ is an open covering of A which, by virtue of the compactness of A, contains a finite subcovering. Specifically, A contains finitely many points $a_1, \ldots, a_n$ such that

$$A \subset \bigcup_{i=1}^{n} M(a_i; \tfrac{1}{2}\delta(a_i))$$

and Theorem 18.13 tells us that this covering has a Lebesgue number $\delta > 0$ such that if x and y in A satisfy the inequality $d(x, y) < \delta$, then these points are members of some set $M(a_k; \frac{1}{2}\delta(a_k))$ $(1 \leq k \leq n)$. For such points, however

$$d(f(x), f(y)) \leq d(f(x), f(a_k)) + d(f(y), f(a_k)) < \tfrac{1}{2}\epsilon + \tfrac{1}{2}\epsilon = \epsilon,$$

and the uniform continuity of f is therefore proved.

It is worthwhile to notice the essential role which compactness played in this proof; namely, without it the existence of a Lebesgue number could not be assured.

27.5 / Theorem If f is continuous on the compact set A, then there are points a and b in A such that

$$f(a) = \operatorname*{lub}_{x \in A} \{ f(x) \}$$

and

$$f(b) = \operatorname*{glb}_{x \in A} \{ f(x) \}.$$

Proof According to the last two theorems the set $f(A)$ is bounded, and as such it has a lub and a glb, say, α and β, respectively. Now, for each positive integer n there is a point $x_n \in A$ such that

$$\alpha - \frac{1}{n} < f(x_n) \leq \alpha.$$

If the set of points x_n is finite then there is nothing to prove; otherwise, the Bolzano–Weierstrass Theorem 17.5 tells us that there is a convergent subsequence $\{x_{k_n}\}$. Let its limit be a. Then, owing to the compactness of A, $a \in A$ and hence

$$\lim_{n \to \infty} f(x_{k_n}) = f(a) = \alpha.$$

The second assertion in the theorem is established in a like manner.

When this theorem is coupled with Theorem 18.5 it tells us that the image of a compact interval is a compact interval. The extension of this result to arbitrary compact sets is the topic of the next theorem:

27.6 / Theorem If f is continuous on the compact set A, then also $f(A)$ is compact.

Proof Knowing already that the set $f(A)$ is bounded we have only to establish that it is closed. Thus, let y be a limit point of $f(A)$. Then for each positive integer n there is a point $x_n \in A$ such that $d(f(x_n), y) < 1/n$, and the reasoning employed in the proof above shows that $y \in f(A)$. Hence the set is compact.

The indispensible role of compactness is displayed also in this problem: Given a set A and a continuous one-to-one function $f: A \to \mathcal{R}$, then this f has an inverse $f^{-1}: f(A) \to A$. When is the inverse function continuous?

27.7 / *Example* Consider the set $A = [0, 1) \cup [2, 3]$ and the function

$$f(x) = \begin{cases} x & \text{for } x \in [0, 1) \\ x - 1 & \text{for } x \in [2, 3] \end{cases}$$

(See Fig. 33). This function is a one-to-one continuous mapping of A

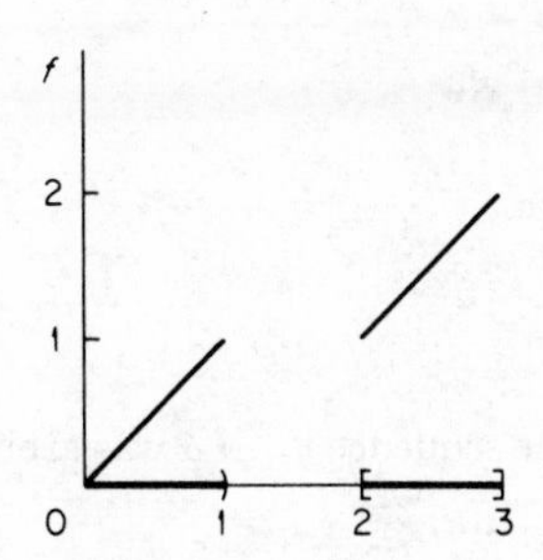

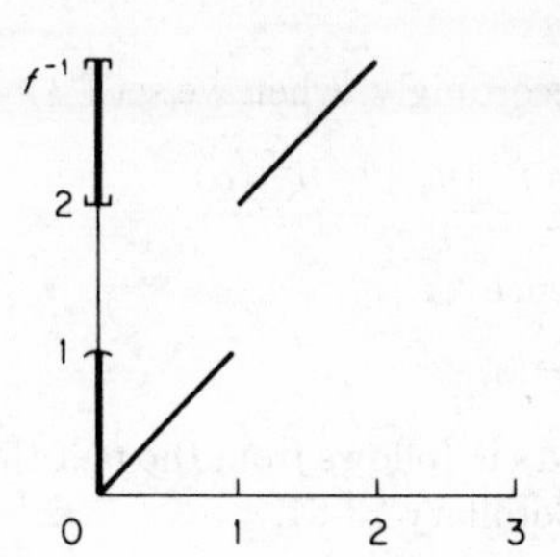

Figure 33

onto the interval $[0, 2]$. Its inverse function is

$$f^{-1}(x) = \begin{cases} x & \text{for} \quad x \in [0, 1) \\ x + 1 & \text{for} \quad x \in [1, 2] \end{cases}$$

which has a simple discontinuity at $x = 1$.

Continuity is preserved when the domain is compact:

27.8 / Theorem If $f: A \to \mathcal{R}$ is a continuous one-to-one function and if A is compact, then the inverse function $f^{-1}: f(A) \to \mathcal{R}$ exists and is continuous.

Proof The compactness of $f(A)$ (Theorem 27.6) and the one-to-one character of f guarantees the existence of the inverse function; let us, therefore, concentrate on its continuity.

Let $\eta \in f(A)$ be arbitrary. If η is an isolated point of $f(A)$ then f^{-1} is automatically continuous there. If η is a limit point of $f(A)$, then it has to be shown that

$$\lim_{y \to \eta} f^{-1}(y) = f^{-1}(\eta). \tag{i}$$

To do this, let $y_n \to \eta$ be an arbitrary sequence with range in $f(A)$. This sequence uniquely determines a sequence $\{x_n\}$ with range in A. Since A is bounded it contains a convergent subsequence $\{x_{n_k}\}$ whose limit, say ξ, belongs to A by virtue of the fact that A is also closed. Hence

$$\lim_{k \to \infty} (f^{-1} \circ f)(x_{n_k}) = (f^{-1} \circ f)(\xi).$$

But the continuity of f says that under the circumstances

$$\lim_{k \to \infty} f(x_{n_k}) = f(\xi).$$

Accordingly, when we set $f(\xi) = \alpha$, then

$$\lim_{k\to\infty} f^{-1}(y_{n_k}) = f^{-1}(\alpha).$$

Hence

$$\alpha = \eta,$$

and (i) follows from the fact that the sequence $y_n \to \alpha$ was arbitrary (Corollary 23.6).

EXERCISES

27-1 Let A be an arbitrary bounded, noncompact set. Construct continuous functions f, g, and h according to the following specifications:
(**1**) f is unbounded on A;
(**2**) g is bounded but attains neither its supremum nor its infimum;
(**3**) h is not uniformly continuous on A.

27-2 Construct a continuous function $f: [0, 1] \cup [2, 3] \to \mathbb{R}$ which is not monotonic on its domain, but whose inverse exists and is continuous.

27-3 If $f: A \to \mathbb{R}$ and $g: A \to \mathbb{R}$ are uniformly continuous, then so is their sum $f + g$: Prove or disprove this statement.

27-4 Uniform continuity is *not* preserved under multiplication. The functions $f(x) = \sin x$ and $g(x) = x$ are uniformly continuous on $\mathbb{R}$, yet $h(x) = f(x) \cdot g(x)$ is not: Prove all statements.

27-5 Show that if f is continuous on $(0, 1)$ but unbounded there, then f is not uniformly continuous.

27-6 Let f have a continuous inverse f^{-1}: If f is uniformly continuous, can we conclude that also f^{-1} is uniformly continuous?

27-7 The function f is *Hölder continuous* if there are constants α and β, $0 < \beta < 1$, such that

$$d(f(x), f(y)) \leq \alpha d(x, y)^\beta$$

for all points x and y in its domain (See Exercise 24-1). Does Hölder continuity imply uniform continuity? What about the converse?

27-8 Let f be uniformly continuous on the interval I. Show that for each $\epsilon > 0$ there is a step function γ_ϵ (consisting of finitely many steps) such that $\| f - \gamma_\epsilon \| < \epsilon$.

27-9 Prove: When f in Exercise 27-8 is merely continuous, then γ_ϵ still exists but it may consist of countably many steps.

27-10 Let A be a given compact set, and let the sequence $\{ f_n \}$ of continuous functions on A be endowed with the following properties:
(a) for each $x \in A$ and integer n, $f_n(x) \geq f_{n+1}(x)$;
(b) the sequence converges pointwise to 0.
Show that the sequence converges uniformly on A to zero.

SEVEN

DIFFERENTIABILITY

The definition of the derivative at a point as the limit of numerical sequences of certain difference quotients is well known from calculus: it is also repeated below. Some conversation should preface a discussion of this numerical process, for the latter beclouds the very important fact that differentiability means the existence of specific linear approximations.† To wit, let f be defined on $\mathfrak{R}$. This function is *continuous* at the point $x = a$ if for each number $\epsilon > 0$ there is a number $\delta > 0$ such that $d(f(x), f(a)) < \epsilon$ when $d(x, a) < \delta$. The function f is *differentiable* at $x = a$ if there is a linear function $\lambda(x) = \theta(x - a) + f(a)$ with the property that for each $\epsilon > 0$ there is a $\delta > 0$ such that $d(f(x), \lambda(x)) < \epsilon d(x, a)$ when $d(x, a) < \delta$. Geometrically this means that, when $x \in N(a; \delta)$, the function f lies between two lines passing through the point $(a, f(a))$ and making an angle ϵ with the line $\lambda(x)$ (see Figure 34).

28 / THE DERIVATIVE AT A POINT

We begin with the definition:

† The reader should also be forewarned that the so very natural identification of the derivative with a number is a peculiarity associated only with functions of one real variable.

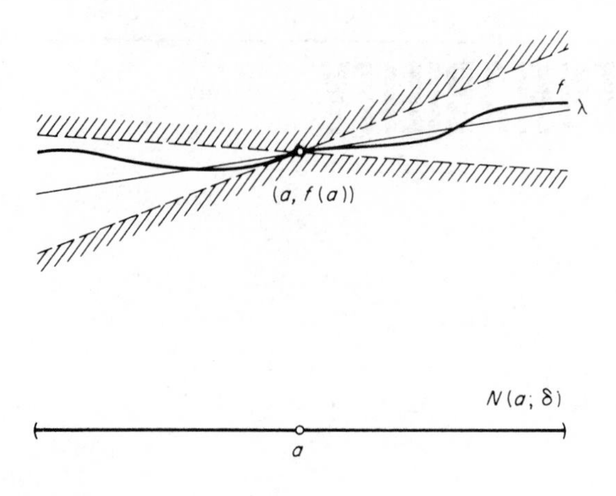

Figure 34

28.1 / ***Definition*** Let the function f be defined on the open interval I and let $a \in I$ be fixed. Let φ be defined on the deleted interval $I / \{a\}$ by

$$\varphi(x) = \frac{f(x) - f(a)}{x - a}. \qquad \text{(i)}$$

If the function φ has a limit at a, then we write

$$\lim_{x \to a} \varphi(x) = f'(a)$$

and call $f'(a)$ the *derivative* of f at a. The function f is said to be *differentiable* at a.

It should, perhaps, be emphasized that the limit called for is that specified in Definitions 23.1. When f is defined on the closed interval $[a, b]$, then f has a *right derivative* at $x = a$ if φ has a right limit there. We write

$$\lim_{x \to a+} \varphi(x) = f'(a+). \qquad \text{(ii)}$$

The *left derivative* of f at b is defined when φ has a left limit at $x = b$:

$$\lim_{x \to b-} \varphi(x) = f'(b-). \qquad \text{(iii)}$$

To avoid the fuss associated with one-sided derivatives we agree once and for all to dispense with the designations in (ii) and (iii) and to include in the term *derivative* left and right derivatives when such are called for. All functions in this section will be assumed to be defined on an interval I.

Let us proceed to connect the ideas developed in the introductory remarks of this chapter with Definition 28.1. This is done via the following statements:

28.2 / Theorem Let $f: I \to \mathcal{R}$ be given, $a \in I$ fixed. If there is a linear function

$$\lambda(x) = \theta(x - a) + f(a), \tag{iv}$$

θ being some constant, for which

$$\frac{d(f(x), \lambda(x))}{d(x, a)} \leq \text{constant}$$

on *some* deleted neighborhood $N^*(a; \delta)$ of a, then f is continuous at $x = a$.

Proof Multiply both sides of the inequality by $d(x, a)$, let $x \to a$ and appeal to Theorem 24.2.

28.3 / Theorem For each f there is at most one linear function λ such that

$$\lim_{x \to a} \frac{d(f(x), \lambda(x))}{d(x, a)} = 0 \qquad (x \neq a).$$

Proof Let also the linear function γ satisfy the theorem. Then, like λ, this function must pass through the point $(a, f(a))$ and hence has a representation of the form

$$\gamma(x) = \theta'(x - a) + f(a).$$

A simple calculation shows that for each admitted value of x

$$0 \leq d(\theta, \theta') = \frac{d(\theta, \theta') \cdot d(x, a)}{d(x, a)} = \frac{d(\lambda(x), \gamma(x))}{d(x, a)}$$

$$\leq \frac{d(f(x), \lambda(x))}{d(x, a)} + \frac{d(f(x), \gamma(x))}{d(x, a)}$$

and the theorem follows.

28.4 / Theorem Let $f: I \to \mathcal{R}$ be given. Let λ and φ be as in formulas (iv) and (i), respectively. The following statements are equivalent:

(1) $\displaystyle\lim_{x \to a} \frac{d(f(x), \lambda(x))}{d(x, a)} = 0,$

(2) $\displaystyle\lim_{x \to a} d(\varphi(x), \theta) = 0.$

Proof For each $x \in N^*(a; \delta) \cap I$,

$$\frac{d(f(x), \lambda(x))}{d(x, a)} = \frac{|f(x) - f(a) - \theta(x-a)|}{|x-a|} = \left|\frac{f(x) - f(a)}{x-a} - \theta\right|$$

$$= |\varphi(x) - \theta| = d(\varphi(x), \theta)$$

and the equivalence of the limits is immediate.

The limit in 28.4(1) always exists when $f(x)$ is a linear function, for then we set $\lambda(x) = f(x)$. It follows that every function $f(x) = \theta x + \beta$ is differentiable and for each point a in its domain $f'(a) = \theta$. In particular, the derivative of the constant function exists and is zero at each point. As in the case of limits and continuity, differentiability at a point is preserved under algebraic operations. Specifically one has:

28.5 / Theorem If f and g are differentiable at the point $a \in I$, then so are the functions $f + g$, $f \cdot g$, and f/g (provided $g(a) \neq 0$, $g'(a) \neq 0$, and $g(x) \neq 0$ in some neighborhood of a). In fact,

(1) $(f + g)'(a) = f'(a) + g'(a)$;

(2) $(fg)'(a) = f'(a)g(a) + f(a)g'(a)$;

(3) $\left(\dfrac{f}{g}\right)'(a) = \dfrac{g(a)f'(a) - g'(a)f(a)}{[g(a)]^2}$.

Proof The first of these statements needs no further justification by virtue of Exercise 23-1 and Definition 28.1. For the verification of (2) it is observed that for each $x \in N^*(a; \delta) \cap I$

$$\frac{(f \cdot g)(x) - (f \cdot g)(a)}{x-a} = \frac{f(x) - f(a)}{x-a} g(a) + f(x) \frac{g(x) - g(a)}{x-a},$$

and an application of Exercise 23-1 yields the result. Finally, (3) is derived from the formula

$$\frac{(f/g)(x) - (f/g)(a)}{x-a}$$

$$= \frac{1}{g(a)g(x)} \left[g(x) \frac{f(x) - f(a)}{x-a} - f(x) \frac{g(x) - g(a)}{x-a} \right]$$

$$(x \in N^*(a; \delta) \cap I)$$

in the same manner.

The differentiation of composite functions is subject to the so-called *chain-rule*:

28.6 / Theorem Let f be continuous on I, differentiable at the point $a \in I$; let g be defined on $f(I)$, differentiable at $f(a)$. Set

$$h(x) = (g \circ f)(x) \qquad (x \in I).$$

Then h is differentiable at the point $x = a$ and

$$h'(a) = (g' \circ f)(a) \cdot f'(a)$$

Remark: It should be noted that $g' \circ f$ designates the derivative of g with respect to the argument $\xi = f$ and not with respect to x.

Proof Let a be fixed, $f(a) = \alpha$. From Definition 28.1 we deduce that when ξ is sufficiently close to α, then

$$g(\xi) - g(\alpha) = (\xi - \alpha)[g'(\alpha) + \eta(\xi)],$$

where $\eta(\xi) \to 0$ as $\xi \to \alpha$. With $\xi = f(x)$ one now finds that, for $x \in I \setminus \{a\}$,

$$\frac{h(x) - h(a)}{x - a} = \frac{(g \circ f)(x) - (g \circ f)(a)}{x - a}$$

$$= \frac{f(x) - f(a)}{x - a}[(g' \circ f)(a) + \eta \circ f(x)].$$

The continuity of f is now required to assure that $f(x) \to f(a)$ as $x \to a$: This, in turn, shows that $\eta \circ f(x) \to 0$ as $x \to a$ and the theorem follows.

28.7 / *Examples*

(1) Consider the function

$$f(x) = \begin{cases} x \sin \dfrac{1}{x} & \text{for} \quad x \neq 0 \\ 0 & \text{at} \quad x = 0 \end{cases}$$

(see Figure 35). This function is Lipschitz continuous yet not differentiable at the point $x = 0$, since for any linear function $\lambda(x) = \theta x$ we have

$$\frac{d(f(x), \lambda(x))}{d(x, 0)} = \frac{d(x \sin \frac{1}{x}, \theta x)}{|x|} = d\left(\sin \frac{1}{x}, \theta\right)$$

but $\sin \dfrac{1}{x}$ has no limit at $x = 0$. On the other hand, the function is differentiable at any other point, and an application of Theorems

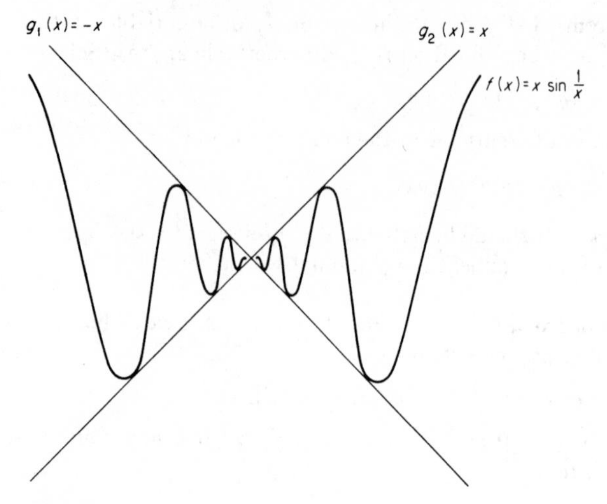

Figure 35

28.5(2) and 28.6 shows that

$$f'(x) = \sin \frac{1}{x} - \frac{1}{x} \cos \frac{1}{x}$$

whenever $x \neq 0$. In this connection, consider the function $f(x) = x^{2/3}$ $(-1 \leq x \leq 1)$. The linear function $\lambda(x) = f(0)$ best approximates f in a neighborhood of the origin, in the sense that if $\gamma(x) = \theta x + f(0)$ is any other line through 0, then for each $\gamma > 0$

$$\operatorname*{lub}_{x \in N(0;\, r)} d(f(x), \lambda(x)) < \operatorname*{lub}_{x \in N(0;\, r)} d(f(x), \gamma(x))$$

(see Figure 36). It is noteworthy that for each value of θ, $f(x) > |\theta x|$ for $x \in N(0; \theta^{-3})$. The function f, in other words, is not Lipschitz continuous in a neighborhood of the origin (see Exercise

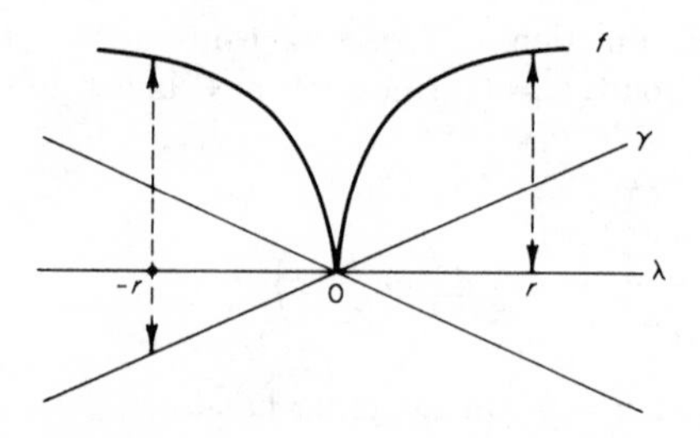

Figure 36

24-1). Indeed, better linear approximations than those for arbitrary continuous functions are available in the class of Lipschitz continuous functions, for if f is such a function, then

$$d(f(x), f(a)) < \epsilon d(x, a)$$

for some *fixed* $\epsilon > 0$. We recognize at once, however, that this recipe is merely a special case of the one described in Theorem 28.2.

(**2**) Now let $\alpha > 0$ be an arbitrary fixed number and let

$$f(x) = \begin{cases} x^{1+\alpha} \sin \dfrac{1}{x} & (x \neq 0) \\ 0 & (x = 0). \end{cases}$$

When x is different from zero we again apply Theorems 28.5(2) and 28.6 to find that

$$f'(x) = (1 + \alpha)x^{\alpha} \sin \frac{1}{x} - x^{\alpha-1} \cos \frac{1}{x} \qquad (x \neq 0).$$

This function is also differentiable at $x = 0$, and, in fact, $f'(0) = 0$, since with $\lambda(x) = 0$,

$$0 \leq \frac{d(f(x), \lambda(x))}{d(x, 0)} = \left| x^{\alpha} \sin \frac{1}{x} \right| \leq | x^{\alpha} |$$

for $x \neq 0$ and Theorem 28.4 is seen to be satisfied.

It should be observed that the function f' is defined for all $x \in \mathcal{R}$ and has a discontinuity of the second kind at $x = 0$ if $0 < \alpha \leq 1$.

EXERCISES

28-1 Consider the following function:

$$f(x) = \begin{cases} \dfrac{1}{4^{n-1}} - x & \left(\dfrac{1}{2 \cdot 4^{n-1}} \leq x \leq \dfrac{1}{4^{n-1}}\right) \\ 2\left(x - \dfrac{1}{4^{n}}\right) & \left(\dfrac{1}{4^{n}} \leq x \leq \dfrac{1}{2 \cdot 4^{n-1}}\right) \\ 0 & (x = 0) \\ f(-x) & (-1 \leq x < 0) \end{cases} \qquad (n \in \mathfrak{N})$$

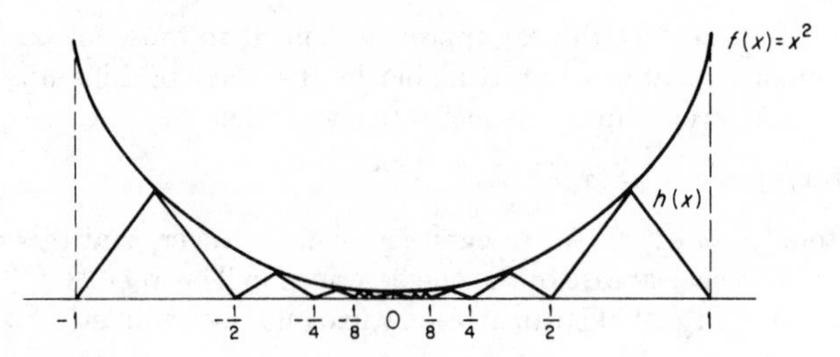

Figure 37

(a) Graph the function.
(b) Discuss the differentiability of f at the point $x = 0$.

28-2 Consider the function h depicted in Figure 37: Is this function differentiable at $x = 0$?

28-3 Discuss the differentiability of the following functions at $x = 0$:

(**1**) $a(x) = |x|$;

(**2**) $b(x) = [x]$, the brackets standing for the integral part of x;

(**3**) $$c(x) = \frac{\alpha_0 + \alpha_1 x + \cdots + \alpha_m x^m}{\beta_0 + \beta_1 x + \cdots + \beta_n x^n}.$$

28-4 Consider an arbitrary enumeration of the rational numbers, say $r_1, r_2, r_3, \ldots$. Discuss the continuity and differentiability of the function

$$f(x) = \sum_{k=1}^{\infty} \frac{|x - r_k|}{2^k}.$$

28-5 Recalling the definition of right and left limits (see Definition 23.1), we define the *generalized right derivative* of f at the point x by the formula

$$f'(x+) = \lim_{t \to x+} \frac{f(t) - f(x+)}{t - x}$$

when the indicated limit exists; the *generalized left derivative* of f at x is

$$f'(x-) = \lim_{t \to x-} \frac{f(t) - f(x-)}{t - x}$$

when the limit exists.

(**1**) Show that $f'(x+) = f'(x-) = f'(x)$ when $f'(x)$ exists.

(**2**) Is there a continuous function f for which $f'(x+) = f'(x-)$, yet $f'(x)$ does not exist?

(**3**) Can you construct a function $f:[0, 1] \to \mathcal{R}$, such that $f'\left(\frac{1}{n}+\right)$ and $f'\left(\frac{1}{n}-\right)$ exist but are not equal for $n = 1, 2, \ldots$?

29 / A CONTINUOUS NOWHERE DIFFERENTIABLE FUNCTION

The function $f: A \to \mathcal{R}$ is differentiable on A if it possesses a derivative at each point thereof. All functions one ordinarily encounters are differentiable save possibly for some exceptional points. The functions in Exercises 28-1 and 28-2, for instance, fail to be differentiable on a countable set. Just as we completely described the nature of the sets of discontinuities of arbitrary functions, it is natural to seek a classification of the sets on which a continuous function may not be differentiable. Attention to this problem was drawn by Ampère who, in 1806, attempted without success to prove that every function is somewhere differentiable. Since piecewise monotonic functions are known to be differentiable except possibly on a set of measure zero† (i.e., such functions are differentiable almost everywhere), it is clear that a nowhere differentiable function, should one exist, must be very pathological. A function $f: \mathcal{R} \to \mathcal{R}$ which is everywhere continuous yet nowhere differentiable was published by Weierstrass in 1872. It seems that another example was known to Bolzano as early as 1830, but it was not published for political reasons. Since that time several such functions have been constructed and the example presented below is essentially due to Van der Waerden (1930).

Consider the following functions: Let $f_0(x) = |x|$ for $|x| \leq \frac{1}{2}$, $f_0(x + n) = f_0(x)$ for each (positive or negative) integer n. By induction define for each $n \in \mathfrak{N}$ the functions

$$f_n(x) = \frac{1}{4^n} f_0(4^n x)$$

(see Figure 38). Each of these functions is piecewise linear and continuous on $\mathcal{R}$. Also, f_n is periodic with period $1/4^n$, and

$$0 \leq f_n(x) \leq \frac{1}{2 \cdot 4^n}. \tag{i}$$

† This fundamental result will not be verified here, but the proof is readily available in Béla Sz.-Nagy's "Introduction to Real Functions and Orthogonal Expansions."

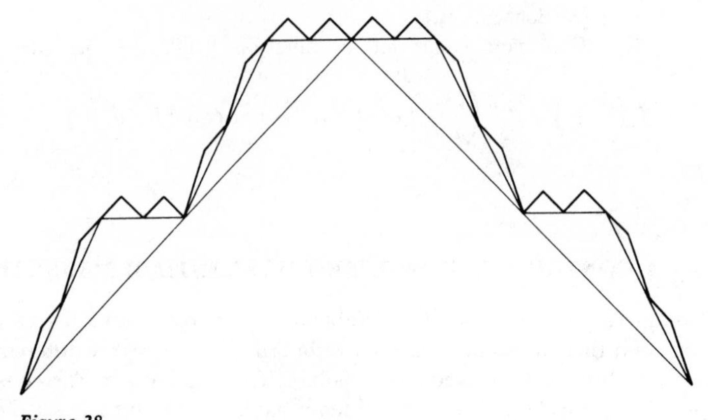

Figure 38

Setting

$$f(x) = \sum_{n=0}^{\infty} f_n(x) = \sum_{n=0}^{\infty} \frac{1}{4^n} f_0(4^n x), \tag{ii}$$

it is asserted that f is a continuous nowhere differentiable function (One sees at once that f fails to have derivatives at all points of the form $k/2^m$, where k and $m \geq 0$ are integers). The first two approximations to f are depicted in Figure 39.

Continuity is easily verified since for arbitrary points x and y and integer p,

$$\begin{aligned} d(f(x), f(y)) &\leq d\left(f(x), \sum_{n=0}^{p} f_n(x)\right) + d\left(f(y), \sum_{n=0}^{p} f_n(y)\right) \\ &\quad + d\left(\sum_{n=0}^{p} f_n(x), \sum_{n=0}^{p} f_n(y)\right) \\ &\leq \left| \sum_{n=p+1}^{\infty} f_n(x) \right| + \left| \sum_{n=p+1}^{\infty} f_n(y) \right| + \sum_{n=0}^{p} d(f_n(x), f_n(y)) \\ &= \alpha + \beta + \gamma. \end{aligned}$$

Owing to (i) we find that $\alpha \leq \dfrac{1}{6 \cdot 4^p}$ and $\beta \leq \dfrac{1}{6 \cdot 4^p}$. Furthermore, given an arbitrary $\epsilon > 0$ we can first select an integer p so large that $\dfrac{1}{6 \cdot 4^p} < \dfrac{1}{3} \epsilon$, and then a $\delta > 0$ so small, that the inequality

$$d(f_n(x), f_n(y)) < \frac{\epsilon}{3(p+1)}$$

holds for all values $n = 0, 1, \ldots, p$ whenever $d(x, y) < \delta$. Consequently

$$0 \leq d(f(x), f(y)) < \frac{\epsilon}{3} + \frac{\epsilon}{3} + \frac{\epsilon}{3} = \epsilon$$

whenever $d(x, y) < \delta$.

Let us now proceed to investigate the differentiability of f. For this let $m \in \mathfrak{N}$ be fixed. Then the periodicity of f_m tells us that for an arbitrary $x \in \mathfrak{R}$,

$$f_m\left(x \pm \frac{1}{4^m}\right) - f_m(x) = 0.$$

But for each $n \in \mathfrak{N}$

$$f_n\left(x \pm \frac{k}{4^n}\right) - f_n(x) = 0 \qquad (k = 1, 2, 3, \ldots).$$

Thus, when $n \geq m$ we write

$$\frac{1}{4^m} = \frac{4^{n-m}}{4^n} = \frac{k}{4^n}$$

and find that

$$f_n\left(x \pm \frac{1}{4^m}\right) - f_n(x) = 0 \qquad (n \geq m) \tag{iii}$$

(see Figure 38). On the other hand, suppose $n \leq m - 1$. Then at least one

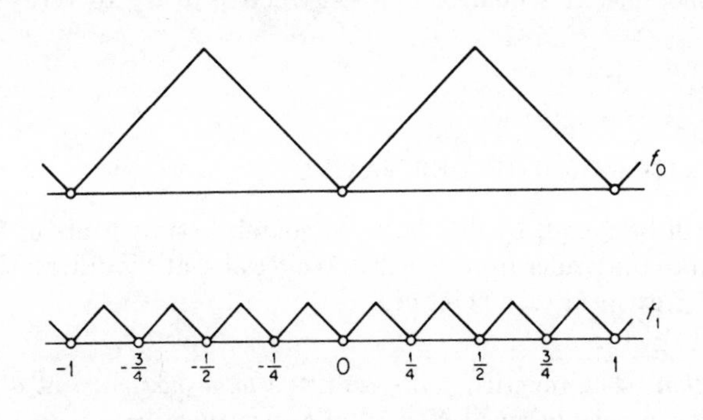

Figure 39

of the relations

$$f_n\left(x + \frac{1}{4^m}\right) - f_n(x) = \pm \frac{1}{4^m} \tag{iv}$$

or

$$f_n(x) - f_n\left(x - \frac{1}{4^m}\right) = \pm \frac{1}{4^m} \tag{v}$$

holds. But whichever the case, the relation which is true for one value $n \le m - 1$ holds for all such values. When, say, (iv) holds, and the information in (iii)–(iv) is applied to formula (ii) we find that

$$\frac{f\left(x + \frac{1}{4^m}\right) - f(x)}{\frac{1}{4^m}}$$

$$= 4^m \left\{\sum_{n=0}^{m-1} \left[f_n\left(x + \frac{1}{4^m}\right) - f_n(x)\right] + \sum_{n=m}^{\infty} \left[f_n\left(x + \frac{1}{4^m}\right) - f_n(x)\right]\right\}$$

$$= 4^m \sum_{n=0}^{m-1} \left[f_n\left(x + \frac{1}{4^m}\right) - f_n(x)\right] = \sum_{n=0}^{m-1} \pm 1$$

so that the difference quotient is an integer whose parity depends on m: it is even when m is even, odd when m is odd. But this means, of course, that the limits of the difference quotients do not exist and f is seen to be differentiable nowhere.

The reader should note that the function f constructed above is nowhere monotonic, which is to say that no point x has a neighborhood in which f is monotonic. It should be quite instructive to try to verify this fact!

30 / PROPERTIES OF THE DERIVATIVE

This section will be found, by and large, to contain restatements of facts already known to the reader from calculus. We recall that f is differentiable on the set A if f' exists at each point of A.

30.1 / *Definition* The function $f: A \to \mathbb{R}$ has a *local maximum* at $a \in A$ if there is a neighborhood $N(a; r)$ of a such that

$$f(x) \le f(a) \qquad (x \in N(a; r) \cap A);$$

f has a *local minimum* at a when

$$f(x) \geq f(a) \qquad (x \in N(a;r) \cap A).$$

A very simple but basic result is:

30.2 / Rolle's Theorem Let the function f meet the following specifications:

(a) f is continuous on $[a, b]$;
(b) f is differentiable on (a, b);
(c) $f(a) = f(b) = 0$.

Then there is a point $\xi \in (a, b)$ such that

$$f'(\xi) = 0.$$

Proof When f is identically zero there is nothing to prove. Otherwise we know from Theorem 27.5 that f attains a nonzero maximum or minimum value. Owing to (c) this value is attained in the interior of the interval. For the sake of argument, then, suppose f has a maximum at the point $\xi \in (a, b)$. Then for a sufficiently small $\delta > 0$,

$$\frac{f(\xi) - f(x)}{\xi - x} \geq 0 \qquad (x \in N^-(\xi;\delta))$$

whereas

$$\frac{f(\xi) - f(x)}{\xi - x} \leq 0 \qquad (x \in N^+(\xi;\delta)).$$

Taking right and left limits as $x \to \xi$ we find that $f'(\xi) \geq 0$ and concurrently $f'(\xi) \leq 0$. Since f is differentiable at ξ, it follows that $f'(\xi) = 0$. A similar argument is used when f is assumed to attain a minimum value.

This leads at once to the next theorem:

30.3 / Generalized Mean Value Theorem Let f and g be defined on $[a, b]$. Suppose:

(a) f and g are continuous on $[a, b]$;
(b) f and g are differentiable on (a, b).

Then a point $\xi \in (a, b)$ can be found such that

$$[f(b) - f(a)]g'(\xi) = [g(b) - g(a)]f'(\xi).$$

Proof Consider the function

$$h(x) = [f(b) - f(a)][g(x) - g(a)] - [g(b) - g(a)][f(x) - f(a)].$$

Owing to Theorem 28.5 this function is differentiable on (a, b) and a simple calculation shows that $h(a) = h(b) = 0$. Hence, there is a point $\xi \in (a, b)$ at which $h'(\xi) = 0$, but this yields the desired formula.

30.4 / Mean Value Theorem If f is continuous on $[a, b]$ and differentiable on (a, b), then there is a point $\xi \in (a, b)$ such that

$$f(b) - f(a) = (b - a)f'(\xi).$$

Proof Simply take the special case $g(x) = x$ in the preceding theorem.

We have just dashed through a sequence of theorems with a recurrent theme: continuity on a compact interval, differentiability on its interior. It is natural to ask if these restrictions can be relaxed. The reader should construct examples to show that the restrictions are necessary. The final theorem gives us important information about the behavior of differentiable functions.

30.5 / Theorem Let f be differentiable on the open interval (a, b). Then

(**1**) f is nondecreasing when $f' \geq 0$,
(**2**) f is nonincreasing when $f' \leq 0$,
(**3**) f is constant when $f' = 0$.

Proof Consider arbitrary points $a \leq x < y \leq b$. Then Theorem 30.4 tells us that

$$f(y) - f(x) = d(x, y)f'(\xi) \tag{i}$$

for some point $x < \xi < y$. The theorem is deduced from the observation that $d(x, y) > 0$.

30.6 / *Remark* It is evident from formula (i) that f is increasing when $f' > 0$, decreasing when $f' < 0$ on (a, b). The strict inequality, however, is sufficient but not necessary, a simple example being $f(x) = x^3$ $(-1 < x < 1)$.

When we examine an arbitrary step function we find that it cannot be a derivative. In Example 28.7(2), on the other hand, we encountered a function with a discontinuity of the second kind which was a derivative.

This brings up the question of possible structure of functions which are derivatives. The characterizing property of derivatives is their intermediate value property (see Theorem 24.9).

30.7 / Intermediate Value Theorem for Derivatives Let f be differentiable on the closed interval $[a, b]$. If $f'(a) < f'(b)$, then for each point $f'(a) < \eta < f'(b)$ there is an $a < \xi < b$ such that $f'(\xi) = \eta$. An analogous result holds when $f'(a) > f'(b)$.

Proof Consult Figure 40. Setting $c = \frac{1}{2}(a + b)$ we introduce the functions

$$\varphi(x) = \begin{cases} a & (a \leq x \leq c) \\ 2x - b & (c \leq x \leq b) \end{cases}$$

and

$$\psi(x) = \begin{cases} 2x - a & (a \leq x \leq c) \\ b & (c \leq x \leq b). \end{cases}$$

These functions are evidently continuous and they are subject to the inequalities

$$a \leq \varphi(x) < \psi(x) \leq b \qquad (a < x < b).$$

The function

$$g(x) = \frac{(f \circ \psi)(x) - (f \circ \varphi)(x)}{\psi(x) - \varphi(x)} \qquad (a < x < b)$$

is accordingly continuous. Moreover,

$$\lim_{x \to a+} g(x) = \lim_{x \to a+} \frac{f(2x - a) - f(a)}{2(x - a)}$$

$$= \lim_{x \to a+} \frac{f(2(x - a) + a) - f(a)}{2(x - a)} = f'(a)$$

and likewise

$$\lim_{x \to b-} g(x) = f'(b).$$

By virtue of Theorem 24.9, therefore, there is a point $a < \xi_0 < b$ such that

$$g(\xi_0) = \eta.$$

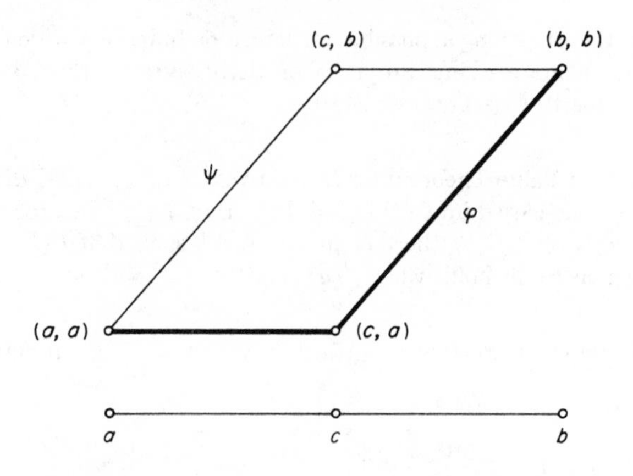

Figure 40

But now we appeal to Theorem 30.4. Transplanted into the present setting the theorem says that there is a point ξ, $\varphi(\xi_0) < \xi < \psi(\xi_0)$, such that

$$(f \circ \psi)(\xi_0) - (f \circ \varphi)(\xi_0) = [\psi(\xi_0) - \varphi(\xi_0)] \cdot f'(\xi),$$

i.e.,

$$\eta = g(\xi_0) = \frac{(f \circ \psi)(\xi_0) - (f \circ \varphi)(\xi_0)}{\psi(\xi_0) - \varphi(\xi_0)} = f'(\xi),$$

as was to be shown.

Now, a function with simple discontinuities does not have the intermediate value property (why?). Hence we have:

30.8 / Corollary If f is differentiable on the interval $[a, b]$, then f' cannot have simple discontinuities.

The reader should attempt to determine the status of the theorems in this section when the underlying sets are no longer assumed to be intervals.

A very useful application of the Generalized Mean Value Theorem concerns the evaluation of limits $f(x)/g(x)$ as $x \to a$. Our previous results in this direction were subject to the existence of the limits of f and g as $x \to a$, plus the restriction $g(x) \neq 0$ in a neighborhood of a, but in many cases such conditions do not prevail. We establish the following:

30.9 / L'Hospital's Rule Let the functions f and g be differentiable on the open interval (a, b) and suppose $g' \neq 0$. If

$$\lim_{x \to a} \frac{f'(x)}{g'(x)} = \eta, \tag{i}$$

then also

$$\lim_{x \to a} \frac{f(x)}{g(x)} = \eta \tag{ii}$$

when *one* of the following conditions is met:

(1) $\lim_{x \to a} f(x) = 0$ and $\lim_{x \to a} g(x) = 0$;

(2) $\lim_{x \to a} g(x) = \infty$.

It should be noted that the limit of f at a is irrelevant in (2).

Proof Consider condition (i). For each $\epsilon > 0$ there is a $\delta > 0$ such that

$$\frac{f'(x)}{g'(x)} \in N\left(\eta; \frac{\epsilon}{2}\right) \qquad \text{when} \quad x \in N^+(a; \delta).$$

Holding ϵ and δ fixed, let $z < x$ be momentarily fixed points in $N^+(a; \delta)$. Then by virtue of Theorem 30.3,

$$\frac{f(z) - f(x)}{g(z) - g(x)} = \frac{f'(\xi)}{g'(\xi)} \in N\left(\eta; \frac{\epsilon}{2}\right) \tag{iii}$$

for *some* $\xi \in N^+(a; \delta)$. Thus, if (1) is the case, then

$$\lim_{z \to a} \frac{f(z) - f(x)}{g(z) - g(x)} = \frac{f(x)}{g(x)} \in N(\eta; \epsilon) \qquad \text{for} \quad x \in N^+(a; \delta),$$

but this clearly implies (ii).

Now let (2) be the case. When the function f is bounded in a neighborhood of a, then the formula in (iii) shows that the left side in (i) vanishes and clearly so does η in (ii). Thus, suppose $f(x) \to \infty$ as $x \to a$ (the case $f(x) \to -\infty$ as $x \to a$ is completely analogous to the present one). Let ϵ and δ be still fixed, and let also z be held fast. Then we can write the left side in (iii) as

$$\frac{f(x)}{g(x)} \frac{\dfrac{f(z)}{f(x)} - 1}{\dfrac{g(z)}{g(x)} - 1} = \frac{f(x)}{g(x)} (1 + \theta),$$

where $\theta \to 0$ as $x \to a$. But this means that for a suitable value of $\delta > 0$,

$$\frac{f(x)}{g(x)} \in N(\eta; \epsilon) \qquad \text{when} \quad x \in N^+(a; \delta)$$

and once more (ii) follows from (i).

EXERCISES

30-1 Imagine a function $f: [a, b] \to \mathcal{R}$ such that

$$d(f(x), f(y)) \leq c \cdot d(x, y)^\alpha \qquad (x, y \in [a, b]),$$

where $c > 0$ and $\alpha > 1$ are fixed constants. What can you say about f?

30-2 Suppose the polynomial p is such that $p(a) = 0$ and $p'(a) = 0$. Then find a polynomial q such that $p(x) = (x - a)^2 q(x)$.

30-3 Let the coefficients of the polynomial equation

$$a_0 + a_1 x + a_2 x^2 + \cdots + a_n x^n = 0$$

be related through the equation

$$a_0 + \frac{a_1}{2} + \frac{a_2}{3} + \cdots + \frac{a_n}{n+1} = 0.$$

Find the zero in the open interval $(0, 1)$ of the polynomial equation.

30-4 Give geometric interpretations to the Generalized Mean Value Theorem and the Mean Value Theorem.

30-5 Prove the Mean Value Theorem directly from Rolle's Theorem.

30-6 Let f be a differentiable function on the open interval $(0, 1)$ and suppose $f' > 0$. Then show that:

(a) f has an inverse f^{-1};
(b) f^{-1} is itself differentiable;
(c) $(f^{-1})' \circ f = \dfrac{1}{f'}$.

30-7 Show that L'Hospital's Rule remains valid in the case

$$-\infty \leq a < b \leq \infty.$$

30-8 Evaluate the following limits:

(a) $\lim_{x \to \infty} x(\sqrt{x^2 + 1} - x)$
(b) $\lim_{x \to 0} x^x$
(c) $\lim_{x \to 0} x \ln x$
(d) $\lim_{x \to \infty} x^{1/x}$
(e) $\lim_{x \to \infty} \dfrac{\ln(1 + e^x)}{x}$.

30-9 Let f be differentiable at the point x, and let $\{h_n\}$ and $\{k_n\}$ be monotonic null sequences of positive terms. Show that

$$f'(x) = \lim_{n \to \infty} \frac{f(x + h_n) - f(x - k_n)}{h_n + k_n}.$$

30-10 Does the conclusion in the preceding problem hold when the sequence $\{k_n\}$ is replaced with the sequence $\{-k_n\}$, where $h_n - k_n \neq 0$ for all values of n?

31 / TAYLOR'S THEOREM

It is clear that an ascending smoothness scale is established when we consider successive classes of functions whose derivatives are themselves differentiable.

31.1 / *Definition* Suppose $f:[a, b] \to \mathcal{R}$ is differentiable on $[a, b]$. If f' is itself differentiable, then we designate its derivative with the symbol f'', called the *second derivative* of f. Inductively, if $f^{(n)}$ is the nth derivative of f $(n \geq 1)$ and if $f^{(n)}$ is differentiable, then the function $f^{(n+1)}$ is the $(n + 1)$st derivative of f. For convenience we set $f^{(0)} = f$.

Let $f: (a, b) \to \mathcal{R}$ be given and suppose $f^{(n)}(\xi)$ exists for some value $a < \xi < b$. Then there is a neighborhood $N(\xi; \delta)$ of ξ such that each of the functions $f^{(k)}$ $(0 \leq k \leq n - 2)$ is differentiable on $N(\xi; \delta)$. We now come to a very important generalization of the Mean Value Theorem:

31.2 / Taylor's Theorem Let there be given an integer $n \in \mathfrak{N}$ and a function $f:[a, b] \to \mathcal{R}$ such that:

(a) the functions $f, f^{(1)}, \ldots, f^{(n-1)}$ are continuous on $[a, b]$;
(b) the function $f^{(n)}$ exists on (a, b).

Then for an arbitrary interval $[\alpha, \beta] \subset [a, b]$ there is a number ξ, $\alpha < \xi < \beta$, such that

$$f(\beta) = f(\alpha) + \frac{f'(\alpha)}{1!}(\beta - \alpha) + \frac{f''(\alpha)}{2!}(\beta - \alpha)^2 + \cdots + \frac{f^{(n-1)}(\alpha)}{(n-1)!}(\beta - \alpha)^{n-1} + R_n, \qquad \text{(i)}$$

where

$$R_n = \frac{f^{(n)}(\xi)}{n!}(\beta - \alpha)^n. \qquad \text{(ii)}$$

Note that for the case $n = 1$ the above theorem reduces to the Mean Value Theorem. The term R_n is called the *Lagrange remainder*. The formula in (i) specifies a polynomial which approximates the n times differentiable function f at β, by means of the value of f and its first $n - 1$

derivatives at α, to within an error R_n. Just as in the Mean Value Theorem, the value of the intermediate point ξ is not known. Estimating the size of R_n is, therefore, a delicate matter.

Proof Let the interval $[\alpha, \beta]$ be fixed. To estimate the difference between $f(\beta)$ and the indicated polynomial we define the number θ through the equation

$$\frac{(\beta-\alpha)^n}{n!}\theta = f(\beta) - \left[f(\alpha) + \frac{f'(\alpha)}{1!}(\beta-\alpha) + \cdots + \frac{f^{(n-1)}(\alpha)}{(n-1)!}(\beta-\alpha)^{n-1}\right] \quad \text{(iii)}$$

and introduce the function φ as follows:

$$\varphi(x) = f(\beta) - \left[f(x) + \frac{f'(x)}{1!}(\beta-x) + \cdots + \frac{f^{(n-1)}(x)}{(n-1)!}(\beta-x)^{n-1} + \frac{\theta}{n!}(\beta-x)^n\right].$$

By virtue of condition (a), φ is continuous on the interval $[a, b]$, and (b) says that φ is differentiable on the interval (a, b). Furthermore, $\varphi(\beta) = 0$ and the choice of θ was such that also $\varphi(\alpha) = 0$. The function φ is thus seen to meet the specifications set forth in the statement of Rolle's Theorem. Accordingly there is a number ξ, $\alpha < \xi < \beta$, such that $\varphi'(\xi) = 0$. Upon differentiation we obtain the following:

$$\varphi'(x) = -\left[f'(x) - f'(x) + \frac{f''(x)}{1!}(\beta-x) - \frac{f''(x)}{1!}(\beta-x) + \cdots + (-1)\frac{f^{(n-1)}(x)}{(n-2)!}(\beta-x)^{n-2} + \frac{f^{(n)}(x)}{(n-1)!}(\beta-x)^{n-1} - \frac{\theta}{(n-1)!}(\beta-x)^{n-1}\right] = \frac{\theta - f^{(n)}(x)}{(n-1)!}(\beta-x)^{n-1}.$$

Since $\varphi'(\xi) = 0$, it follows that $\theta = f^{(n)}(\xi)$ and a substitution into formula (iii) yields the desired result.

Generalizations as well as applications of these results are taken up in Section 34.

EXERCISES

31-1 Consider the function $f(x) = \sin x$ ($x \in \mathcal{R}$). Referring to formula (ii), show that for $\alpha = 0, \beta = x$,

$$| R_n(x) | \leq \frac{| x |^n}{n!}.$$

31-2 Discuss the applicability of Taylor's Theorem to the function $f(x) = (1 - x)^{1/2}$.

31-3 Obtain the *Leibnitz formula* for the nth derivative of the product function $f \cdot g$:

$$(f \cdot g)^{(n)}(x) = \sum_{k=0}^{n} \binom{n}{k} (f^{(k)} \cdot g^{(n-k)})(x).$$

31-4 If $f''(\xi)$ exists, then

$$f''(\xi) = \lim_{h \to 0} \frac{f(\xi + h) - 2f(\xi) + f(\xi - h)}{h^2}.$$

Prove this statement. Give an example of a function f for which the right side limit exists yet $f''(\xi)$ does not exist.

EIGHT

SPACES OF CONTINUOUS FUNCTIONS

32 / THE PROBLEMS OF SEPARABILITY AND CONVERGENCE

Compactness, it will be recalled, played a key role in establishing many of the important properties of continuous functions. This concept also lies at the heart of the discussion here. Thus, let there be given a fixed nonempty compact set $A \subset \mathcal{R}$ and let $\mathcal{C}(A)$ stand for the set of all continuous functions $f: A \to \mathcal{R}$. With the distance function

$$\| f - g \| = \max_{x \in A} d(f(x), g(x)),$$

which was introduced in Example 3.13, $\mathcal{C}(A)$ is a *metric space*† and it is therefore meaningful to inquire if the metric attributes of $\mathcal{R}$ have their analog in $\mathcal{C}(A)$. To show that the above formula indeed specifies a metric is simple. Owing to Theorems 27.3 and 27.4 the right side in the formula is finite. The statements

(a) $\| f - g \| \geq 0$, equality holding if and only if $f = g$,

(b) $\| f - g \| = \| g - f \|$,

require no further justification. The *triangle inequality*

(c) $\| f - g \| \leq \| f - h \| + \| h - g \|$

† Again we do not distinguish notationally between the *set* $\mathcal{C}(A)$ and the corresponding *metric space* $\mathcal{C}(A)$, but this should not lead to any confusion.

is verified as follows: For each fixed x,

$$d(f(x), g(x)) \leq d(f(x), h(x)) + d(h(x), g(x))$$

(see Example 3.9). In particular, this inequality is not disturbed when in the right side we replace each summand with its maximum:

$$d(f(x), g(x)) \leq \| f - h \| + \| h - g \|.$$

Being true for each admitted value of x, this inequality now leads to (c). While $\mathcal{C}(A)$ is the only function space discussed here, it should always by borne in mind that it is but one of many. $\mathcal{C}(A)$ is, of course, a *linear space* over $\mathcal{R}$, but the linear structure of $\mathcal{C}(A)$ is not the subject of study here. Rather, our concern involves the metric $\| \ \|$.

The central metric themes in the architecture of $\mathcal{R}$ were its separability and completeness (convergence): These topics occupy a prominent position also in this discussion. Consider the first of these. The existence of dense countable sets enabled us to approximate arbitrary members of $\mathcal{R}$ to within a predetermined error in a finite number of steps: the importance of this cannot be overstated. The separability of $\mathcal{C}(A)$ is a consequence of the following facts:

The set $P(A)$ of polynomials on A with coefficients in $\mathcal{R}$ is dense in $\mathcal{C}(A)$: Given an arbitrary function $f \in \mathcal{C}(A)$ and a number $\epsilon > 0$, then there is a polynomial $p \in P(A)$ with the property that $\| f - p \| < \epsilon$.

The set $P^*(A)$ of polynomials with coefficients in $\mathcal{R}^*$ is dense in $P(A)$.

The set $P^*(A)$ is countable.

The first of these statements is Weierstrass' approximation theorem which is proved later in this chapter. The second statement is verified as follows: Let the polynomial $p(x) = \sum_{k=0}^{n} a_k x^k$ ($a_k \in \mathcal{R}$) and a number $\epsilon > 0$ be given; let $m = \max\{x^k \mid x \in A, 0 \leq k \leq n\}$. If we select rational numbers a_k^* such that $| a_k - a_k^* | < \epsilon/(n + 1)m$ for $k = 0, 1, \ldots, n$, then the polynomial $p^*(x) = \sum_{k=0}^{n} a_k^* x^k$ is endowed with the property that $\| p - p^* \| < \epsilon$. A simple generalization of Theorem 17.7 shows $P^*(A)$ to be dense in $\mathcal{C}(A)$ when $P(A)$ is dense in $\mathcal{C}(A)$. Owing to Exercise 4–8 the set $P^*(A)$ is countable.

Let us digress here momentarily to point to some very challenging problems emerging out of the Weierstrass approximation theorem. One considers various classes of functions $T \subset \mathcal{C}(A)$, such as the class of polynomials whose degree does not exceed a given integer n, relative to which questions such as these are posed: If $f \in \mathcal{C}(A)$ is given, then

(a) Is there a member $p \in T$ which best approximates f, in the sense that for each $q \in T$, $\| f - p \| \leq \| f - q \|$?

(b) Is the best approximation unique?

(c) How can the best approximation be constructed?

These questions are among those dealt with in approximation theory. Can you answer them when $f = |x - \frac{1}{2}|$ $(0 \leq x \leq 1)$ and T is restricted to the class of polynomials of degree ≤ 2?

The question of convergence is more delicate since the Cauchy criterion for convergence can be extended from $\mathcal{R}$ to $\mathcal{C}(A)$ in two distinct ways. Given a sequence $\{f_n\}$ of functions of $\mathcal{C}(A)$, then each fixed point $a \in A$ specifies a numerical sequence $\{f_n(a)\}$: The statement that the function sequence $\{f_n\}$ converges when every numerical sequence $\{f_n(x)\}$ $(x \in A)$ converges is sensible. Formally we state the following definition:

32.1 / *Definition* The sequence $\{f_n\}$ $(f_n \in \mathcal{C}(A))$ converges *pointwise* to the function φ if

$$\lim_{n\to\infty} f_n(x) = \varphi(x)$$

for *each* $x \in A$.

To show that $\mathcal{C}(A)$ is not necessarily complete under pointwise convergence it suffices to demonstrate that φ may be discontinuous. This is done in the next example:

32.2 / *Example* Let $A = [0, 1]$ and for each $n \in \mathfrak{N}$ let $f_n(x) = x^n$. Then by Theorem 11.9:

$$\lim_{n\to\infty} f_n(x) = \begin{cases} 0 & \text{for } 0 \leq x < 1 \\ 1 & \text{at } x = 1, \end{cases}$$

that is, the function sequence $\{f_n\}$ converges pointwise to a dis-

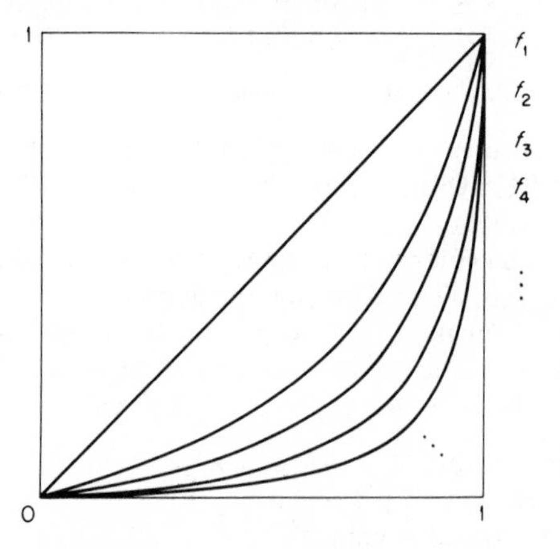

Figure 41

continuous function (see Figure 41): All subsequences thereof are readily seen to have a common limit.

We note that when the sequence $\{f_n\}$ converges pointwise to φ, then for each point $x \in A$ and number $\epsilon > 0$ there is a positive integer $k = k(x; \epsilon)$ such that $d(f_n(x), \varphi(x)) < \epsilon$ for all values $n > k$. The failure of the function sequence in Example 32.2 to converge to a continuous function stems from the fact that there is no fixed integer k which works for all x. Thus, when x is confined to the interval $2^{-1/k} < x < 1$, then $d(x^k, \varphi(x)) = x^k > \frac{1}{2}$. How, then, can the continuity of the limit function be guaranteed? Combining Definition 32.1 and Theorem 24.2, we find that for φ to be continuous at $x = a$, it is necessary that

$$\varphi(a) = \lim_{x \to a} \varphi(x) = \lim_{x \to a} \lim_{n \to \infty} f_n(x)$$

while, at the same time,

$$\varphi(a) = \lim_{n \to \infty} f_n(a) = \lim_{n \to \infty} \lim_{x \to a} f_n(x);$$

that is, the order in which the iterated limits are taken must be interchangeable:

$$\lim_{x \to a} \lim_{n \to \infty} f_n(x) = \lim_{n \to \infty} \lim_{x \to a} f_n(x).$$

In Example 32.2, for instance,

$$\lim_{x \to 1} \lim_{n \to \infty} x^n = 0,$$

whereas

$$\lim_{n \to \infty} \lim_{x \to 1} x^n = 1.$$

The answer to the problem at hand is given in the next section, where it is shown that uniform convergence preserves continuity:

32.3 / *Definition* The function sequence $\{f_n\}$ $(f_n \in \mathcal{C}(A))$ converges *uniformly* on A to the limit function φ if to every $\epsilon > 0$ there corresponds an integer n_ϵ such that

$$\| f_n - \varphi \| < \epsilon$$

for all values $n > n_\epsilon$.

Uniform convergence clearly implies pointwise convergence.

Another fundamental property of $\mathcal{R}$ which has no simple analog in $\mathcal{C}(A)$ is the Bolzano–Weierstrass Theorem which asserts that every infinite bounded subset of $\mathcal{R}$ has a limit point: The set of functions $\{x^n\}$ of Example 32.2 has no limit point in $\mathcal{C}(A)$. This problem is the subject of the Arzelà–Ascoli Theorem which is proved later in this chapter. Here we make only this observation: Assuming momentarily that uniform convergence does, indeed, guarantee the continuity of the limit function, the question posed here is this: How can we characterize the subsets of $\mathcal{C}(A)$ from which a uniformly convergent sequence can be extracted?

The remainder of this section is devoted to some examples.

32.4 / *Examples*

(**1**) The set of constant functions $f_n(x) = (-1)^n$ has two limit points, $a(x) = -1$ and $b(x) = 1$.

(**2**) Consider the functions $f_n(x) = \dfrac{x}{n}$ $(n \in \mathfrak{N})$ which are continuous on $\mathcal{R}$. On every bounded set the sequence $\{f_n\}$ tends uniformly to the continuous limit function $\varphi(x) = 0$, yet this fails to be true on $\mathcal{R}$. This is easy to see, for given an arbitrary integer $k > 0$, then $d\left(\dfrac{x}{k}, \varphi(x)\right) = \dfrac{x}{k} > 1$, when $k < x < \infty$.

(**3**) Just as in the case of numerical sequences, we can associate with each sequence $\{f_n\}$ the series $\sum f_n$ which is said to converge pointwise to φ when

$$\lim_{k\to\infty} \sum_{n=0}^{k} f_n(x) = \varphi(x)$$

for each $x \in A$. Thus, consider the functions

$$f_n(x) = \frac{x^2}{(1+x^2)^n} \qquad (n \in \mathfrak{N})$$

with domain $A = [-1, 1]$, and put

$$\varphi(x) = \sum_{n=0}^{\infty} \frac{x^2}{(1+x^2)^n}.$$

Then a simple argument shows that

$$\varphi(x) = \begin{cases} 0 & \text{at} \quad x = 0 \\ 1 + x^2 & \text{when} \quad x \neq 0 \end{cases}$$

(The reader recognizes, no doubt, that we have here a geometric series for each x). Hence, the pointwise limit of the series is discontinuous.

(**4**) In the situations depicted thus far, the limit function was continuous except at some points. Let us now show that a pointwise convergent sequence may have an everywhere discontinuous limit. For each value of m and n consider, namely, the "saw tooth" functions $f_{mn}: \mathcal{R} \to \mathcal{R}$ defined piecewise to be

$$f_{mn}(x) = |\, 2m!\cdot x - 2k + 1 \,|^n$$

$$(k \leq m!\cdot x \leq k + 1; \quad k = 0, \pm 1, \pm 2, \ldots),$$

m and n being positive integers. It is easily verified that for fixed m and n, $0 \leq f_{mn}(x) \leq 1$; in fact, $f_{mn}(x) = 1$ when $m!x$ is an integer, and $f_{mn}(x) < 1$ otherwise. Hence, when we set for each value of m

$$f_m(x) = \lim_{n\to\infty} f_{mn}(x),$$

then by Theorem 11.9

$$f_m(x) = \begin{cases} 1 & \text{when} \quad m!x \quad \text{is integral} \\ 0 & \text{elsewhere.} \end{cases}$$

Now set

$$f(x) = \lim_{m\to\infty} f_m(x).$$

Since for each m, $f_m(x) = 0$ when x is irrational it follows that also $f(x) = 0$ in this case. On the other hand, when x is rational, say, $x = \dfrac{p}{q}$, with p and q relatively prime integers, q is a factor of $m!$ when $m \geq q$, and hence $f_m(x) = 1$. Thus also $f(x) = 1$. Consequently

$$f(x) = \begin{cases} 0 & \text{at irrational} \quad x \\ 1 & \text{at rational} \quad x, \end{cases}$$

i.e., the pointwise limit function exists but is discontinuous everywhere.

EXERCISES

32-1 A set with the binary operations $+$ and $\cdot$ which satisfy the axioms 2.2(1) through (9) is a *ring* (more precisely, a commutative ring with an identity). The elements $a \neq 0$ and $b \neq 0$ are *zero divisors* of the ring when $ab = 0$. Show that $\mathcal{C}(A)$ is a ring with zero divisors over $\mathcal{R}$.

32-2 Is $\mathcal{C}(A)$ a metric space under the specified distance function when A is noncompact?

32-3 Let $L \subset \mathcal{C}(A)$ be the set of constant functions $f(x) = c$ $(c \in \mathcal{R})$. Show that for each $f \in \mathcal{C}(A)$ there is a $\lambda \in L$ such that

$$\| f - \lambda \| = \inf_{\gamma \in L} \| f - \gamma \|.$$

32-4 Can you prove that the above exercise remains correct when L is replaced with the set of linear functions of $\mathcal{C}(A)$?

32-5 Can you extend the notion of continuity to the metric function $\| \, \| : \mathcal{C}(A) \times \mathcal{C}(A) \to \mathcal{R}$.

32-6 Prove or disprove: Consider a sequence of functions $f_n \in \mathcal{C}(A)$ such that $\| f_n - f_m \| \to 0$ as $m, n \to \infty$. If the subsequence $\{ f_{n_k} \}$ converges uniformly to the function f then also the sequence $\{ f_n \}$ converges to f.

32-7 Referring to Exercise 24-1, consider the class of Lipschitz continuous functions on $[0, 1]$ with constant $\alpha = 1$:

$$| f(x) - f(y) | \leq | x - y |$$

for all admitted points, and such that $\| f \| \leq 10$. Let $\mathcal{H}$ be that subclass of these functions which satisfy the requirement $\| f - g \| \geq 1$. How many members does $\mathcal{H}$ contain?

33 / UNIFORM CONVERGENCE

Our purpose in this section is to examine the various facets of uniform convergence and to investigate the preservation of continuity and differentiability under this limit process. We recall Example 32.4(2) where it was shown that uniform convergence is not necessary to assure the continuity of the limit function; the sufficiency of the condition is exhibited below, but first we establish another characterization of uniform convergence.

It is only natural that the Cauchy convergence criterion (Definition 9.2), which played the leading role in the completeness of $\mathcal{R}$ should have its analog here. Indeed, we have:

33.1 / Theorem The following statements are equivalent:

(1) The sequence $\{ f_n \}$ converges uniformly on A;

(2) For each number $\epsilon > 0$ there is an integer k_ϵ such that

$$\| f_m - f_n \| < \epsilon$$

whenever $m, n > k_\epsilon$.

Proof Suppose f is the uniform limit of the sequence $\{f_n\}$. Then owing to Definition 32.3 there is an integer k_ϵ such that $\|f_n - f\| < \frac{1}{2}\epsilon$ for all values $n > k_\epsilon$ and hence the triangle inequality tells us that

$$\|f_n - f_m\| \leq \|f_n - f\| + \|f_m - f\| < \frac{\epsilon}{2} + \frac{\epsilon}{2} = \epsilon$$

for $m, n > k_\epsilon$. It follows that the sequence is a Cauchy sequence.

Now suppose (2) is valid. Since, for each $x \in A$,

$$d(f_n(x), f_m(x)) \leq \|f_n - f_m\| < \epsilon,$$

it follows from the completeness property of the real numbers that there is a function $f(x)$ which is the pointwise limit of $\{f_n(x)\}$ (The uniqueness of the limit guarantees that f is, indeed, a function). Hence, it must be shown that this f is the uniform limit of the sequence $\{f_n\}$. Look at the formula in (2). It implies that $|f_m(x) - f_n(x)| < \epsilon$ for all $x \in A$ and integers $m, n > k_\epsilon$. Thus, if we take the pointwise limit $\lim_{m\to\infty} f_m(x) = f(x)$, then we obtain the inequalities $|f_n(x) - f(x)| \leq \epsilon$ for all $x \in A$ and integers $n > k_\epsilon$. Consequently,

$$\|f_n - f\| \leq \epsilon$$

for all $n > k_\epsilon$, as was to be shown.

It should be explicitly understood that no claim is made in this theorem regarding the continuity of the limit function.

33.2 / Theorem If the sequence $\{f_n\}$ ($f_n \in \mathcal{C}(A)$) converges uniformly on A to f, then $f \in \mathcal{C}(A)$.

Proof Let $a \in A$ be an arbitrary point. Given an $\epsilon > 0$, there is an integer k such that

$$\|f_n - f\| < \frac{\epsilon}{3}$$

for all values $n > k$. The continuity of f_n provides for a $\delta > 0$ with the property that

$$d(f_n(x), f_n(a)) < \frac{\epsilon}{3}$$

when $x \in N(a;\delta) \cap A$. These two estimates show that

$$\begin{aligned} d(f(x), f(a)) &\leq d(f_n(x), f(x)) + d(f_n(a), f(a)) + d(f_n(x), f_n(a)) \\ &\leq 2\|f_n - f\| + d(f_n(x), f_n(a)) = \frac{2\epsilon}{3} + \frac{\epsilon}{3} = \epsilon \end{aligned}$$

for the indicated values of x, and the theorem follows.

Thus, if we agree to call a uniformly convergent sequence of continuous functions a Cauchy sequence, then we can state a completeness theorem which is analogous to the one stated for $\mathcal{R}$:

33.3 / Theorem If A is compact, then $\mathcal{C}(A)$ is complete.

Without further justification we have:

33.4 / Corollary If the sequence $\{f_n\}$ converges uniformly on A, then

$$\lim_{x\to a}\lim_{n\to\infty} f_n(x) = \lim_{n\to\infty}\lim_{x\to a} f_n(x)$$

for each $a \in A$.

Let us now determine conditions under which pointwise convergence implies uniform convergence. Thus, let the sequence $\{f_n\}$ converge pointwise on A to f. Examples 32.2 and 32.4(2) suggest four conditions for consideration:

(a) The f_n are continuous;
(b) The sequence $\{f_n\}$ is monotone: either $f_n(x) \leq f_{n+1}(x)$ or $f_{n+1}(x) \leq f_n(x)$ for all $x \in A$ and all $n \in \mathfrak{N}$;
(c) f is continuous;
(d) A is compact.

Example 32.2 offers an instance in which the requirements in (a), (b), and (d) are met, yet the convergence is not uniform; this conclusion applies also to Example 32.4(2) in which conditions (a), (b), and (c) are satisfied. In addition to these, regard the following examples:

33.5 / *Examples*

(**1**) For each $n \in \mathfrak{N}$ let

$$f_n(x) = \begin{cases} 1 & \text{when} \quad 0 < x < \dfrac{1}{n} \\ 0 & \text{when} \quad x = 0 \quad \text{or} \quad \dfrac{1}{n} \leq x \leq 1. \end{cases}$$

This sequence of discontinuous functions converges pointwise to the continuous function $f(x) = 0$. To see this, select an arbitrary $x \in [0, 1]$. If $x = 0$, then there is nothing to prove. Otherwise, select k so large, that $1/k < x$. Then $f_n(x) = 0$ for all integers $n \geq k$ and the assertion follows.

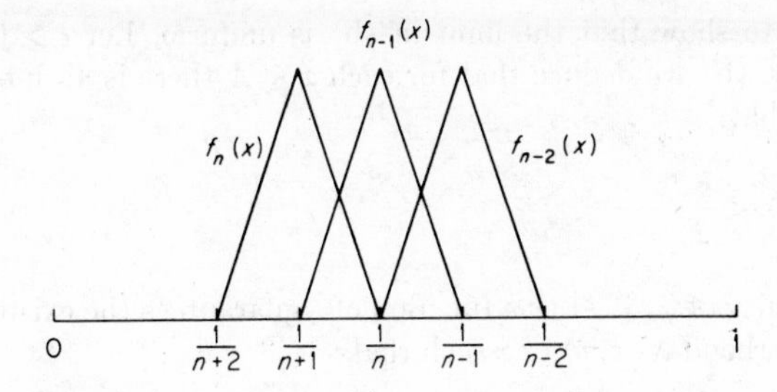

Figure 42

Specifically, depicted here is a situation in which the demands in (b), (c), and (d) are fulfilled, yet the convergence is evidently not uniform.

(**2**) The only combination which was not ruled out is (a), (c), and (d). This is corrected with the following sequence: For each $n \in \mathfrak{N}$ set

$$f_n(x) = \begin{cases} (n+1)[(n+2)x - 1] & \left(\dfrac{1}{n+2} \leq x \leq \dfrac{1}{n+1}\right) \\ (n+1)[1 - nx] & \left(\dfrac{1}{n+1} \leq x \leq \dfrac{1}{n}\right) \\ 0 & \text{elsewhere on } [0, 1] \end{cases}$$

(see Figure 42).

One shows as in the above example that the pointwise limit of this sequence is $f(x) = 0$. Hence, conditions (a), (c), and (d) are satisfied without implying uniform convergence. That the four conditions together imply uniform convergence is proved in the next theorem:

33.6 / Dini's Theorem If the sequence $\{f_n\}$ ($f_n \in \mathcal{C}(A)$) converges pointwise on the compact set A to the continuous limit function f, and if for each $x \in A$

$$f_1(x) \leq f_2(x) \leq f_3(x) \leq \cdots,$$

then $\{f_n\}$ converges uniformly on A to f.

Proof Setting $\varphi_n = f - f_n$ for each $n \in \mathfrak{N}$ we conclude that:

(a) Each function φ_n is continuous;

(b) For each $x \in A$, $\lim_{n\to\infty} \varphi_n(x) = 0$;

(c) $\varphi_1(x) \geq \varphi_2(x) \geq \varphi_3(x) \geq \cdots$ for all $x \in A$.

It remains to show that the limit in (b) is uniform. Let $\epsilon > 0$ be given. From (b) we deduce that for each $x \in A$ there is an integer $n(x)$ such that

$$0 \leq \varphi_{n(x)}(x) < \frac{\epsilon}{2}.$$

The continuity of $\varphi_{n(x)}(\xi)$ as a function of ξ guarantees the existence of a neighborhood $N(x; r)$ of x such that

$$0 \leq \varphi_{n(x)}(\xi) < \epsilon$$

for all points $\xi \in N(x; r) \cap A$. The family of neighborhoods $N(x; r)$ is clearly an open covering of A:

$$A \subset \bigcup_{x \in A} N(x; r).$$

Since A is compact, there is a finite set of points $x_1, x_2, \ldots, x_k$, such that

$$A \subset \bigcup_{j=1}^{k} N(x_j; r).$$

Set

$$n = \max\{n(x_1), \ldots, n(x_k)\}. \tag{i}$$

If $\xi \in A$ is arbitrary, then $\xi \in N(x_m; r)$ for some integer $1 \leq m \leq k$ and consequently

$$\varphi_{n(x_m)}(\xi) < \epsilon.$$

The relations in (c) and (i) now show that for each $p > n$

$$0 \leq \varphi_p(\xi) \leq \varphi_n(\xi) \leq \varphi_{n(x_m)}(\xi) < \epsilon$$

for all $\xi \in A$, and hence the convergence is uniform as asserted.

Remark: The preceding theorem remains true when the inequalities in its formulation are reversed. In this case one would simply replace the φ_n with the functions $\eta_n = f_n - f$.

Regarding differentiation, it is important to know if a convergent sequence of differentiable functions admits term by term differentiation; that is, we ask if

$$\lim_{n \to \infty} \lim_{x \to a} \frac{f_n(x) - f_n(a)}{x - a} = \lim_{x \to a} \lim_{n \to \infty} \frac{f_n(x) - f_n(a)}{x - a}.$$

By way of motivation consider:

33.7 / *Example* The sequence

$$f_n(x) = \frac{\sin nx}{\sqrt{n}}$$

whose members are continuous on $\mathcal{R}$, is easily seen to converge uniformly to the limit function $f(x) = 0$; namely, given an arbitrary $\epsilon > 0$, then

$$\left\| \frac{\sin nx}{\sqrt{n}} - 0 \right\| = \frac{1}{\sqrt{n}} \| \sin nx \| \leq \frac{1}{\sqrt{n}} < \epsilon$$

when $n > 1/\epsilon^2$. Hence $f'(x) = 0$. The sequence $\{ f'_n \}$, on the other hand, diverges everywhere. Since $f'_n(x) = \sqrt{n} \cos nx$, it suffices to show that for each fixed point x, there are arbitrarily large values of n for which $| \cos nx | \geq \frac{1}{2}$. This is trivial when $x = 0$; otherwise, when $x \neq 0$ is fixed and $| \cos nx | < \frac{1}{2}$, then

$$| \cos 2nx | = | 2 \cos^2 nx - 1 | = 1 - 2 \cos^2 nx \geq \tfrac{1}{2},$$

as was to be shown.

In summary, we have here a sequence of continuously differentiable functions which converges uniformly to zero, but whose derived sequence diverges everywhere.

33.8 / *Example* Let there be given the sequence

$$f_n(x) = \frac{x}{1 + n^2x^2} \qquad (-1 \leq x \leq 1).$$

Then the uniform limit exists and is $f(x) = 0$. But

$$\lim_{n\to\infty} f'_n(x) = \lim_{n\to\infty} \frac{1 - n^2x^2}{(1 + n^2x^2)^2} = \begin{cases} 1 & \text{at} \quad x = 0 \\ 0 & \text{when} \quad 0 < | x | \leq 1. \end{cases}$$

In this case the derived sequence has a limit at each admitted value of x, but the limit at $x = 0$ does not coincide with the derivative of the limit of $\{ f_n \}$.

Next we prove the following:

33.9 / Theorem Given a sequence $\{ f_n \}$ $(f_n \in \mathcal{C}(A))$ with domain $A = [a, b]$ such that:

(**1**) There is a point $y \in A$ for which $\{ f_n(y) \}$ converges;
(**2**) Each function f_n is differentiable on A;
(**3**) The sequence $\{ f'_n \}$ converges uniformly on A.

Then the sequence $\{f_n\}$ converges uniformly on A to a differentiable function f, such that

$$f' = \lim_{n\to\infty} f'_n.$$

It should be noted that the functions f'_n are permitted to be discontinuous.

Proof Consider the point y in (1). Given an $\epsilon > 0$, then we are guaranteed the existence of an integer n_1 such that

$$d(f_m(y), f_n(y)) < \frac{\epsilon}{3} \tag{ii}$$

whenever $m, n > n_1$. For this same ϵ the condition in (3) lets us specify an integer n_2 with the property that

$$\|f'_m - f'_n\| < \frac{\epsilon}{3d(b, a)} \tag{iii}$$

when $m, n > n_2$: put $k = \max\{n_1, n_2\}$.

Consider the restriction of the function $f_m - f_n$ to an interval $[x, y] \subset A$. For this interval the Mean Value Theorem and the estimate in (iii) show that, for some point $x < \xi < y$ and $m, n > k$,

$$d(f_m(x) - f_n(x), f_m(y) - f_n(y)) = d(x, y)d(f'_m(\xi), f'_n(\xi))$$

$$\leq d(x, y)\,\|f'_m - f'_n\| < \frac{\epsilon}{3}. \tag{iv}$$

In particular, (ii) and (iv) admit the estimate

$$d(f_m(x), f_n(x)) \leq d(f_m(x) - f_n(x), f_m(y) - f_n(y))$$

$$+ d(f_m(y), f_n(y)) < \frac{\epsilon}{3} + \frac{\epsilon}{3} = \frac{2\epsilon}{3} < \epsilon$$

whenever $m, n > k$. Being true for each $x \in A$, it follows that also $\|f_m - f_n\| < \epsilon$ for all $m, n > k$, thereby showing that the sequence $\{f_n\}$ converges uniformly on A: let its limit be f.

Fixing an arbitrary point $\xi \in A$, let us now show that f is differentiable there. We again apply the Mean Value Theorem to the function $f_m - f_n$, this time taking an interval with endpoints x and ξ. As in (iv), we deduce that

$$d\left(\frac{f_m(\xi) - f_m(x)}{\xi - x}, \frac{f_n(\xi) - f_n(x)}{\xi - x}\right) < \frac{\epsilon}{3}$$

for all integers $m, n > k$. Holding n fixed but letting $m \to \infty$, we arrive at the inequality

$$d\left(\frac{f(\xi) - f(x)}{\xi - x}, \frac{f_n(\xi) - f_n(x)}{\xi - x}\right) \leq \frac{\epsilon}{3} \qquad (\xi \neq x) \tag{v}$$

when $n > k$.

Let φ stand for the uniform limit of the sequence $\{f_n'\}$. Then for the given ϵ there is an integer k' such that

$$d(f_n'(\xi), \varphi(\xi)) < \epsilon \tag{vi}$$

whenever $n > k'$, and we set $p = \max\{k, k'\}$. First, we deduce from (2) that there is a $\delta > 0$ such that

$$d\left(\frac{f_p(\xi) - f_p(x)}{\xi - x}, f_p'(\xi)\right) < \frac{\epsilon}{3} \qquad (\xi \neq x) \tag{vii}$$

for all points $x \in N(\xi; \delta) \cap A$. Next, we infer from (v), (vi), and (vii) that

$$d\left(\frac{f(\xi) - f(x)}{\xi - x}, \varphi(\xi)\right)$$

$$\leq d\left(\frac{f(\xi) - f(x)}{\xi - x}, \frac{f_p(\xi) - f_p(x)}{\xi - x}\right) + d\left(\frac{f_p(\xi) - f_p(x)}{\xi - x}, f_p'(\xi)\right)$$

$$+ d(f_p'(\xi), \varphi(\xi)) < \frac{\epsilon}{3} + \frac{\epsilon}{3} + \frac{\epsilon}{3} = \epsilon \qquad (\xi \neq x)$$

for all $x \in N(\xi; \delta) \cap A$, showing that $f'(\xi)$ exists and is as asserted.

33.10 / ***Example*** The function f in Example 28.7(2) was shown to be everywhere differentiable, with f' discontinuous at $x = 0$. Let us modify this example in a trivial manner in order to illustrate the last theorem. Specifically, we take $A = [-1, 1]$ and set for each $n \in \mathfrak{N}$

$$f_n(x) = \begin{cases} \dfrac{n-1}{n} x^2 \sin \dfrac{1}{x} & \text{when} \quad x \neq 0 \\ 0 & \text{at} \quad x = 0 \end{cases}$$

and then

$$f_n'(x) = \begin{cases} \dfrac{n-1}{n}\left(2x \sin \dfrac{1}{x} - \cos \dfrac{1}{x}\right) & \text{when} \quad x \neq 0 \\ 0 & \text{at} \quad x = 0. \end{cases}$$

Despite its discontinuity (of the second kind) at $x = 0$, the sequence $\{f'_n\}$ converges uniformly to the function

$$f'(x) = \begin{cases} 2x \sin \frac{1}{x} - \cos \frac{1}{x} & (x \neq 0) \\ 0 & (x = 0), \end{cases}$$

since, for each $\epsilon > 0$,

$$\|f'_n - f'\| = \frac{1}{n} \|f'\| \leq \frac{3}{n} < \epsilon$$

when $n > \frac{3}{\epsilon}$. Hence, the conditions of Theorem 33.9 are met.

As a final topic of this section we take up the question of convergence of series of functions. To have a formal statement of the problem we lay down:

33.11 / ***Definition*** Let $\sum f_n$ be a given series of members of $\mathcal{C}(A)$ with partial sums

$$\sigma_n = \sum_{k=0}^{n} f_k.$$

Then the series $\sum f_k$ converges *uniformly* on A when the sequence $\{\sigma_n\}$ so converges.

It is clear that the limit function is continuous in this case. In the case of numerical sequences we amassed a wide selection of convergence criteria. Hence the task of determining the convergence of series is facilitated with the following:

33.12 / **Weierstrass M-Test** Let $\{f_n\}$ be a sequence of functions and, for each n, let M_n be a constant such that

$$\|f_n\| \leq M_n.$$

Then $\sum f_n$ converges uniformly when $\sum M_n$ converges.

Proof The convergence of the series $\sum M_n$ implies that for each $\epsilon > 0$ there is an integer n_ϵ with the property

$$\|\sigma_n - \sigma_m\| = \left\| \sum_{k=m+1}^{n} f_k \right\| \leq \sum_{k=m+1}^{n} \|f_k\| \leq \sum_{k=m+1}^{n} M_k < \epsilon,$$

for all integers $m > n > n_\epsilon$.

While power series are treated in the next section, we present here one result involving them as an application of the M-test.

33.13 / Theorem If the power series $\sum a_n x^n$ converges for some point $x = x_0$, then it converges uniformly on every interval $[-\xi, \xi]$, where $0 < \xi < |x_0|$.

Proof We know that the series converges absolutely for x, where $|x| < |x_0|$, when $\sum a_n x_0{}^n$ converges. Hence $\sum a_n \xi^n$ converges absolutely; that is, $\sum |a_n| \xi^n < \infty$. But for $M_n = |a_n| \xi^n$ we find that, for all $-\xi \leq x \leq \xi$, $\| a_n x^n \| \leq M_n$ and the theorem follows. (This theorem is already known to the reader who did Exercise 16-6.)

EXERCISES

33-1 Demonstrate the converse of Theorem 33.2 to be false.

33-2 Show that the sequence $\{f_n\}$ fails to converge uniformly on A to f if and only if there is some $\epsilon > 0$ such that for no integer k is the inequality $\| f_n - f \| < \epsilon$ correct for all $n > k$.

33-3 Prove the following consequence of Theorem 33.6: Consider a series $\sum f_n$ of continuous functions with a compact domain A such that $f_n \geq 0$ for each n. If the series converges on A to the continuous function f, then it converges uniformly on A to f.

33-4 State and prove the analog of Theorem 33.9 in the case of series (i.e., replace the f_n with partial sums σ_n).

33-5 Can a continuous function be the uniform limit of a sequence of discontinuous functions?

33-6 For each $n \in \mathfrak{N}$ let

$$f_n(x) = \begin{cases} 1 - nx & \left(0 \leq x \leq \dfrac{1}{n}\right) \\ 0 & \left(\dfrac{1}{n} \leq x \leq 1\right). \end{cases}$$

Then each function f_n is continuous on $[0, 1]$. Discuss the convergence of the sequence $\{f_n\}$.

33-7 Prove or disprove: If $\{f_n\}$ and $\{g_n\}$ are uniformly convergent, then the sequence $\{f_n g_n\}$ is convergent (at least pointwise).

33-8 Construct sequences $\{f_n\}$ and $\{g_n\}$ of continuous functions which converge uniformly on $\mathfrak{R}$, but whose sequence of products $\{f_n g_n\}$ does *not* converge uniformly on $\mathfrak{R}$.

33-9 Use the Root Test and Theorem 33.12 to prove the following: If $\sum_{n=0}^{\infty} a_n x^n$ converges uniformly on some interval $(-x_0, x_0)$, $x_0 > 0$, to the function f, then f is differentiable and moreover

$$f'(x) = \sum_{n=1}^{\infty} n a_n x^{n-1}.$$

33-10 Use Exercise 33-9 to deduce that if $f(x) = \sum_{n=0}^{\infty} a_n x^n$ on an open interval $(-x_0, x_0)$ then, for each $p \in \mathfrak{N}$,

$$f^{(p)}(x) = \sum_{n=p}^{\infty} n(n-1)\cdots(n-p+1)a_n x^{n-p},$$

that is, the function specified by the power series has derivatives of all orders in the given domain.

33-11 Show that the series $\sum x^n e^{-nx}$ converges uniformly in the interval $[0, 2]$.

33-12 Prove that

$$\lim_{x\to 1} \sum_{n=1}^{\infty} \frac{nx^2}{n^3 + x^3} = \sum_{n=1}^{\infty} \frac{n}{n^3+1}.$$

33-13 Consider the series

$$\sum_{k=2}^{\infty} \frac{\sin kx}{\ln k}$$

(see Exercise 16-13). Discuss the pointwise and uniform convergence of this series.

34 / POWER SERIES

Of particular interest among series of functions $\sum f_n$ are the series in which $f_n(x) = a_n(x-c)^n$, $a_n \in \mathfrak{R}$, c being a fixed point. As in the case of sequences, pronounced differences exist between function series and numerical series. Formally we state:

34.1 / *Definition* Given a sequence $\{a_n\}$ $(a_n \in \mathfrak{R})$, then the series

$$\sum_{n=0}^{\infty} a_n(x-c)^n$$

is a *power series* in $\xi = x - c$, where c is a fixed number.

Theorem 33.13 establishes the fundamental property that the convergence of the series above at a point $x \neq c$ implies uniform convergence in a neighborhood of c; the Root Test 16.11 tells us when there is convergence. Specifically, one has:

34.2 / Theorem If

$$\frac{1}{\rho} = \limsup_{n\to\infty} \sqrt[n]{|a_n|} = \alpha,$$

where $\rho = 0$ when $\alpha = \infty$ and $\rho = \infty$ when $\alpha = 0$, then the series in 34.1 converges when $|x - c| < \rho$ and diverges when $|x - c| > \rho$.

The number ρ which is attached to each series is called its *radius of convergence*; $(c - \rho, c + \rho)$ is the *interval of convergence*. The function

$$f(x) = \sum a_n(x - c)^n$$

is said to be *analytic*; the right side is called the *expansion of f in a power series about the point $x = c$*, or simply the *power series* of f about $x = c$. For brevity of notation one often substitutes for the above series

$$\varphi(x) = f(x + c) = \sum a_n x^n,$$

which represents an expansion of φ about the origin. Some of the situations one encounters are depicted below:

34.3 / *Examples*

(1) The series

$$f(x) = \sum n^n x^n$$

converges only in the trivial case $x = 0$, i.e., $\rho = 0$.

(2) Consider the series

$$g(x) = \sum_{n=1}^{\infty} \frac{x^n}{n},$$

where $\rho = 1$. From Chapter Four we know that the series diverges at $x = 1$ yet converges at $x = -1$.

(3) If we let

$$h(x) = \sum_{n=1}^{\infty} \frac{x^n}{n^2},$$

then again $\rho = 1$, but this series converges on $[-1, 1]$.

(4) When

$$k(x) = \sum_{n=1}^{\infty} \frac{x^{2n}}{n},$$

then the series is seen to diverge at both end points, $x = -1$ and $x = 1$, of the interval of convergence.

The preceding examples show that the behavior of a series at the endpoints of the interval of convergence is not easily described.

Power series were used in introducing the Cauchy product of two numerical series (Definition 16.12). Indeed, we have:

34.4 / Theorem If the series $\sum a_n x^n$ and $\sum b_n x^n$ converge on the interval $(-\rho, \rho)$ to the functions $f(x)$ and $g(x)$, respectively, then the product series $\sum c_n x^n$ converges on this interval to $(f \cdot g)(x)$, each c_n being given by the formula

$$c_n = \sum_{j=0}^{n} a_j b_{n-j}.$$

Proof According to Theorem 34.2 the series in question converge absolutely for $|x| < \rho$. All we need do, therefore, is apply Theorem 16.14 to the series $\sum \alpha_n$ and $\sum \beta_n$, where $\alpha_n = a_n x^n$ and $\beta_n = b_n x^n$.

Thus far no claim was made regarding the continuity of a series at an endpoint of its interval of convergence. It was shown by Abel that if a given series converges at an endpoint, then the convergence of the series is uniform up to this point. Specifically, we prove:

34.5 / Abel's Theorem If the power series $\sum a_n x^n$ converges for $|x| < \rho$ and if $\sum a_n \rho^n = \alpha$, then the series converges uniformly on the interval $[0, \rho]$ and

$$\lim_{x \to \rho-} \sum a_n x^n = \alpha.$$

Proof Without loss of generality we may suppose that $\rho = 1$, for otherwise we would consider the series in $\xi = x/\rho$. With this agreement, let $f(x) = \sum a_n x^n$; consider the function $g(x) = 1/(1-x)$: For $|x| < 1$ this function has the power series expansion $\sum x^n$ about the point $x = 0$. Theorem 34.4 shows that

$$(f \cdot g)(x) = \sum c_n x^n$$

for $|x| < 1$, where

$$c_n = \sum_{j=0}^{n} a_j$$

for each admitted value of n. Let $\epsilon > 0$ be given. Since, by hypothesis, $\sum a_n = \alpha$, we can fix an integer k so large, that

$$d(c_n, \alpha) < \frac{\epsilon}{2} \qquad (n > k).$$

Since $|x^n| \leq 1$ for all values of n, we have the estimate

$$(1-x)\left|\sum_{n=0}^{k}(c_n-\alpha)x^n\right| \leq (k+1)\cdot \max_{0\leq n\leq k}\{|c_n-\alpha|\}(1-x)$$
$$= \beta(1-x).$$

The inequality $\beta(1-x) < \frac{\epsilon}{2}$ will be valid when $1 - \frac{\epsilon}{2\beta} < x < 1$.

Thus, when we put $\delta = \frac{\epsilon}{2\beta}$, we find that

$$(1-x)\left|\sum_{n=0}^{k}(c_n-\alpha)x^n\right| < \frac{\epsilon}{2}$$

for all points $x \in (1-\delta, 1)$. From these two estimates and the identity $(1-x)g(x) = 1$ $(|x| < 1)$ we deduce that, for each x such that $1-\delta < x < 1$,

$$d(f(x), \alpha) = (1-x)\,|f(x)g(x) - \alpha g(x)|$$
$$= (1-x)\left|\sum_{n=0}^{\infty}(c_n-\alpha)x^n\right|$$
$$\leq (1-x)\left|\sum_{n=0}^{k}(c_n-\alpha)x^n\right| + (1-x)\sum_{n=k+1}^{\infty}|c_n-\alpha|\,x^n$$
$$< \frac{\epsilon}{2} + (1-x)\frac{\epsilon}{2}\sum_{n=k+1}^{\infty}x^n = \frac{\epsilon}{2} + \frac{\epsilon}{2}x^{k+1} < \epsilon.$$

Hence, $d(f(x), \alpha) < \epsilon$ for the specified values of x and the theorem follows.

It was demonstrated in Exercises 33-9 and 33-10 that if the series in Definition 34.1 converges in some interval $(c-\rho,\ c+\rho)$, then f has derivatives of all orders in that interval. In particular, upon setting $x = c$ in the formula of Exercise 33-10 for the nth derivative of f we find that

$$a_n = \frac{f^{(n)}(c)}{n!}, \tag{i}$$

thus showing that the coefficients of the expansion of f in a power series about a point $x = c$ are uniquely determined by the values of f and its derivatives at that point. That is,

$$f(x) = \sum_{n=0}^{\infty}\frac{f^{(n)}(c)}{n!}(x-c)^n. \tag{ii}$$

This form of the series is called a *Taylor series*. For the sake of completeness we formalize the following uniqueness statement:

34.6 / Theorem Suppose the series $f(x) = \sum a_n x^n$ and $g(x) = \sum b_n x^n$ converge on an interval $(-\rho, \rho)$. If $f = g$, then $a_n = b_n$ for all values of n.

Thus, since the coefficients a_n can be read-off from (i) when the successive derivatives of f are known, one is led to consider this question: If f has derivatives of all orders in an interval $(c - \rho, c + \rho)$, $\rho > 0$, does the function have a power series expansion about the point $x = c$? Equivalently stated, does the series in the right side of (ii) necessarily converge to f on the interval in question? Regretfully the answer is negative, as shown in the next example:

34.7 / *Example* Consider the function (see Exercises 34-5–34-8)

$$f(x) = \begin{cases} e^{-1/x^2} & (x \neq 0) \\ 0 & (x = 0). \end{cases}$$

The student should require no assistance to prove that f has derivatives of all orders in $\mathcal{R}$ with the property that $f^{(n)}(0) = 0$ for all $n \in \mathfrak{N}$. This function f has, therefore, no power series expansion about the point $x = 0$, since the Taylor series does not converge to f.

The final arguments in this section will concern conditions which guarantee the Taylor series expansion of a given function. The first result in this direction is:

34.8 / Taylor's Theorem Let the series $f(x) = \sum a_n x^n$ converge for $|x| < \rho$. If the point c is such that $-\rho < c < \rho$, then f has a power series expansion also about the point $x = c$:

$$f(x) = \sum_{n=0}^{\infty} \frac{f^{(n)}(c)}{n!} (x - c)^n,$$

and the series converges for $|x - c| < \rho - |c|$.

Proof Let x be fixed. From the identities $x^n = [(x - c) + c]^n$ and the binomial theorem we deduce that

$$f(x) = \sum_{n=0}^{\infty} a_n x^n = \sum_{n=0}^{\infty} a_n \sum_{m=0}^{n} \binom{n}{m} c^{n-m}(x-c)^m$$

$$= \sum_{n=0}^{\infty} \sum_{m=0}^{n} a_n \binom{n}{m} c^{n-m}(x-c)^m$$

$$= \sum_{n=0}^{\infty} \sum_{m=0}^{\infty} b_{mn}(x),$$

where we have set $b_{mn}(x) = 0$ for all integers $m > n$. Owing to Theorem 16.9 we may interchange the order of summation in the double series whenever $\sum_n \sum_m | b_{mn}(x) |$ converges. But for each value of n,

$$\sum_{m=0}^{\infty} | b_{mn}(x) | = \sum_{m=0}^{n} | a_n | \binom{n}{m} | c |^{n-m} | x - c |^m$$

$$= | a_n | \sum_{m=0}^{n} \binom{n}{m} | c |^{n-m} | x - c |^m = | a_n | (| x - c | + | c |)^n$$

and we know that the series $\sum_n | a_n | (| x - c | + | c |)^n$ converges when $| x - c | + | c | < \rho$. Hence, for values of x confined to the specified domain, f may be expressed in the form

$$f(x) = \sum_{m=0}^{\infty} \left[\sum_{n=m}^{\infty} \binom{n}{m} a_n c^{n-m} \right] (x-c)^m.$$

That the coefficients are, indeed, of the desired variety follows, of course, from the uniqueness of theorem 34.6.

When f is not known to have a power series representation, then the following criterion may be useful:

34.9 / Theorem If f has derivatives of all orders on a closed interval I and if there is a constant γ such that

$$|| f^{(n)} || \leq \gamma$$

for all values of n, then f has a Taylor series expansion about each point $c \in I$.

Proof Owing to Theorem 31.2 this function f has a representation

$$f(x) = f(c) + f'(c)(x - c) + \frac{f''(c)}{2!}(x - c)^2 + \cdots$$

$$+ \frac{f^{(n)}(c)}{n!}(x - c)^n + R_n(x),$$

for each $a \in I$ and each n. Using the Lagrange form of the remainder R_n we find that, for each value of n,

$$\left\| f(x) - \sum_{k=0}^{n} \frac{f^{(k)}(c)}{k!}(x - c)^k \right\| = \left\| \frac{f^{(n+1)}(\xi)}{(n+1)!}(x - c)^{n+1} \right\| \leq \frac{\gamma d^{n+1}}{(n+1)!}.$$

where d stands for the length of I and $\xi \in I$ depends on x. The proof can now be completed by the student.

EXERCISES

34-1 Establish the following counterpart of Exercise 13-2: If $f(x) = \sum a_n x^n$ and $g(x) = \sum b_n x^n$, then $(f + g)(x) = \sum (a_n + b_n) x^n$.

34-2 Complete the proof of Theorem 34.9.

34-3 We say that the numerical series $\sum a_n$ is *Abel Summable* to α if

$$\lim_{x \to 1-} \sum a_n x^n = \alpha,$$

the series being assumed to converge on $[0, 1)$. Show that the divergent series $\sum (-1)^n$ is Abel Summable to $\frac{1}{2}$.

34-4 Discuss the Abel summability of the divergent series $1 - 2 + 3 - 4 + \cdots$.
(*Hint*: You may find Exercise 33-9 useful here.)

34-5 Defining

$$E(x) = \sum_{n=0}^{\infty} \frac{x^n}{n!},$$

show that:

(**1**) The series converges on $\mathcal{R}$; does it converge uniformly there?

(**2**) $E(x)E(y) = E(x + y)$. From this relation we deduce that $E(x) \neq 0$ for all values of x, since $E(x)E(-x) = E(x - x) = E(0) = 1$. Hence show that, in fact, $E(x) > 0$ for all $x \in \mathcal{R}$;

(**3**) $E'(x) = E(x)$;

(**4**) For each $n \in \mathfrak{N}$, $E(n) = e^n$, where $e = \sum_{n=0}^{\infty} \frac{1}{n!}$;

(**5**) $E(r) = e^r$ for each rational number r.

34-6 For each $x \in \mathcal{R}$ set

$$e^x = \operatorname*{lub}_{r<x} e^r \qquad (r \in \mathcal{R}^*)$$

and show that

$$E(x) = e^x.$$

34-7 Show that

$$\lim_{x\to\infty} e^x = \infty$$

and

$$\lim_{x\to-\infty} e^x = 0.$$

34-8 Demonstrate that for each $n \in \mathfrak{N}$

$$\lim_{x\to\infty} x^n e^{-x} = 0.$$

35 / THE APPROXIMATION OF FUNCTIONS

Let there now be given a compact interval $A = [a, b]$ and the corresponding space $\mathcal{C}(A)$ of continuous functions. Two naive approximation theorems were formulated in Exercises 27-8 and 27-9, and a refinement of these is given in Exercise 35-1. In the year 1885 Weierstrass established the following far reaching proposition:

35.1 / Weierstrass Approximation Theorem Given an arbitrary function $f \in \mathcal{C}(A)$ and a number $\epsilon > 0$, then there is a polynomial $p \in \mathcal{C}(A)$ for which $\| f - p \| < \epsilon$.

An equivalent formulation of this important result concerns the separability of $\mathcal{C}(A)$:

35.2 / Corollary The set of polynomials with rational coefficients is dense in $\mathcal{C}(A)$.

It is clear that Theorem 35.1 can be formulated for continuous functions whose domain is an arbitrary compact subset of $\mathcal{R}$: this follows an application of the Tietze Extension Theorem. Of the several available proofs of the Weierstrass theorem we feature here the beautiful proof of Bernstein. To this end, let $f(t) \in \mathcal{C}(A)$ be fixed. The transformation

$$t = a + (b - a)x,$$

which relates $[a, b]$ in a one-to-one manner to the interval $I = [0, 1]$, renders the function $g(x) = f(t)$ continuous on I. If p is a polynomial such that $\| g - p \| < \epsilon$ on I, then $p(x) = p\left(\frac{t-a}{b-a}\right) = q(t)$ is a polynomial with the property $\| f - q \| < \epsilon$ on $[a, b]$. This argument shows that it suffices to construct a proof of the theorem for the case $[a, b] = [0, 1]$ and the stage is now set for Bernstein's proof. This entails the construction of specific polynomials which converge uniformly to the given function, the polynomials being appropriately called Bernstein polynomials:

35.3 / *Definition* Given a function $f: I \to \mathcal{R}$, then the polynomial

$$B_n(x; f) = \sum_{k=0}^{n} f\left(\frac{k}{n}\right)\binom{n}{k} x^k (1-x)^{n-k} \qquad (n \in \mathfrak{N})$$

is the nth *Bernstein polynomial* of f.

These polynomials should be carefully examined by the reader. In particular it should be noted that $B_n(x; 1) = (x + (1-x))^n = 1$ for each $n \in \mathfrak{N}$. We have the theorem:

35.4 / Bernstein's Approximation Theorem If $f: I \to \mathcal{R}$ is continuous, then

$$\lim_{n\to\infty} \| f - B_n(x; f) \| = 0;$$

that is, f is the uniform limit of the associated Bernstein polynomials.

The proof of this theorem rests on the following:

35.5 / Lemma For each $x \in I$ and $n \in \mathfrak{N}$ we have the inequality

$$\sum_{k=0}^{n} \left(x - \frac{k}{n}\right)^2 \binom{n}{k} x^k (1-x)^{n-k} \le \frac{1}{4n}.$$

Proof According to the binomial theorem

$$\sum_{k=0}^{n} \binom{n}{k} x^k a^{n-k} = (x + a)^n \qquad (n \in \mathfrak{N}). \tag{i}$$

Differentiating both sides of this equation with respect to x and then multiplying the resulting equation by x yields the formula

$$\sum_{k=0}^{n} \binom{n}{k} k x^k a^{n-k} = nx(x + a)^{n-1}, \tag{ii}$$

where we have used the fact that the term corresponding to $k = 0$ vanishes. Applying the same scheme to the last equation results in the formula

$$\sum_{k=0}^{n} \binom{n}{k} k^2 x^k a^{n-k} = n(n-1)x^2(x+a)^{n-2} + nx(x+a)^{n-1}. \tag{iii}$$

When we now replace a by $1 - x$ in (i), (ii), and (iii) we arrive at the relations

$$\sum_{k=0}^{n} \binom{n}{k} x^k (1-x)^{n-k} = 1, \tag{iv}$$

$$\sum_{k=0}^{n} k \binom{n}{k} x^k (1-x)^{n-k} = nx,$$

and

$$\sum_{k=0}^{n} k^2 \binom{n}{k} x^k (1-x)^{n-k} = n(n-1)x^2 + nx.$$

These equations are multiplied, in turn, by n^2x^2, $-2nx$, and 1, and then added. The formula

$$\sum_{k=0}^{n} \left(x - \frac{k}{n}\right)^2 \binom{n}{k} x^k (1-x)^{n-k} = \frac{x(1-x)}{n}$$

follows a trivial simplification, and the desired estimate is derived from the fact that $x(1-x) \leq 1$ on I.

Let us now turn to the proof of Theorem 35.4.

Proof of 35.4 By virtue of (iv) above, f may be written in the form

$$f(x) = \sum_{k=0}^{n} f(x) \binom{n}{k} x^k (1-x)^{n-k}$$

and hence

$$\begin{aligned} d(f(x), B_n(x;f)) &= \left| \sum_{k=0}^{n} \left[f(x) - f\left(\frac{k}{n}\right) \right] \binom{n}{k} x^k (1-x)^{n-k} \right| \\ &\leq \sum_{k=0}^{n} d\left(f(x), f\left(\frac{k}{n}\right)\right) \binom{n}{k} x^k (1-x)^{n-k}. \end{aligned} \tag{v}$$

Being continuous on a compact interval, $\|f\|$ is finite. Also, given an arbitrary $\epsilon > 0$ there is a $\delta > 0$ such that $d(f(x), f(y)) < \epsilon/2$

whenever $d(x, y) < \delta$: Let such a δ be fixed, and select an integer n so large that

$$n \geq \max\left\{\frac{1}{\delta^4}, \frac{\|f\|^2}{\epsilon^2}\right\}. \tag{vi}$$

Let $x \in I$ be arbitrary but fixed. Consider in the right side of (v) those values of k (if any) for which $d\left(x, \frac{k}{n}\right) < n^{-1/4} < \delta$. For these values of k we have the estimate:

$$\sum_k d\left(f(x), f\left(\frac{k}{n}\right)\right)\binom{n}{k} x^k(1-x)^{n-k} \leq \frac{\epsilon}{2}\sum_k \binom{n}{k} x^k(1-x)^{n-k}$$

$$\leq \frac{\epsilon}{2}\sum_{k=0}^{n}\binom{n}{k} x^k(1-x)^{n-k} = \frac{\epsilon}{2}.$$

When $d\left(x, \frac{k}{n}\right) \geq n^{-1/4}$ we write this in the form $\left(x - \frac{k}{n}\right)^2 \geq n^{-1/2}$ and make use of the lemma, as follows: Owing to the triangle inequality $d\left(f(x), f\left(\frac{k}{n}\right)\right) \leq 2\|f\|$ and hence

$$\sum_k d\left(f(x), f\left(\frac{k}{n}\right)\right)\binom{n}{k} x^k(1-x)^{n-k}$$

$$\leq 2\|f\| \sum_k \frac{\left[d\left(x, \frac{k}{n}\right)\right]^2}{\left[d\left(x, \frac{k}{n}\right)\right]^2}\binom{n}{k} x^k(1-x)^{n-k}$$

$$\leq 2\|f\|\sqrt{n}\sum_{k=0}^{n} d\left[\left(x, \frac{k}{n}\right)\right]^2\binom{n}{k} x^k(1-x)^{n-k} \leq \frac{\|f\|}{2\sqrt{n}} < \epsilon,$$

the last inequality being a consequence of (vi). Thus we have shown that

$$d(f(x), B_n(x; f)) < \epsilon$$

and being true for each $x \in I$, it follows that, in particular,

$$\|f(x) - B_n(x; f)\| < \epsilon$$

when n is as demanded. This completes the proof of Bernstein's Theorem.

By now it is expected of course that polynomials are not the only family of functions which admit an approximation theorem in the spirit of Theorem 35.1. To recognize such families it is expedient to discover those properties of the set of polynomials on which the theorem depends. A superficial examination already reveals that the set of polynomials is closed under addition, multiplication, and scalar multiplication, and this observation motivates us to introduce the concept of an algebra:

35.6 / ***Definition*** Given an arbitrary set A, then the family $\mathfrak{A}$ of functions $f: A \to \mathcal{R}$ is an *algebra* if

$$f + g \in \mathfrak{A},$$

$$f \cdot g \in \mathfrak{A},$$

$$c \cdot f \in \mathfrak{A} \qquad (c \in \mathcal{R})$$

whenever f and g are members of $\mathfrak{A}$. The algebra is *uniformly closed* when it contains the limits of all its uniformly convergent sequences, that is, when $f_n \in \mathfrak{A}$ and $\lim \| f_n - f \| = 0$ imply that $f \in \mathfrak{A}$.

The set $\mathcal{P}$ of polynomials is evidently an algebra which, according to Theorem 35.1, is not uniformly closed. In fact, if we call the set $\mathfrak{A}^*$ of uniform limits of an algebra $\mathfrak{A}$ its *uniform closure*, then $\mathfrak{A}^* \supset \mathfrak{A}$ and the Weierstrass theorem can be rephrased as follows:

35.7 / **Corollary** The space $\mathcal{C}(A)$ of continuous functions on a compact set is the uniform closure of its algebra of polynomials.

It is interesting to note that the set of even polynomials† is an algebra. The even polynomials, however, lack the very important capability of separating points:

35.8 / ***Definition*** The family $\mathfrak{A}$ of functions $f: A \to \mathcal{R}$ *separates* the points of A if for each pair of points, $x \neq y$, of A there is a function $f \in \mathfrak{A}$ such that $f(x) \neq f(y)$.

The significance of this condition is the following: Let $A = [-1, 1]$ and let $\mathfrak{Q}$ be the algebra of even polynomials. Then for each $p \in \mathfrak{Q}$, $p(x) = p(-x)$ for all $x \in A$, and consequently, when $| 1 - p(1) | < 1$ then $| -1 - p(-1) | = | 1 + p(-1) | \geq 1$. This implies that $\| x - p(x) \| \geq 1$ for each $p \in \mathfrak{Q}$, so that

$$\inf_{p \in Q} \| x - p(x) \| \geq 1,$$

† That is, polynomials of the form $a_0 + a_1 x^2 + a_2 x^4 + \cdots + a_n x^{2n}$.

which means that the function $f(x) = x$ cannot be uniformly approximated on A with the members of $\mathfrak{Q}$ (see also Exercise 35-10).

A point separating algebra may yet fail to qualify for our purpose. Namely, if $\mathfrak{A}$ meets the demands in Definitions 35.6 and 35.8 but for some point $a \in A$, $p(a) = 0$ for all functions $p \in \mathfrak{A}$, then f cannot be closely approximated with members of $\mathfrak{A}$ if $f(a) = 1$. In short, we are led to introduce:

35.9 / *Definition* The family $\mathfrak{A}$ of functions $f: A \to \mathfrak{R}$ is said to *vanish at no point of A* if for each $x \in A$ there is a member $f \in \mathfrak{A}$ such that $f(x) \neq 0$.

An intriguing algebra is described next:

35.10 / *Example* Let $\mathfrak{S}$ stand for the family of step functions $f: [a, b] \to \mathfrak{R}$. Let it be stated once and for all that $\varphi \in \mathfrak{S}$, if it can be written in the form

$$\varphi = \sum_{p=1}^{n} a_p \chi_{I_p},$$

where the a_p are constants, and the I_p are intervals such that

$$[a, b] = \bigcup_{p=1}^{n} I_p \qquad \text{(vii)}$$

and

$$I_p \cap I_q = \varnothing \qquad (p \neq q), \qquad \text{(viii)}$$

χ_{I_p} being the characteristic function of I_p (Example 3.7).

We claim that the family $\mathfrak{S}$ satisfies the conditions in Definitions 35.6, 35.8, and 35.9. Since the last two definitions are trivially accommodated by the family in question, we shall only show that $\mathfrak{S}$ is an algebra. To show this suppose also $\psi \in \mathfrak{S}$, say,

$$\psi = \sum_{q=1}^{m} b_q \chi_{J_q},$$

where the J_q are intervals as in (vii) and (viii). Setting

$$H_{pq} = I_p \cap J_q,$$

then evidently the new intervals are mutually exclusive,

$$[a, b] = \bigcup_{p,q} H_{pq},$$

and a cursory examination reveals that

$$\varphi + \psi = \sum_{p,q} (a_p + b_q)\chi_{H_{pq}},$$

$$\varphi \cdot \psi = \sum_{p,q} a_p b_q \chi_{H_{pq}}$$

and

$$c \cdot \varphi = \sum_{p,q} (ca_p)\chi_{I_p} \qquad (c \in \mathfrak{R}).$$

Hence, the sum, product, and scalar product of step functions are again step functions, thereby showing that $\mathfrak{S}$ is an algebra. The reader is urged to illustrate the previous argument with suitable diagrams.

According to Exercise 27-8, $\mathfrak{C}(A)$ is contained in the uniform closure $\mathfrak{S}^*$ of $\mathfrak{S}$, but so is $\mathfrak{S}$ itself. In fact, it is evident that $\mathfrak{S}^*$ contains discontinuous functions other than step functions. It would be very instructive to try to describe $\mathfrak{S}^*$ more completely.

The foregoing considerations lead to Stone's generalization of the Weierstrass Approximation Theorem. Prefacing the statement of this result are the following consequences of the concepts just introduced: The family $\mathfrak{A}$ will tacitly be assumed to be an algebra of continuous functions on a compact set A which meets the conditions in Definitions 35.8 and 35.9.

35.11 / Lemma Given a pair of points $\xi \neq \eta$ in A and arbitrary constants a and b, then there is a function $f \in \mathfrak{A}$ with the property that $f(\xi) = a$ and $f(\eta) = b$.

Proof Let the points $\xi \neq \eta$ be given. According to Definitions 35.8 and 35.9 there are functions $u \in \mathfrak{A}$ and $v \in \mathfrak{A}$ such that $u(\xi) \neq u(\eta)$ and $v(\xi) \neq 0$. Let the function u_1 be defined as follows:

$$u_1 = \begin{cases} u & \text{if } u(\xi) \neq 0 \\ v & \text{if } u(\xi) = 0 \quad \text{and} \quad v(\xi) \neq v(\eta) \\ u + v & \text{if } u(\xi) = 0 \quad \text{and} \quad v(\xi) = v(\eta) \end{cases}.$$

Let a function u_2 be defined in the same manner, but with ξ replaced by η. If we set

$$w_1(x) = u_1^2(x) - u_1(\eta)u_1(x)$$

$$w_2(x) = u_2^2(x) - u_2(\xi)u_2(x),$$

then the resulting functions w_1 and w_2 are endowed with the properties

$$w_1(\xi) \neq 0, \qquad w_1(\eta) = 0, \qquad w_2(\xi) = 0, \qquad w_2(\eta) \neq 0.$$

This enables us to write the function

$$f(x) = \frac{a}{w_1(\xi)} w_1(x) + \frac{b}{w_2(\eta)} w_2(x)$$

which is as demanded.

35.12 / Lemma If $f \in \mathfrak{A}^*$, then also $|f| \in \mathfrak{A}^*$.

Proof In proving this result we use the special case of the Weierstrass Approximation Theorem formulated in Exercise 35-4. The significance of this exercise lies in the fact that $|x|$ is uniformly approximated with polynomials having no constant term.

Thus, let $\epsilon > 0$ be given. The compactness of A guarantees $\|f\|$ to be finite and the cited exercise provides a polynomial $p(\xi) = \sum_{k=1}^{n} a_k \xi^k$ such that

$$d(p(\xi), |\xi|) < \epsilon \qquad \text{for} \quad |\xi| \leq \|f\|.$$

If we put $\xi = f(x)$, $\xi^k = [f(x)]^k$, then

$$(p \circ f)(x) = \sum_{k=1}^{n} a_k [f(x)]^k$$

is, of course, a member of $\mathfrak{A}^*$ with the property

$$d((p \circ f)(x), |f(x)|) < \epsilon \qquad (x \in A).$$

Hence, $\| p \circ f - |f| \| < \epsilon$. Since $\mathfrak{A}^*$ is the uniform closure of $\mathfrak{A}$, it follows that $|f| \in \mathfrak{A}^*$, as was to be shown.

The next lemma requires some prior terminology. Namely, given continuous functions f and g on some set A, then the function h, which is defined pointwise as

$$h(x) = \max\{f(x), g(x)\}, \tag{ix}$$

is also continuous on A; the function

$$k(x) = \min\{f(x), g(x)\} \tag{x}$$

is likewise continuous on A. To verify these statements it suffices to observe that

$$h = \frac{f+g}{2} + \frac{|f-g|}{2} \tag{xi}$$

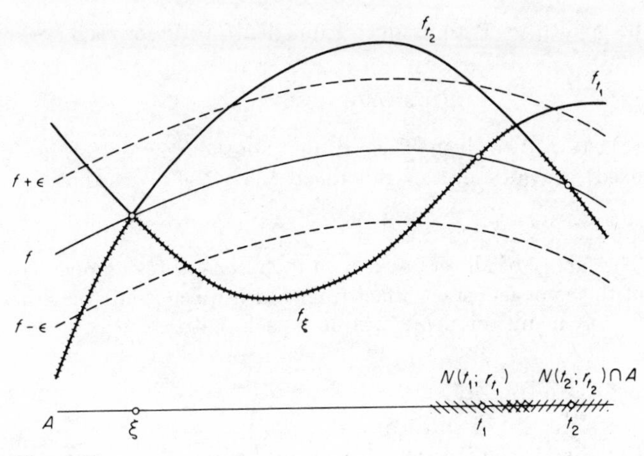

Figure 43

and

$$k = \frac{f+g}{2} - \frac{|f-g|}{2}. \tag{xii}$$

35.13 / Lemma If f and g belong to $\mathfrak{A}^*$, then so do the functions $\max\{f, g\}$ and $\min\{f, g\}$.

The proof of this lemma is a consequence of the preceding lemma and the identities in (xi) and (xii).

This brings us now to the anticipated theorem:

35.14 / Stone–Weierstrass Theorem Let $\mathfrak{A}$ be a given subalgebra of $\mathcal{C}(A)$. Suppose further that $\mathfrak{A}$ separates the points of A and that it vanishes at no point thereof. Then $\mathfrak{A}^* = \mathcal{C}(A)$.

Proof The relation $\mathfrak{A}^* \subset \mathcal{C}(A)$ being *a priori* known, it suffices to show that if $f \in \mathcal{C}(A)$ is arbitrary, then for each $\epsilon > 0$ there is a function $p \in \mathfrak{A}^*$ such that $\| p - f \| < \epsilon$: We are within easy reach of this conclusion.

Let the function $f \in \mathcal{C}(A)$ and a number $\epsilon > 0$ be fixed. To begin with, let us show that for each $\xi \in A$ there is a function $f_\xi \in \mathfrak{A}^*$ with the properties:

(a) $f_\xi(\xi) = f(\xi)$,
(b) $f_\xi(x) < f(x) + \epsilon \qquad (x \in A)$

(see Figure 43). From the inclusion $\mathfrak{A} \subset \mathfrak{A}^*$ we deduce that $\mathfrak{A}^*$ itself satisfies Definitions 35.8 and 35.9. Lemma 35.11 can, therefore, be applied to $\mathfrak{A}^*$ to produce a function $f_t \in \mathfrak{A}^*$ for each $t \in A$ such that

$$f_t(\xi) = f(\xi), \qquad f_t(t) = f(t).$$

Let $\epsilon > 0$ be given. The continuity of the function f_t (t being held fixed) provides for a neighborhood $N(t; r_t)$ of t such that

$$f_t(x) < f(x) + \epsilon \qquad (x \in N(t; r_t) \cap A). \tag{xiii}$$

The family of all sets $N(t; r_t)$, $t \in A$, constitutes an open covering of the compact set A, a fact from which we conclude that there are finitely many points $t_1, \ldots, t_n$ in A such that

$$A \subset \bigcup_{i=1}^{n} N(t_i; r_{t_i}).$$

Considering the functions f_{t_i} corresponding to these points, put

$$f_\xi = \min\{ f_{t_1}, f_{t_2}, \ldots, f_{t_n}\}. \tag{xiv}$$

A trivial extension of Lemma 35.13 shows that f_ξ belongs to $\mathfrak{A}^*$.

Since, for each $t \in A$, $f_t(\xi) = f(\xi)$, it follows that, in particular, $f_\xi(\xi) = f(\xi)$. Furthermore, given an arbitrary $x \in A$ then there is a point $t_i \in A$ such that $x \in N(t_i; r_{t_i})$: on this set $f_{t_i}(x) < f(x) + \epsilon$ so that $f(x)$ also satisfies this inequality on $N(t_i; r_{t_i})$. Hence f_ξ has the stipulated properties.

Consulting Figure 43, one is easily satisfied that although the inequality in (xiii) can be replaced by $d(f_t(x), f(x)) < \epsilon$, this cannot be done in (b). To complete the proof of the theorem we apply a modified version of the above argument to the functions f_ξ constructed in (xiv). Namely, suppose we constructed a function as in (xiv) for each point $\xi \in A$. Using the same $\epsilon > 0$ as above, we use the continuity of f_ξ to obtain neighborhood $N(\xi; r_\xi)$ of the fixed point ξ such that

$$f_\xi(x) > f(x) - \epsilon \qquad (x \in N(\xi; r_\xi) \cap A).$$

The compactness of A again permits us to specify finitely many points $\xi_1, \ldots, \xi_k$ such that

$$A \subset \bigcup_{i=1}^{k} N(\xi_i; r_{\xi_i})$$

and, corresponding to (xiv), we set

$$p = \max\{ f_{\xi_1}, f_{\xi_2}, \ldots, f_{\xi_k}\}.$$

We can now conclude that

$$p(x) > f(x) - \epsilon \qquad (x \in A),$$

whereas it follows from (b) that

$$p(x) < f(x) + \epsilon \qquad (x \in A),$$

and hence $\| p - f \| < \epsilon$, as was to be demonstrated.

Analogously to Corollary 35.7, the Stone–Weierstrass Theorem can be stated in the language of sets as follows:

35.15 / Corollary If A is compact, then a subalgebra of $\mathcal{C}(A)$ which separates the points of A but vanishes at no point thereof is dense in $\mathcal{C}(A)$.

The reader will note that the dependence of Theorem 35.14 on the Weierstrass Approximation Theorem was confined to the application of Exercise 35-4 in the proof of Lemma 35.12; this lemma, in turn, was needed only to establish Lemma 35.13. Therefore, if in the statement of Theorem 35.14 we add the latter lemma as a hypothesis, then the resulting theorem can be proved independently of Theorem 35.1: this is known as *Stone's Theorem.*

EXERCISES

35-1 If $f: [a, b] \to \mathcal{R}$ is continuous, then it can be uniformly approximated with continuous piecewise linear functions. Prove this fact.

35-2 Sketch the polynomials $B_n(x; x)$ for $n = 1, 2, 3$.

35-3 Modify the Bernstein polynomials to obtain a uniform approximation to f on $[-a, a]$.

35-4 Using the polynomials obtained in the previous exercise, construct polynomials $p_n(x)$ which approximate $f(x) = |x|$ uniformly on $[-a, a]$ and such that $p_n(0) = 0$ for each $n \in \mathfrak{N}$ (Explicitly stated, the polynomials p_n have zero constant terms).

35-5 The partial sums $\sigma_n(x)$ of a power series are polynomials. Discuss the major differences between uniformly convergent sequences of polynomials and the uniformly convergent sequence $\{\sigma_n\}$.

35-6 Which of the following families of functions are algebras?

(a) The set P of polynomials p on $[0, 1]$ such that $\| p \| \leq 1$.
(b) The collection W of power series on $[0, 1]$.
(c) The set L of Lipschitz continuous functions on $[a, b]$ (see Exercise 24-1).

35-7 Referring to Exercise 35-6, what can you say about the uniform closure of each of the depicted families?

35-8 If $\mathfrak{A}^*$ is the uniform closure of an algebra $\mathfrak{A}$ of bounded functions $f: A \to \mathfrak{R}$, then $\mathfrak{A}^*$ is a uniformly closed algebra. Can the boundedness restriction be lifted?

35-9 Regarding Theorem 35.14, suppose the hypothesis that $\mathfrak{A}$ vanishes at no point of A is replaced by the assumption that the functions $h(x) = 1$ and $g(x) = x$ are in $\mathfrak{A}$. Show that the new theorem implies the former; is the converse true?

35-10 Consider an arbitrary interval $[a, b]$. When is there a continuous function f with domain $[a, b]$ such that if p is an arbitrary polynomial of even degree, then $\| f - p \| \geq 1$?

35-11 Verify the relations in (xi) and (xii).

35-12 Given continuous functions f and g with domain A, then the function $h = \max\{ f, g\}$ is also continuous on A. Deduce therefrom that also $k = \min\{ f, g\}$ is continuous on A.

35-13 Referring to the algebra $\mathfrak{S}$ of step functions in Example 35.10, is it true that if $f \in \mathfrak{S}^*$, then f has at most countably many discontinuities?

35-14 The function $f: [a, b] \to \mathfrak{R}$ is of class *Baire-1* when it is the pointwise limit of a sequence of continuous functions. Show that every function of $\mathfrak{S}^*$ is of class *Baire-1*.

35-15 Consider the function

$$f(x) = \begin{cases} \sin \dfrac{1}{x} & (0 < x \leq 1) \\ 0 & (x = 0). \end{cases}$$

Show that f is of class Baire-1. Is this function the uniform limit of step functions?

36 / EQUICONTINUITY

"Every infinite bounded subset of $\mathfrak{R}$ has a limit point": This is the statement of the famous Bolzano–Weierstrass Theorem which, together with the corresponding property of numerical sequences (Exercise 9-12), was shown earlier in this chapter to be false in $\mathcal{C}(A)$. The determination of the additional hypotheses which will admit such a theorem in $\mathcal{C}(A)$ is the goal of this section.

When examining the concept of boundedness of sets of functions, it is observed at once that, as in the case of convergence, two avenues of approach are open to us, the first leading to the following:

36.1 / *Definition* The family $\mathfrak{F}$ of functions $f: A \to \mathcal{R}$ is *pointwise bounded* when, for each $x \in A$, the numerical set $\{f(x)\}$ $(f \in \mathfrak{F})$ is bounded. Equivalently stated, there is a function φ such that

$$|f(x)| < \varphi(x) \qquad (x \in A;\ \ f \in \mathfrak{F}). \tag{i}$$

The function φ is evidently nonunique. Although this definition presents a seemingly natural extension of boundedness in $\mathcal{R}$, the two situations are not analogous. The reason for this is that a family such as $\mathfrak{F} = \{1/x\}$, $0 < x < 1$, is pointwise bounded, yet there is no bounded function φ for which the inequality in (i) holds. To remedy this dilemma we offer:

36.2 / *Definition* The family $\mathfrak{F}$ of functions $f: A \to \mathcal{R}$ is *uniformly bounded* when there is a constant α such that

$$\|f\| < \alpha \qquad (f \in \mathfrak{F}).$$

The statement that the property in 36.2 implies that of 36.1 but not conversely requires no further justification. Definitions similar to the above can be formulated for sequences of functions. Rather than doing that let us examine minutely a uniformly convergent sequence of functions in an attempt to assess the common features of its constitutents.

Suppose A is a compact set and $\{f_n\}$ is a uniformly convergent sequence of continuous functions on A. Then there is an integer k such that $\|f_k - f_n\| < 1$ for all integers $n > k$. From the compactness of A we deduce that each function f_n is bounded and we let

$$m = \max\{\|f_1\|, \ldots, \|f_k\|\} + 1.$$

Then for an arbitrary integer $n > k$,

$$\|f_n\| \le \|f_k\| + \|f_n - f_k\| < \|f_k\| + 1 < m,$$

showing the given sequence to be uniformly bounded.

Now let an arbitrary $\epsilon > 0$ be given. Again there is an integer n for which $\|f_k - f_n\| < \frac{1}{3}\epsilon$. Also, to each $n \in \mathfrak{N}$ there corresponds a number $\delta_n > 0$ such that $d(x, y) < \delta_n$ $(x, y \in A)$ implies that $d(f_n(x), f_n(y)) < \frac{1}{3}\epsilon$. If we let

$$\delta = \min\{\delta_1, \ldots, \delta_k\},$$

then $\delta > 0$, and we observe that if x and y belong to A, $d(x, y) < \delta$ and $n \le k$, then $d(f_n(x), f_n(y)) < \frac{1}{3}\epsilon$. On the other hand, when $n > k$, then

$$\begin{aligned} d(f_n(x), f_n(y)) &\le d(f_n(x), f_k(x)) + d(f_k(x), f_k(y)) \\ &\quad + d(f_n(y), f_k(y)) < \frac{\epsilon}{3} + \frac{\epsilon}{3} + \frac{\epsilon}{3} = \epsilon. \end{aligned} \tag{ii}$$

The conclusion is that for each $\epsilon > 0$ there is a $\delta > 0$ *which depends only on* ϵ such that (ii) holds when $d(x, y) < \delta$. This suggests the following:

36.3 / *Definition* The family $\mathfrak{F}$ of functions on a given set $A \subset \mathcal{R}$ is *equicontinuous* on A if, for each $\epsilon > 0$ there is a $\delta > 0$ such that whenever $x, y \in A$ and $d(x, y) < \delta$, then $d(f(x), f(y)) < \epsilon$ for all members of $f \in \mathfrak{F}$.

The argument leading to this definition can be summarized as follows:

36.4 / Theorem Let A be a given compact set. If the sequence $\{f_n\}$ of continuous functions is uniformly convergent on A, then it is equicontinuous and uniformly bounded on A.

The important converse of this proposition is established in the next theorem:

36.5 / Theorem Every uniformly bounded equicontinuous sequence of functions on a compact set A has a uniformly convergent subsequence.

Proof In this proof we encounter once more the Cantor diagonal method which was a key tool in Chapters One and Three.

Let $\{f_n\}$ satisfy the hypotheses of the theorem; the first of these provides for a constant α such that

$$||f_n|| < \alpha \qquad (n \in \mathfrak{N}).$$

Let $\{b_n\}$ be an arbitrary but fixed sequence whose range B is dense in A (for example, the rational numbers of A arranged in an arbitrary sequence would qualify). Consider the numerical sequence

$$f_1(b_1), f_2(b_1), f_3(b_1), \ldots.$$

Being bounded, this sequence has a convergent subsequence

$$f_{1,1}(b_1), f_{1,2}(b_1), f_{1,3}(b_1), \ldots.$$

The last sequence gives rise to the bounded sequence

$$f_{1,1}(b_2), f_{1,2}(b_2), f_{1,3}(b_2), \ldots,$$

which, in turn, also contains a convergent subsequence, say,

$$f_{2,1}(b_2), f_{2,2}(b_2), f_{2,3}(b_2), \ldots.$$

Continuing in this manner, we arrive at a family of sequences

$$\sigma_1(x): \quad f_{1,1}(x), f_{1,2}(x), f_{1,3}(x), \ldots$$

$$\sigma_2(x): \quad f_{2,1}(x), f_{2,2}(x), f_{2,3}(x), \ldots$$

$$\sigma_3(x): \quad f_{3,1}(x), f_{3,2}(x), f_{3,3}(x), \ldots$$

$$\vdots$$

with the following properties:

(a) For each $n \in \mathfrak{N}$, $\sigma_{n+1}(x)$ is a subsequence of $\sigma_n(x)$;
(b) $\sigma_n(b_n) = \{ f_{nm}(b_n) \}$ converges as $m \to \infty$, n being held fixed;
(c) the order of the functions in each sequence is the same;
which is to say that two functions, once they stand in a given relation to each other in one sequence, remain so related in every succeeding sequence until one or the other of the two functions no longer appears.

From the above matrix we now select the main diagonal:

$$\sigma(x): f_{1,1}(x), f_{2,2}(x), f_{3,3}(x), \ldots.$$

According to (c), the sequence

$$f_{n,n}(x), f_{n+1,n+1}(x), f_{n+2,n+2}(x), \ldots$$

is a subsequence of $\sigma_n(x)$ $(n \in \mathfrak{N})$, and hence (b) tells us that $\sigma(x)$ converges pointwise on the set $B \subset A$. This is as far as the boundedness assumption carries us, and now we appeal to the stipulated equicontinuity of the original sequence $\{ f_n \}$ to show that the sequence σ converges uniformly on A.

Let $\epsilon > 0$ be given. Then there is a $\delta > 0$ such that

$$d(f_n(x), f_n(y)) < \frac{\epsilon}{3} \qquad (n \in \mathfrak{N})$$

whenever $d(x, y) < \delta$, the points being assumed to be in A. With each point $b \in B$ associate the neighborhood $N(b; \delta)$. Then from the fact that A is compact and B is dense in A we deduce that there are finitely many points $b_1, \ldots, b_r$ in B for which

$$A \subset \bigcup_{j=1}^{r} N(b_j; \delta).$$

Regarding these points we can select an integer k such that

$$d(f_{m,m}(b_j), f_{n,n}(b_j)) < \frac{\epsilon}{3} \qquad (1 \le j \le r)$$

for all integers $m, n > k$.

Now, if $x \in A$ is arbitrary, then $x \in N(b_j; \delta)$ for some admitted value of j and hence

$$d(f_{m,m}(x), f_{n,n}(x)) \le d(f_{m,m}(x), f_{m,m}(b_j)) + d(f_{m,m}(b_j), f_{n,n}(b_j))$$
$$+ d(f_{n,n}(b_j), f_{n,n}(x)) < \frac{\epsilon}{3} + \frac{\epsilon}{3} + \frac{\epsilon}{3} = \epsilon$$

for all integers m, $n > k$. Since this estimate is independent of the choice of x, it follows that

$$\| f_{m,m} - f_{n,n} \| < \epsilon,$$

and the desired uniform convergence is thereby established.

The reader has, no doubt, noted that the uniform boundedness of $\{f_n\}$ in the above theorem could have been replaced with pointwise boundedness. In fact, we have:

36.6 / Lemma If $\{f_n\}$ is pointwise bounded and equicontinuous on the compact set A, then the sequence is uniformly bounded on A.

Proof Fix an $\epsilon > 0$ and, using the equicontinuity of the sequence, let $\delta > 0$ be such that

$$d(f_n(x), f_n(y)) < \epsilon$$

when $d(x, y) < \delta$, $x, y \in A$. For each $x \in A$, let

$$\varphi(x) = \operatorname*{lub}_{n \in \mathfrak{N}} | f_n(x) |.$$

We shall show that φ is continuous on A.

Let the points x and y be held fixed, and suppose

$$| f_n(y) | < | f_n(x) | + \epsilon.$$

Then, for each fixed value of n, $| f_n(y) | \le \varphi(x) + \epsilon$ and consequently

$$\varphi(y) \le \varphi(x) + \epsilon.$$

Similarly it follows that

$$| f_n(x) | < | f_n(y) | + \epsilon$$

implies the inequality

$$\varphi(x) \le \varphi(y) + \epsilon;$$

the conclusion being that

$$d(\varphi(x), \varphi(y)) \leq \epsilon$$

when $d(x, y) < \delta$. Since φ is continuous on a compact set, it is bounded there and consequently the sequence $\{f_n\}$ is uniformly bounded.

All of these results are combined in the ensuing theorem:

36.7 / Arzelà–Ascoli Theorem Let $A \subset \mathfrak{R}$ be compact and let $\mathfrak{F}$ be a family of continuous functions with domain A. Then the following statements are equivalent:
(1) The family $\mathfrak{F}$ is pointwise bounded and equicontinuous on A;
(2) Every sequence from $\mathfrak{F}$ contains a uniformly convergent subsequence.

Proof Theorem 36.5 and Lemma 36.6 prove that the statement in (1) implies that in (2). To prove the converse implication we shall show that if (1) is false, then also (2) is false. That is, we show that if the family $\mathfrak{F}$ is not pointwise bounded on A or is not equicontinuous on A, then it contains a sequence with no convergent subsequence.

If $\mathfrak{F}$ is not pointwise bounded, then there is a sequence $\{f_n\}$, $f_n \in \mathfrak{F}$, such that $\|f_n\| \geq n$ for $n = 1, 2, 3, \ldots$. Evidently no subsequence of $\{f_n\}$ is uniformly bounded, and Theorem 36.4 tells us as a consequence that $\{f_n\}$ can have no uniformly convergent subsequences.

Now suppose $\mathfrak{F}$ is not equicontinuous on A. Then for some $\epsilon > 0$ there is a sequence $\{f_n\}$, $f_n \in \mathfrak{F}$, which has a subsequence $\{f_{n_k}\}$ corresponding to each $\delta > 0$ with the property that $d(f_{n_k}(x), f_{n_k}(y)) \geq \epsilon$ when $d(x, y) = \delta$. The sequence $\{f_{n_k}\}$ does not contain a uniformly convergent subsequence, and this completes the proof of the theorem.

EXERCISES

36-1 Let $\{f_n\}$ be a differentiable sequence of functions $f_n:[0, 1] \to \mathfrak{R}$, such that the sequence $\{f_n(x_0)\}$ converges for some x_0. Show that if the derived sequence $\{f_n'\}$ is uniformly bounded on $[0, 1]$, then $\{f_n\}$ contains a convergent subsequence.

36-2 Which of the following families is equicontinuous on the indicated sets?

(a) The Lipschitz continuous functions on $[0, 1]$ (see Exercise 24-1);
(b) the continuous functions $f:[a, b] \to \mathcal{R}$ satisfying $\|f\| \leq 1$;
(c) the set $\{\sin nx\}$, $n \in \mathfrak{N}$, $0 \leq x \leq \pi$.

36-3 Discuss the behavior of the sequence of functions
$f_n(x) = (-1)^n(x + 1)^n$ on $[0, 1]$.

37 / SUMMARY

Having traversed in this chapter some rugged and varied terrain, let us pause momentarily to review in retrospect its major landmarks.

The space $\mathcal{C}(A)$ of continuous functions $f:A \to \mathcal{R}$ was treated as a metric space, the metric being derived from the uniform norm $\| \ \|$. While discussion was essentially limited to the case in which A is compact, it is true that $\mathcal{C}(A)$ is a metric space when A is noncompact but membership in $\mathcal{C}(A)$ is restricted to bounded continuous functions (Exercise 32-2). The following facts are now known:

(1) The Cauchy sequences of $\mathcal{C}(A)$ are its uniformly convergent sequences; according to Theorem 33.2 the space $\mathcal{C}(A)$ is complete.
(2) Like $\mathcal{R}$, the space $\mathcal{C}(A)$ is separable, and the existence of countable dense sets in $\mathcal{C}$ is a consequence of the Weierstrass Approximation Theorem 35.1.
(3) The "bounded" sets of $\mathcal{C}(A)$ are its uniformly bounded sets (Definition 36.2); infinite bounded subsets of $\mathcal{C}(A)$ have a limit point if and only if they are equicontinuous (Arzelà–Ascoli Theorem 36.7).

Our approach in this chapter represents a departure from the earlier arguments in the sense that collections of functions were studied (much the same as point sets) rather than individual functions. This fruitful and far reaching line of investigation is pursued in the branch of mathematics called "Functional Analysis." For example, with an arbitrary function $f \in \mathcal{C}(A)$ we can associate the set

$$N(f; \delta) = \{g \in \mathcal{C}(A) \mid \|f - g\| < \delta\};$$

this set may be justifiably called a neighborhood of f. With this, one can proceed as in Chapter Five to construct a *topology* on $\mathcal{C}(A)$ by defining open sets, closed sets, etc. The reader should attempt, in fact, to determine how many of the concepts and theorems on Chapter Five remain meaningful in $\mathcal{C}(A)$.

NINE

MEASURE AND INTEGRATION

Consider the set A of irrational numbers in $E = [0, 1]$ and its characteristic function χ_A:

$$\chi_A(x) = \begin{cases} 1 & (x \in A) \\ 0 & (x \notin A). \end{cases}$$

Imagine a given number $\epsilon > 0$ and some enumeration $r_1, r_2, r_3, \ldots$ of the rational numbers in E. For each $n \in \mathfrak{N}$ enclose the vertical line segment

$$\lambda(r_n) = \{(r_n, y) \mid 0 \leq y \leq 1\} \text{ in the } \epsilon\text{-strip} \left(r_n - \frac{\epsilon}{2^{n+1}}, r_n + \frac{\epsilon}{2^{n+1}}\right) \times E:$$

each such strip having area $\epsilon \cdot 2^{-n}$ implies that the total area of the union of all these ϵ-strips does not exceed $\sum \epsilon \cdot 2^{-n} = \epsilon$. Since ϵ is arbitrary we are led to conclude that the union of the line segments $\lambda(r_n)$ has measure zero (see Remark 17.15), which is to say that the only meaningful "area" of the set is zero; the set $A \times E$, therefore, has measure 1. To calculate this as well as areas in similar situations, our definition of the integral should be

$$\int_0^1 \chi_A(x)\, dx = 1.$$

This, however, need not be so for all definitions of integral.

The *Riemann integral* of a function $f:E \to R$ is the limit of sums

$$\sum_{i=1}^{n} f(\xi_i)\ \Delta x_i$$

in which $\{\Delta x_i\}$ represents a partition of E into subintervals, and $\xi_i \in \Delta x_i$ (see Figure 44). This scheme of area approximation rests on the assumption that within each interval Δx_i, $f(\xi_i)$ approximates f: this is far from being the case in the above situation. If, for example, we persistently select rational points $\xi_i \in \Delta x_i$, then

$$\sum_{i=1}^{n} \chi_A(\xi_i)\ \Delta x_i = 0$$

and hence also the limit is zero as $n \to \infty$; an exclusive choice of irrational points ξ_i at each stage renders the sums

$$\sum_{i=1}^{n} \chi_A(\xi_i)\ \Delta x_i = \sum_{i=1}^{n} \Delta x_i = 1$$

and hence the limit is one. In short, no meaningful area can be assigned to the set $A \times E$ with approximations in which the "nearness" of values of the function in question on small intervals Δx_i is crucial.

The preceding argument shows that a new strategy is required in cases such as the one depicted. The innovation introduced by Lebesgue was to partition the *range* of the function f to be integrated rather than its domain: in this scheme the "nearness" of values of f is automatic. However, this leads to the following problem:

Let $f:E \to \mathcal{R}$ be given. To approximate the area corresponding to the values $y_i \leq f \leq y_{i+1}$ (see Figure 45) it must be possible to assign a measure (length) to the sets

$$A_i = \{x \in E \mid y_{i-1} \leq f(x) \leq y_i\} \qquad (i = 1, 2, \ldots, n).$$

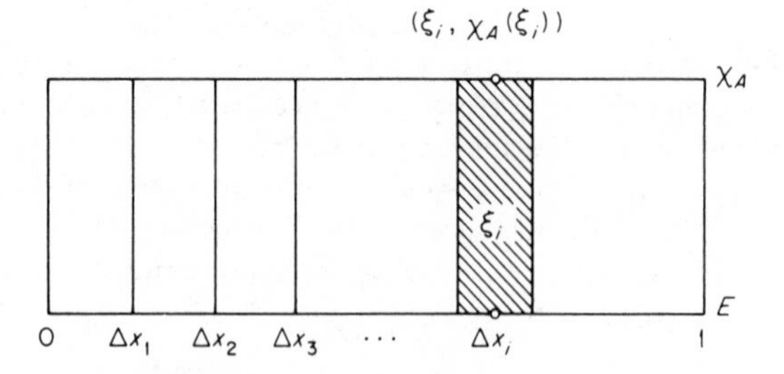

Figure 44

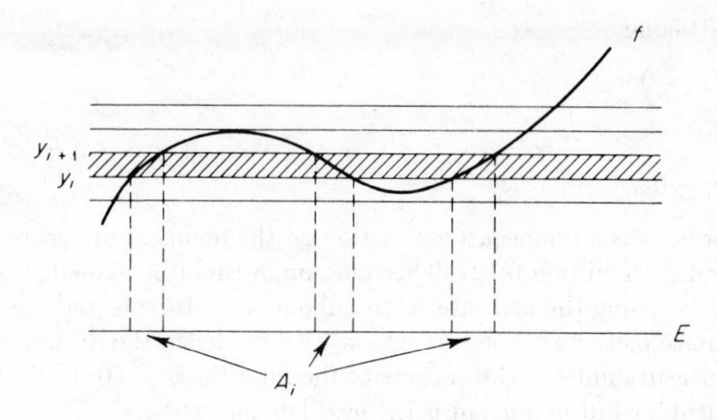

Figure 45

The measure that we seek must be commensurate with the length, which is nonnegative and additive. The function m, in other words, should have these properties:

(a) $m(A) \geq 0$;
(b) $m(A) = d(a, b)$ when A is an interval with endpoints a and b;
(c) if $A_1, A_2, A_3, \ldots$ are pairwise disjoint sets, and $A = \bigcup_{i=1}^{\infty} A_i$, then

$$m(A) = \sum_{i=1}^{\infty} m(A_i).$$

The Lebesgue approach to integration suggests, therefore, that the concept of length be extended to the widest possible class of sets in $\mathcal{R}$. One such extension will be considered in the following pages. It will be shown in Section 39 that a measure which satisfies the property in (c) cannot extend to all subsets of $\mathcal{R}$, in fact, not even to the class of all bounded subsets of $\mathcal{R}$.

38 / MEASURABLE SETS

The measure of an interval is well defined through its length:

38.1 / *Definition* If I stands for one of the intervals (a, b), $(a, b]$, $[a, b)$ or $[a, b]$, then the number

$$m(I) = d(a, b)$$

is the *measure* of I.

In particular

$$m(\varnothing) = 0$$

and otherwise

$$m(I) \geq 0.$$

In other words, m is a nonnegative function on the bounded intervals of $\mathcal{R}$ into $\mathcal{R}$ (see also Definition 9.12). The function m could be extended to all intervals by assigning the measure ∞ to unbounded intervals, but we wish to avoid infinite measures. For this reason we consider in the discussion to follow the measurability of the subsets of the interval $E = [0, 1]$.

Elementary but important is the next lemma:

38.2 / Lemma Let $I_1, I_2, \ldots, I_n$ be given intervals. If

$$\bigcup_{p=1}^{n} I_p \subset E \qquad \text{and} \qquad I_p \cap I_q = \varnothing \qquad \text{whenever} \quad p \neq q$$

then

$$\sum_{p=1}^{n} m(I_p) \leq 1;$$

if

$$\bigcup_{p=1}^{n} I_p \supset E$$

then

$$\sum_{p=1}^{n} m(I_p) \geq 1.$$

The proof of this lemma, while intuitively clear, should nevertheless be written out by the reader.

Consider an arbitrary open set $A \subset E$. This set has a unique decomposition into the union of at most countably many mutually exclusive open intervals I_p. Owing to Lemma 38.2 we can apply the Boundedness Test 14.1 to the series $\sum m(I_p)$ which accordingly converges (absolutely). This fact allows us to extend the set function m to the family of all open sets of E, the extension being again designated by m. Formally we state the following:

38.3 / *Definition* Let A be an arbitrary open set in E with components I_p. Then the measure of A is the number

$$m(A) = \sum_{p=1}^{\infty} m(I_p).$$

It goes without saying that some or all of the intervals I_p may be empty. The function m is said to be *countably additive.* The convergence of the above series assures the existence of an integer n corresponding to each $\epsilon > 0$ such that

$$\sum_{p=n+1}^{\infty} m(I_p) < \epsilon.$$

With these simple means we are ready to extend our measure to more general sets. Let $A \subset E$ be arbitrary, and consider all open sets G containing it: The set $\{m(G)\}$ is bounded below and this allows us to associate with A a unique number, namely, the greatest lower bound of $\{m(G)\}$.

38.4 / *Definition* The *outer measure* $\mu^*(A)$ of the set $A \subset E$ is the number

$$\mu^*(A) = \operatorname*{glb}_{G \supset A} \{m(G)\},$$

where the infimum is taken over all open sets containing A.

At the same time we consider all open sets containing the complement in E of A, and this brings us to the next definition:

38.5 / *Definition* The *inner measure* $\mu_*(A)$ of A is the number

$$\mu_*(A) = 1 - \mu^*(E \setminus A).$$

We observe that for each interval $I \subset E$ we have the relations

$$\mu^*(I) = m(I) = \mu_*(I).$$

Let us, in fact, verify the first equality (the second can be done as an exercise). If G is an arbitrary open set containing I, then I must be contained in one of the component intervals of G (in this connection, see Theorem 22.2). This tells us that $m(I) \leq m(G)$, and accordingly $m(I) \leq \mu^*(I)$. The inequality $\mu^*(I) \leq m(I)$ is clear from Definition 38.4. This shows that no discrepancies between the numbers $\mu_*(A)$ and $\mu^*(A)$ can be tolerated when the measure m is to be extended to A. This is the motive for introducing:

38.6 / *Definition* The set $A \subset E$ is *Lebesgue measurable* or simply *measurable* when

$$\mu_*(A) = \mu^*(A).$$

The common value of the outer and inner measures of A is denoted by the symbol $\mu(A)$ and designated as the *Lebesgue measure* of the set A.

Some of the important relations which exist between the various measures considered above will now be explored. An immediate consequence of the definitions is:

38.7 / Theorem The inequality

$$\mu_*(A) \le \mu^*(A)$$

is universally valid in E.

Proof According to Definition 38.5 the inequality in the theorem is equivalent to the statement

$$\mu^*(A) + \mu^*(E \setminus A) \ge 1,$$

which we proceed to verify. Consider open sets $G \supset A$ and $H \supset E \setminus A$ with components $\{I_p\}$ and $\{J_q\}$, respectively. Since evidently $G \cup H \supset E$ we deduce from Theorem 18.11 that there is a finite subfamily of intervals, say $I_{p1}, \ldots, I_{pr}, J_{q1}, \ldots, J_{qs}$, which forms an open covering of E. By virtue of Lemma 38.2,

$$\sum_{i=1}^{r} m(I_{p_i}) + \sum_{j=1}^{s} m(J_{p_j}) \ge 1.$$

But at the same time

$$m(G) \ge \sum_{i=1}^{r} m(I_{p_i})$$

and

$$m(H) \ge \sum_{j=1}^{s} m(I_{q_j}).$$

Consequently,

$$\mu^*(A) + \mu^*(E \setminus A) = \operatorname*{glb}_{G \supset A} \{m(G)\} + \operatorname*{glb}_{H \supset E \setminus A} \{m(H)\}$$

$$= \operatorname*{glb}_{\substack{G \supset A \\ H \supset E \setminus A}} \{m(G) + m(H)\} \ge 1.$$

A first characterization of measurable sets is easily come by.

38.8 / Theorem The set $A \subset E$ is measurable if and only if

$$\mu^*(A) + \mu^*(E \setminus A) = 1.$$

Proof If A is measurable, then

$$\mu^*(A) = \mu_*(A) = 1 - \mu^*(E \setminus A).$$

On the other hand, when the equation in the theorem holds, then

$$\mu^*(A) = 1 - \mu^*(E \setminus A) = \mu_*(A).$$

38.9 / Corollary If $A \subset E$ is measurable, then so is $E \setminus A$.

This follows at once with the observation that $E \setminus (E \setminus A) = A$.

We recall that the finite unions of pairwise disjoint intervals are *elementary sets* (see Exercise 9-11). This family is enhanced with the property of being closed under the set operations: If A and B are elementary sets, then so are the sets $A \cup B$, $A \cap B$, $A \setminus B$ and $A \triangle B$.
We have the following corollary:

38.10 / Corollary If $A \subset E$ is an elementary set, then A is measurable.

The following characterizations of measurability are useful:

38.11 / Theorem Let $A \subset E$ be given. Then the following statements are equivalent:

(**1**) the set A is measurable;
(**2**) corresponding to each $\epsilon > 0$ there are open sets $G_1 \supset A$ and $G_2 \supset E \setminus A$ such that

$$m(G_1 \cap G_2) < \epsilon;$$

(**3**) for each $\epsilon > 0$ there is an elementary set B such that

$$\mu^*(A \triangle B) < \epsilon.$$

In the proof we rely on the following:

38.12 / Lemma If G_1 and G_2 are bounded open sets, then

$$m(G_1) + m(G_2) = m(G_1 \cup G_2) + m(G_1 \cap G_2).$$

Proof Let an arbitrary number $\epsilon > 0$ be preassigned. Each set G_i $(i = 1, 2)$ can be expressed as the union of two disjoint open sets,

$$G_i = F_i \cup (G_i \setminus F_i),$$

such that F_i is the finite union of open intervals, and

$$m(G_i \setminus F_i) < \frac{\epsilon}{4}.$$

In particular, the inequalities

$$m(F_i) \leq m(G_i) \leq m(F_i) + \frac{\epsilon}{4} \tag{i}$$

are a direct consequence of Exercise 38-1 and Definition 38.3. Also, simple Venn diagrams show that

$$G_1 \cup G_2 \subset (F_1 \cup F_2) \cup (G_1 \setminus F_1) \cup (G_2 \setminus F_2)$$

and

$$G_1 \cap G_2 \subset (F_1 \cap F_2) \cup (G_1 \setminus F_1) \cup (G_2 \setminus F_2).$$

where each of the involved sets is open. Again we appeal to the cited exercise and definition to conclude that

$$m(G_1 \cup G_2) \leq m(F_1 \cup F_2) + \frac{\epsilon}{2}$$

and

$$m(G_1 \cap G_2) \leq m(F_1 \cap F_2) + \frac{\epsilon}{2}.$$

Now, owing to (i), Exercises 38-14 and 38-1,

$$\begin{aligned} m(G_1) + m(G_2) &\leq m(F_1) + m(F_2) + \frac{\epsilon}{2} \\ &= m(F_1 \cup F_2) + m(F_1 \cap F_2) + \frac{\epsilon}{2} \\ &\leq m(G_1 \cup G_2) + m(G_1 \cap G_2) + \frac{\epsilon}{2}. \end{aligned}$$

At the same time we see that

$$\begin{aligned} m(G_1) + m(G_2) \geq m(F_1) + m(F_2) &= m(F_1 \cup F_2) + m(F_1 \cap F_2) \\ &\geq m(G_1 \cup G_2) + m(G_1 \cap G_2) - \epsilon. \end{aligned}$$

When put together, these estimates result in the inequalities

$$\begin{aligned} m(G_1 \cup G_2) + m(G_1 \cap G_2) - \epsilon &\leq m(G_1) + m(G_2) \\ &\leq m(G_1 \cup G_2) + m(G_1 \cap G_2) + \frac{\epsilon}{2} \end{aligned}$$

and the lemma follows from the fact that $\epsilon > 0$ was arbitrary.

Proof of 38.11 We prove Theorem 38.11, first showing that the first statement implies the second. To this end, let $G_1 \supset A$ and $G_2 \supset E \setminus A$

be open sets such that

$$m(G_1) < \mu^*(A) + \frac{\epsilon}{2}$$

and

$$m(G_2) < \mu^*(E \setminus A) + \frac{\epsilon}{2}.$$

From the preceding lemma and Lemma 38.2, we conclude that

$$\begin{aligned} m(G_1 \cap G_2) &= m(G_1) + m(G_2) - m(G_1 \cup G_2) \\ &< \mu^*(A) + \mu^*(E \setminus A) - m(G_1 \cup G_2) + \epsilon \\ &\leq \mu^*(A) + \mu^*(E \setminus A) - 1 + \epsilon, \end{aligned}$$

where the last inequality is true because of the inclusion $E \subset G_1 \cup G_2$. We now make use of the hypothesis in (1): Since A is measurable, $\mu^*(A) = \mu_*(A)$, and Theorem 38.8 tells us that $\mu^*(A) + \mu^*(E \setminus A) - 1 = 0$. Hence $m(G_1 \cap G_2) < \epsilon$ as was to be shown.

We now demonstrate that (2) implies (3). For this, let G_1 and G_2 be as stipulated, except that $m(G_1 \cap G_2) < \frac{1}{2}\epsilon$. Let G_1 be written as the union of two disjoint open sets, $G_1 = G_1' \cup G_1''$, where $m(G_1') < \frac{1}{2}\epsilon$. Then the inclusion $A \subset G_1$ implies the inclusion

$$A \setminus G_1'' \subset G_1'$$

and evidently

$$G_1'' \setminus A \subset G_1'' \cap G_2.$$

This is equivalent to the statement

$$A \triangle G_1'' \subset G_1' \cup (G_1'' \cap G_2).$$

The set $G_1'' \cap G_2$ is open and Lemma 38.12 tells us that

$$m(G_1 \cup (G_1'' \cap G_2)) \leq m(G_1) + m(G_1'' \cap G_2).$$

At the same time we have the estimate $m(G_1') < \frac{1}{2}\epsilon$ and the estimate

$$m(G_1'' \cap G_2) \leq m(G_1 \cap G_2) < \frac{\epsilon}{2}.$$

Hence we conclude that

$$\mu^*(A \triangle G_1'') \leq \frac{\epsilon}{2} + \frac{\epsilon}{2} = \epsilon.$$

It goes without saying, of course, that G_1'' can be taken to be an elementary set.

To complete the proof of the theorem it remains only to show that (3) implies (1). The argument of this case rests on the following fact: Consider arbitrary sets B_1 and B_2 in E such that $\mu^*(B_1) \geq \mu^*(B_2)$. Evidently

$$B_1 \subset B_2 \cup (B_1 \triangle B_2)$$

and if F and G are open sets such that $G_1 \supset B_2$, $G_2 \supset B_1 \triangle B_2$, then

$$\mu^*(B_1) \leq m(G_1 \cup G_2) \leq m(G_1) + m(G_2).$$

It now follows at once that

$$\mu^*(B_1) \leq \mu^*(B_2) + \mu^*(B_1 \triangle B_2).$$

When $\mu^*(B_2) \geq \mu^*(B_1)$ we interchange the roles of B_1 and B_2 in the derivation above, therefrom concluding that

$$| \mu^*(B_1) - \mu^*(B_2) | \leq \mu^*(B_1 \triangle B_2).$$

This result holds for arbitrary sets B_1 and B_2. For the application to the specific situation at hand let $\epsilon > 0$ be given and suppose $\mu^*(A \triangle B) < \frac{1}{2}\epsilon$. Then

$$| \mu^*(A) - \mu^*(B) | \leq \mu^*(A \triangle B) < \frac{\epsilon}{2}.$$

Furthermore, since $A \subset E$ and $B \subset E$ one readily verifies that

$$(E \setminus A) \setminus (E \setminus B) = (E \setminus A) \cap B = B \setminus A$$

and

$$(E \setminus B) \setminus (E \setminus A) = (E \setminus B) \cap A = A \setminus B.$$

But this means that

$$(E \setminus A) \triangle (E \setminus B) = A \triangle B$$

so that also

$$| \mu^*(E \setminus A) - \mu^*(E \setminus B) | < \frac{\epsilon}{2}.$$

But

$$\mu^*(B) + \mu^*(E \setminus B) = 1.$$

Hence, the following derivation is valid:

$$\begin{aligned} &| \mu^*(A) + \mu^*(E \setminus A) - 1 | \\ &\qquad \le | \mu^*(A) - \mu^*(B) | + | \mu^*(E \setminus B) - \mu^*(E \setminus A) | \\ &\qquad < \frac{\epsilon}{2} + \frac{\epsilon}{2} = \epsilon. \end{aligned}$$

The conclusion

$$\mu^*(A) + \mu^*(E \setminus A) - 1 = 0$$

is warranted since $\epsilon > 0$ is arbitrary. Hence A is measurable.

We conclude this section with the following important characterization of measurability of Carathéodory:

38.13 / Theorem The set $A \subset E$ is measurable if and only if

$$\mu^*(X) = \mu^*(X \cap A) + \mu^*(X \setminus A)$$

for each set $X \subset E$.

Proof Let $A \subset E$ be a set for which the above formula holds for each $X \subset E$. Putting $X = E$, we find that

$$\mu^*(E) = 1 = \mu^*(A) + \mu^*(E \setminus A),$$

and Theorem 38.8 tells us that the set A is measurable.

Conversely, let A be measurable and let $X \subset E$ be arbitrary. We take open sets $G_1 \supset X \cap A$ and $G_2 \supset X \setminus A$. Then $G_1 \cup G_2 \supset X$, and according to Lemma 38.12

$$m(G_1 \cup G_2) \le m(G_1) + m(G_2).$$

From Definition 38.4 it now follows that

$$\mu^*(X) \le \mu^*(X \cap A) + \mu^*(X \setminus A).$$

Let us now establish the converse inequality. Given an arbitrary $\epsilon > 0$, there is an open set $G_1 \supset X$ such that

$$m(G_1) < \mu^*(X) + \epsilon.$$

According to Theorem 38.11(2) there are open sets $G_2 \supset A$ and $G_3 \supset E \setminus A$ such that

$$m(G_2 \cap G_3) < \epsilon.$$

The sets $G_1 \cap G_2$ and $G_1 \cap G_3$ are open, $G_1 \cap G_2 \supset X \cap A$, $G_1 \cap G_3 \supset X \setminus A$, and

$$\mu^*(X \cap A) \leq m(G_1 \cap G_2),$$

$$\mu^*(X \setminus A) \leq m(G_1 \cap G_3).$$

In particular,

$$m(G_1) \geq m(G_1 \cap G_2) + m(G_1 \cap G_3) - m(G_2 \cap G_3),$$

and the preceding estimates show that

$$\mu^*(X) + \epsilon \geq \mu^*(X \cap A) + \mu^*(X \setminus A) - \epsilon.$$

The desired inequality follows from the fact that ϵ was arbitrary, and the proof of the theorem is complete.

EXERCISES

38-1 Let G_1 and G_2 be open sets in E. Show that the inclusion $G_1 \subset G_2$ implies the inequality $m(G_1) \leq m(G_2)$.

38-2 Are there open sets $G_1 \subset G_2$ such that $G_1 \neq G_2$ yet $m(G_1) = m(G_2)$?

38-3 Prove or disprove: If the open set G is such that $m(G) = 0$, then $G = \varnothing$.

38-4 Let $\{A_p\}$ $(A_p \subset E, p \in \mathfrak{N})$ be open intervals; let the set $A = \bigcup_{p \in \mathfrak{N}} A_p$ have components I_q. Show that $\sum m(I_q) \leq \sum m(A_p)$.

38-5 Show that if $\mu^*(A) = 0$, then A is measurable. This fact justifies our earlier claims that countable sets have measure zero.

38-6 Prove that every open and every closed set in E is measurable.

38-7 Show that $A \subset E$ is nonmeasurable if and only if

$$\mu^*(A) + \mu^*(E \setminus A) > 1.$$

38-8 Prove or disprove: The set $A \subset E$ is measurable if and only if for each $\epsilon > 0$ there is an open set G such that

$$\mu^*(A \triangle G) < \epsilon.$$

38-9 Show that for all sets $A \subset B \subset E$,

$$\mu_*(A) \leq \mu_*(B),$$

$$\mu^*(A) \leq \mu^*(B).$$

38-10 With every set $A \subset E$ associate its *translate*

$$A + a = \{x + a \mid x \in A\},$$

where $a \in \mathcal{R}$ is fixed. Prove that if A is measurable, then so is $A + a$ for each $a \in \mathcal{R}$. Furthermore,

$$\mu(A) = \mu(A + a).$$

38-11 Show that if H is a closed set, G an open set, $H \subset G$, then $\mu(H) \leq \mu(G)$. Can equality ever hold?

38-12 Let A be an elementary set. Show that $\mu(A) = 0$ if and only if $m(A) = 0$.

38-13 Show that for an arbitrary set $A \subset E$,

$$\mu^*(A) - \mu^*(E \setminus A) = \mu_*(A) - \mu_*(E \setminus A).$$

38-14 Let A and B be each the union of finitely many open intervals in E. Deduce directly from Definition 38.3 that

$$m(A) + m(B) = m(A \cup B) + m(A \cap B).$$

(*Hint*: Use the fact that if $A = (a_1, a_2)$, $B = (a_1, c)$ and $C = (c, a_2)$, $a_1 < c < a_2$, then $m(A) = m(B) + m(C)$.)

38-15 According to Lemma 38.12, $m(G_1 \cup G_2) \leq m(G_1) + m(G_2)$ when G_1 and G_2 are open sets. Establish the following generalization of this inequality: If $\{G_p\}$ is a family of open sets in E, and

$$G = \bigcup_{p \in \mathfrak{N}} G_p,$$

then

$$m(G) \leq \sum_{p=1}^{\infty} m(G_p).$$

(*Hint*: You may first prove the version in which only finitely many of the sets G_p are nonempty.)

38-16 Let $I_1, \ldots, I_r$ and $J_1, \ldots, J_s$ be each a family of pairwise disjoint intervals. Show that if

$$\dot{\bigcup_{p=1}^{r}} I_p = \dot{\bigcup_{q=1}^{s}} J_q,$$

then

$$\sum_{p=1}^{r} m(I_p) = \sum_{q=1}^{s} m(J_q).$$

(*Note*: This result would permit us to extend the function m to the elementary subsets of E.)

39 / PROPERTIES OF MEASURABLE SETS

We already know intuitively that measurability is preserved under the basic set operations, and we shall now verify this. We begin with an important inequality:

39.1 / Lemma Let G represent the countable union of open sets $G_p \subset E$. Then

$$m(G) \leq \sum_{p=1}^{\infty} m(G_p).$$

Proof The lemma is obvious when the series in the right side of the inequality diverges (to $+\infty$). Otherwise, let $\{I_q\}$ be the connected components of G; for each value of p, let I_{pr} be the components of G_p. Then it is clear that either $I_{pr} \cap I_q = \emptyset$ or else $I_{pr} \subset I_q$. In particular, there are intervals $I_{p_i r_j}$ for each value of q such that

$$I_q = \bigcup_{i,j} I_{p_i r_j}.$$

The intervals $I_{p_i r_j}$ constitute a covering of I_q, and a simple extension of Lemma 38.2 shows that

$$m(I_q) \leq \sum_{i,j} m(I_{p_i r_j}).$$

For each p_i, however,

$$\bigcup_j I_{p_i r_j} \subset G_{p_i}$$

and

$$I_{p_i r_j} \cap I_{p_i r_k} = \emptyset \qquad \text{when} \quad j \neq k,$$

so that

$$\sum_j m(I_{p_i r_j}) \leq m(G_{p_i}).$$

Hence,

$$\sum_{i,j} m(I_{p_i r_j}) \leq \sum_i m(G_{p_i}),$$

and accordingly,

$$m(I_q) \leq \sum_i m(G_{p_i}).$$

Using now the absolute convergence of all the series involved, we see that

$$m(G) = \sum_{q=1}^{\infty} m(I_q) \leq \sum_{p=1}^{\infty} m(G_p).$$

39.2 / Theorem If $A_1, \ldots, A_n$ is an arbitrary collection of measurable subsets of E, then also the set

$$A = \bigcup_{p=1}^{n} A_p$$

is measurable.

Proof Let $\epsilon > 0$ be a given number. According to Theorem 38.11 there are open sets $G_p \supset A_p$ and $G'_p \supset E \setminus A_p$ such that

$$m(G_p \cap G'_p) < \frac{\epsilon}{n} \qquad (p = 1, 2, \ldots, n). \tag{i}$$

Clearly $\bigcup_{p=1}^{n} G_p \supset A$, and in view of the relation

$$E \setminus \bigcup_{p=1}^{n} A_p = \bigcap_{p=1}^{n} (E \setminus A_p)$$

we see that

$$\bigcap_{p=1}^{n} G'_p \supset E \setminus A.$$

Put

$$H = \bigcup_{p=1}^{n} G_p \qquad \text{and} \qquad H' = \bigcap_{p=1}^{n} G'_p.$$

We assert that $m(H \cap H') < \epsilon$. Indeed, it is readily seen that

$$G_p \cap \left(\bigcap_{q=1}^{n} G'_q \right) \subset G_p \cap G'_p$$

for each admitted value of p, and hence

$$H \cap H' = \bigcup_{q=1}^{n} \left(G_q \cap \left(\bigcap_{p=1}^{n} G'_p \right) \right) \subset \bigcup_{p=1}^{n} (G_p \cap G'_p)$$

and by virtue of Exercise 38-1, Lemma 39.1 and (i) above

$$m(H \cap H') \leq m\left(\bigcup_{p=1}^{n} (G_p \cap G'_p) \right) \leq \sum_{p=1}^{n} m(G_p \cap G'_p) < \sum_{p=1}^{n} \frac{\epsilon}{n} = \epsilon.$$

The sets H and H' are open, and according to Theorem 38.11 the set A is measurable.

From this theorem we derive:

39.3 / Corollary If $A_1, \ldots, A_n$ is an arbitrary collection of measurable subsets of E, then also the set

$$B = \bigcap_{p=1}^{n} A_p$$

is measurable.

Proof Put $B_p = E \setminus A_p$. Then

$$E \setminus \left(\bigcup_{p=1}^{n} B_p \right) = \bigcap_{p=1}^{n} A_p.$$

The sets A_p are measurable, so are the sets B_p, and owing to Theorem 39.2 so is $\bigcup_{p=1}^{n} B_p$, and the corollary follows.

We also have the following corollary:

39.4 / Corollary If A and B are measurable sets, then so are the sets $A \setminus B$ and $A \triangle B$.

The measurability of $A \setminus B$ follows from the relation $A \setminus B = A \cap (E \setminus B)$ and Corollary 38.9. The fact that $A \triangle B = (A \setminus B) \cup (B \setminus A)$ shows also this set to be as asserted.

Of fundamental importance is the next theorem:

39.5 / Theorem Let the sets A_p ($p \in \mathfrak{N}$) be arbitrary pairwise disjoint measurable subsets of E. Then also the set $A = \bigcup_{p \in \mathfrak{N}} A_p$ is measurable, and moreover

$$\mu(A) = \sum_{p=1}^{\infty} \mu(A_p).$$

This fact follows from the following two lemmas:

39.6 / Lemma Let $\{A_p\}$ ($A_p \subset E$, $p \in \mathfrak{N}$) be a family of measurable sets. If

$$A = \bigcup_{p \in \mathfrak{N}} A_p,$$

then

$$\mu^*(A) \leq \sum_{p=1}^{\infty} \mu(A_p).$$

Proof Let an arbitrary $\epsilon > 0$ be given. For each $p \in \mathfrak{N}$ there are open sets $G_p \supset A_p$ such that

$$m(G_p) < \mu(A_p) + \frac{\epsilon}{2^p}.$$

Then owing to Definition 38.4 and Lemma 39.1,

$$\mu^*(A) \leq m(\bigcup_{p \in \mathfrak{N}} G_p) \leq \sum_{p=1}^{\infty} m(G_p) \leq \sum_{p=1}^{\infty} \mu(A_p) + \epsilon,$$

and the desired inequality follows from the fact that ϵ was arbitrary.

Note: Although we could replace the measure m by μ in the preceding proofs, we prefer not to do so for a while. This way it is easy to see how open sets are used in the approximation of arbitrary measurable sets.

39.7 / Lemma Let $\{A_p\}$ $(A_p \subset E,\ p \in \mathfrak{N})$ be a family of pairwise disjoint measurable sets. If

$$A = \bigcup_{p \in \mathfrak{N}} A_p,$$

then

$$\mu_*(A) \geq \sum_{p=1}^{\infty} \mu(A_p).$$

Proof Pick an integer $n \geq 1$. According to Theorem 38.11 there are open sets, G and H, such that

$$G \supset \bigcup_{p=1}^{n} A_p,$$

$$H \supset E \setminus (\bigcup_{p=1}^{n} A_p),$$

$$m(G \setminus E) < \frac{\epsilon}{4},$$

and

$$m(G \cap H) < \frac{\epsilon}{4};$$

the last estimate owing to the fact that the set $\bigcup A_p$ is measurable.

With the help of Exercise 38-9 we show that

$$\mu_*(A) \geq \mu_*\Big(\bigcup_{p=1}^{n} A_p\Big) = 1 - \mu^*\Big(E \setminus \bigcup_{p=1}^{n} A_p\Big) \geq 1 - \mu^*(H)$$
$$= \mu_*(E \setminus H)$$
$$= \mu(E \setminus H), \qquad \text{(ii)}$$

the last equality being due to the fact that H is measurable (see Corollary 38.9). Lemma 39.1 can now be applied to the decomposition

$$G = (E \setminus H) \cup (G \cap H) \cup (G \setminus E),$$

to yield the inequality

$$m(G) = \mu(G) \leq \mu(E \setminus H) + \mu(G \cap H) + \mu(G \setminus E)$$
$$\leq \mu(E \setminus H) + \frac{\epsilon}{4} + \frac{\epsilon}{4} = \mu(E \setminus H) + \frac{\epsilon}{2}.$$

This estimate is now applied to (ii). The result is that

$$\mu_*(A) \geq \mu(G) - \frac{\epsilon}{2}. \qquad \text{(iii)}$$

A lower estimate for $m(G)$ is obtained as follows: Since the sets A_p are in G, there are open sets G_p, $G \supset G_p \supset A_p$, $p = 1, 2, \ldots, n$, such that

$$m(G_p \cap G_q) < \frac{\epsilon}{n(n-1)} \qquad (p \neq q).$$

This is due to the fact that $A_q \subset E \setminus A_p$ and $A_p \subset E \setminus A_q$, and Theorem 38.11. Now put

$$H_p = \bigcup_{j=p}^{n} G_j.$$

Then Lemma 38.12 tells us that, for each value of p,

$$m(G_p \cup H_{p+1}) = m(G_p) + m(H_{p+1}) - m(G_p \cap H_{p+1}).$$

Hence,

$$m\Big(\bigcup_{p=1}^{n} G_p\Big) = \sum_{p=1}^{n} m(G_p) - \sum_{p=1}^{n-1} m(G_p \cap H_{p+1}).$$

Since

$$G_p \cap H_{p+1} = \bigcup_{q=p+1}^{n} (G_p \cap G_q),$$

it follows that

$$m(G_p \cap H_{p+1}) \leq \sum_{q=p+1}^{n} m(G_p \cap G_q),$$

and consequently

$$m(\bigcup_{p=1}^{n} G_p) \geq \sum_{p=1}^{n} m(G_p) - \sum_{\substack{p<q \\ 2\leq q\leq n}} m(G_p \cap G_q).$$

The last sum contains $(n-1) + (n-2) + \cdots + 1 = n(n-1)/2$ summands, each not exceeding $\epsilon/n(n-1)$. This leads us to conclude that

$$m(G) \geq m(\bigcup_{p=1}^{n} G_p) \geq \sum_{p=1}^{n} m(G_p) - \frac{n(n-1)}{2} \cdot \frac{\epsilon}{n(n-1)}$$

$$= \sum_{p=1}^{n} m(G_p) - \frac{\epsilon}{2}.$$

Combined with the estimate in (iii), this and Exercise 38-9 show that

$$\mu_*(A) \geq \sum_{p=1}^{n} m(G_p) - \epsilon \geq \sum_{p=1}^{n} \mu(A_p) - \epsilon.$$

Since ϵ was arbitrary, it follows that

$$\mu_*(A) \geq \sum_{p=1}^{n} \mu(A_p).$$

This estimate is true for each positive integer, and the lemma follows.

Proof of Theorem 39.5 Let us now examine Theorem 39.5. The inequality $\mu_*(A) \leq \mu^*(A)$, which was established in Theorem 38.7, shows that

$$\sum_{p=1}^{\infty} \mu(A_p) \leq \mu_*(A) \leq \mu^*(A) \leq \sum_{p=1}^{\infty} \mu(A_p).$$

Hence, A is measurable and the asserted equality holds.

Theorem 39.2 and Corollary 39.3 can now be extended to countable families of measurable sets:

39.8 / Theorem If $\{A_p\}$ $(p \in \mathfrak{N})$ is an arbitrary family of measurable subsets of E, then also the sets

$$A = \bigcup_{p\in\mathfrak{N}} A_p$$

and

$$B = \bigcap_{p \in \mathfrak{N}} A_p$$

are measurable.

Proof Let us express the set A in the form

$$A = A_1 \cup [A_2 \setminus A_1] \cup [A_3 \setminus (A_1 \cup A_2)] \cdots.$$

Each set in brackets is measurable according to the facts just established. Moreover, the sets are all pairwise disjoint. Theorem 39.5, therefore, tells us that A is measurable.

The proof of Corollary 39.3 can be repeated without the finiteness restriction, and this proves the second assertion in the theorem.

Because of Theorem 39.5 the measure μ is said to be *σ-additive* or *completely additive*. This fact has far reaching consequences, one of which is the existence of nonmeasurable sets. This is considered next:

39.9 / *Example* Let us associate with each number $x \in [-1, 1]$ the class $A(x)$ of points within $N(x; 1)$ whose distance from it is rational:

$$A(x) = \{x + r \mid r \in \mathfrak{R}^*, d(x, r) \leq 1\}.$$

We introduce the following equivalence relation: $x \sim y$ when $x - y \in \mathfrak{R}^*$. The transitive character of the equivalence tells us that the classes are mutually exclusive: if $x \nsim y$, then $A(x) \cap A(y) = \varnothing$. Clearly each class $A(x)$ is countable, yet their union contains the interval $[-1, 1]$. We conclude, therefore, that there are uncountably many such classes. Using the axiom of choice, let the set A consist of precisely one member from each class $A(x)$. Evidently no two members of A are equivalent and we assert that this set is nonmeasurable.

Since we are going to make use of the translation invariant character of the measure μ (see Exercise 38-10) we shall digress here momentarily to verify this fact. To begin with, it is obvious that if I is an arbitrary interval of E, and if $I + a$ represents its translation by the distance a, then $m(I + a) = m(I)$. That open sets share this property is easily seen, and an application of Theorem 38.11 shows that all measurable sets have this property.

Thus, consider some listing of the rational numbers in the interval $[-2, 2]$ in a sequence, say, $r_1, r_2, r_3, \ldots$. For each r_p consider the translation

$$A + r_p = \{x + r_p \mid x \in A\}.$$

The sets $A + r_p$ are said to be *congruent*. We shall show that

$$(A + r_p) \cap (A + r_q) = \varnothing \qquad (r_p \neq r_q). \tag{iv}$$

Suppose, on the contrary, that z belongs to this intersection. Then this number has two distinct representations, $z = x_p + r_p$ and $z = x_q + r_q$, for which $x_p = z - r_p$ and $x_q = z - r_q$ belong to distinct classes $A(x)$. This implies, of course, that $x_p - x_q \notin \mathcal{R}^*$. At the same time, however, $x_p - x_q = r_q - r_p \in \mathcal{R}^*$, and we are compelled to conclude that the assertion in (iv) was correct after all.

Finally, let it be shown that

$$[-1, 1] \subset \bigcup_{p \in \mathfrak{N}} (A + r_p) \subset [-3, 3]. \tag{v}$$

This is easy, for if $z \in [-1, 1]$, then there is a class $A(x)$ such that $z \in A(x)$, implying the existence of a rational number $r_n \in [-2, 2]$ for which $z - x = r_n$. But this means that $z \in A + r_n$, and the first inclusion is thereby established. The second inclusion is a direct consequence of the construction.

The stage is now set for the concluding argument. If A is assumed to be measurable, then

$$\mu(A + r_p) = \mu(A)$$

for each $p \in \mathfrak{N}$. On account of (iv) we can apply Theorem 39.5 and then (v) to yield the relations

$$2 \leq \mu\Big[\bigcup_{p \in \mathfrak{N}} (A + r_p)\Big] = \sum_{p=1}^{\infty} \mu(A + r_p) = \sum_{p=1}^{\infty} \mu(A) \leq 6.$$

The inequality $2 \leq \sum \mu(A)$ demands that $\mu(A) > 0$, whereas the inequality $\sum \mu(A) \leq 6$ cannot be satisfied unless $\mu(A) = 0$. Having thus arrived at a contradiction, we conclude that A was not measurable.

The final application of Theorem 39.5 to be considered here concerns a certain continuity property with which the function μ is endowed. To wit, we prove:

39.10 / Corollary Consider a nested sequence of measurable sets

$$A_1 \supset A_2 \supset A_3 \supset \cdots;$$

put

$$A = \bigcap_{p \in \mathfrak{N}} A_p.$$

Then

$$\mu(A) = \lim_{p\to\infty} \mu(A_p).$$

Proof Recognizing that

$$A_p = A \cup (A_p \setminus A)$$

for each $p \in \mathfrak{N}$, we note the relation

$$A \cap (A_p \setminus A) = \varnothing$$

and find that

$$\mu(A_p) - \mu(A) = \mu(A_p \setminus A).$$

We put $B_p = A_p \setminus A$ and see that the conclusion of the theorem is equivalent to the statement

$$\lim_{p\to\infty} \mu(B_p) = 0,$$

where

$$\bigcap_{p\in\mathfrak{N}} B_p = \varnothing.$$

Now, the sets $B_p \setminus B_{p+1}$ are mutually exclusive and since

$$B_p = (B_p \setminus B_{p+1}) \cup (B_{p+1} \setminus B_{p+2}) \cup \cdots$$

for each fixed value of p, we conclude from Theorem 39.5 that

$$\mu(B_p) = \sum_{k=p}^{\infty} \mu(B_k \setminus B_{k+1}) \qquad (p \in \mathfrak{N}).$$

But $\mu(B_p)$ may be regarded as the pth remainder of the convergent series

$$\sum_{k=1}^{\infty} \mu(B_k \setminus B_{k+1}).$$

Hence $\mu(B_p) \to 0$ as $p \to \infty$.

EXERCISES

39-1 Show that every measurable set $A \subset E$ of positive measure contains a nonmeasurable set. Where does the argument break down when it is applied to sets of measure zero?

39-2 Prove or disprove: The set $A \subset E$ is measurable if and only if for each $\epsilon > 0$, there is an open set G such that $\mu^*(G \triangle A) < \epsilon$.

39-3 How would you define the Lebesgue measure for unbounded sets?

39-4 Let r be a given number, $0 < r < 1$. Referring to Section 20, construct a nowhere dense perfect set in E with measure r.

39-5 Consider the monotonic increasing sequence

$$A_1 \subset A_2 \subset A_3 \subset \cdots \subset E$$

of measurable sets. Prove: if

$$A = \bigcup_{p \in \mathfrak{N}} A_p,$$

then

$$\mu(A) = \lim_{p \to \infty} \mu(A_p).$$

39-6 The nonempty class $\mathfrak{A}$ of sets is a *ring* if $A \in \mathfrak{A}$ and $B \in \mathfrak{A}$ implies that also $A \cup B \in \mathfrak{A}$ and $A \setminus B \in \mathfrak{A}$. If $\mathfrak{A}$ contains every countable union of its members, then $\mathfrak{A}$ is said to be a *σ-ring*.

(a) Is an algebra of sets (Exercise 2-11) a ring? a σ-ring? what about the converse?

(b) Show that the subsets of E of measure zero form a σ-ring.

(c) Show that the class of all finite and countable subsets of E is a ring.

(d) Do the nonmeasurable subsets of E constitute a ring?

40 / MEASURABLE FUNCTIONS

The characterization of continuity in Theorem 24.6 and Exercise 24-3 deserves a reexamination, for ensuing from it is the important fact that if the function f is continuous on a bounded interval I, then the four sets $I[f < \alpha]$, $I[f > \alpha]$, $I[f \leq \alpha]$, and $I[f \geq \alpha]$ are measurable for each number $\alpha \in \mathfrak{R}$. We know the converse of this statement to be false, but it was indicated in the introduction of this chapter that even an integral of an everywhere discontinuous function f is feasible when the indicated sets are measurable. Against this background we introduce the concept of a measurable function. First, however, we prove:

40.1 / Theorem Let the function $f: A \to \mathfrak{R}$ be given. If one of the sets $A[f < \alpha]$, $A[f > \alpha]$, $A[f \leq \alpha]$, or $A[f \geq \alpha]$ is measurable for some fixed number α, then so are the others.

Proof For each number α the set A can be expressed in the form

$$A = A[f < \alpha] \cup A[f \geq \alpha].$$

Owing to Corollary 38.9, therefore, the measurability of one of the two sets implies that of the other. The same reasoning applies to the pair $A[f > \alpha]$ and $A[f \leq \alpha]$. Consequently, it suffices to verify that the measurability of $A[f > \alpha]$ is both necessary and sufficient to claim the measurability of the set $A[f \geq \alpha]$.

Thus, suppose the set $A[f \geq \alpha]$ is measurable for an arbitrary $\alpha \in \mathcal{R}$. Then so is each of the sets $A\left[f \geq \alpha + \dfrac{1}{p}\right]$, where α is now fixed but p ranges over $\mathfrak{N}$. By virtue of the representation

$$A[f > \alpha] = \bigcup_{p \in \mathfrak{N}} A\left[f \geq \alpha + \frac{1}{p}\right]$$

and Theorem 39.8 also the set $A[f > \alpha]$ is measurable. The converse argument is carried out with the relation

$$A[f \geq \alpha] = \bigcap_{p \in \mathfrak{N}} A\left[f > \alpha - \frac{1}{p}\right].$$

In view of the above theorem we can lay down the following definition:

40.2 / *Definition* The function $f: A \to \mathcal{R}$ is *measurable* if and only if the sets

$$A[f < \alpha] = \{x \in A \mid f(x) < \alpha\}$$

are measurable for each fixed $\alpha \in \mathcal{R}$.

Thus, an arbitrary continuous function is automatically measurable. We also have:

40.3 / Corollary If $f: A \to \mathcal{R}$ is measurable, then each of the sets $A[f = \alpha]$ is measurable.

This result is a direct consequence of the representation

$$A[f = \alpha] = A \setminus \{A[f < \alpha] \cup A[f > \alpha]\}.$$

It is interesting to note that this corollary has no true converse.

40.4 / *Example* Consider the open interval $\mathring{E} = (0, 1)$ and a nonmeasurable set $A \subset \mathring{E}$. Let a function f be defined as follows:

$$f(x) = \begin{cases} x & (x \in A) \\ -x & (x \in \mathring{E} \setminus A). \end{cases}$$

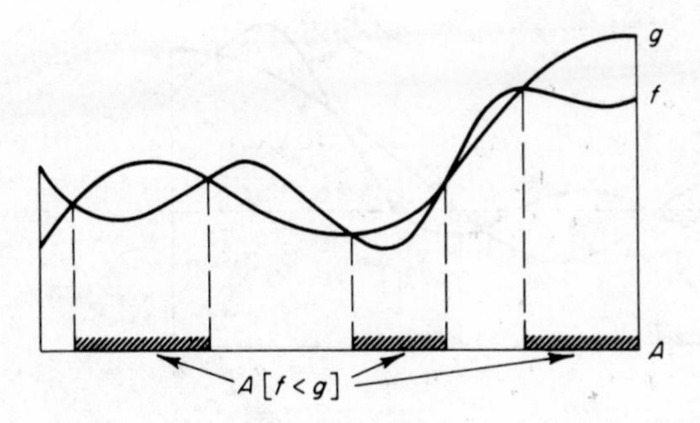

Figure 46

If α is an arbitrary number, then the set $\mathring{E}[f = \alpha]$ is either a singleton or empty, but measurable in any case. Yet the set $\mathring{E}[f > 0] = A$ is not measurable, and so neither is f.

We observe that the number α in Definition 40.2 can be regarded as the constant (measurable) function $g(x) = \alpha$. This leads us to inquire if sets such as $A[f < g]$ are measurable when both f and g are arbitrary measurable functions (see Figure 46). Indeed, we have:

40.5 / Theorem If f and g are measurable functions, then each of the sets $A[f < g]$, $A[f > g]$, $A[f \leq g]$, and $A[f \geq g]$ is measurable.

Proof The verification of all cases except the first is left as an exercise. The proof of the first case is based on the separability of $\mathcal{R}$. Namely, if $x \in A$ is fixed, then $f(x) < g(x)$ if and only if a rational number r can be found such that $f(x) < r < g(x)$. Enumerating the rational numbers in some manner, say, $r_1, r_2, r_3, \ldots$, let us consider the sets

$$A_p = A[f < r_p] \cap A[g > r_p]$$

(see Figure 47). Evidently each set A_p is measurable and since

$$A[f < g] = \bigcup_{p \in \mathfrak{N}} A_p,$$

it follows from Theorem 39.8 that the set $A[f < g]$ is measurable.

The algebraic operations which preserve continuity also preserve measurability. Thus, for example, we have:

40.6 / Theorem If the function f is measurable, then so are the functions $c \cdot f$ for an arbitrary constant c and $|f|$.

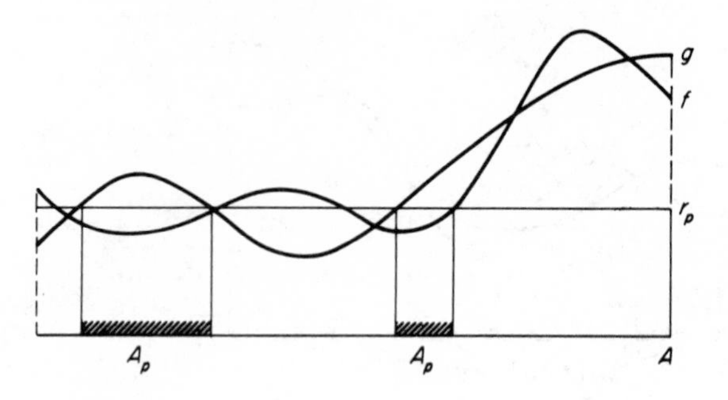

Figure 47

Proof When $c = 0$, then $c \cdot f = 0$ and hence is measurable. Otherwise, we find that $A[cf < \alpha] = A\left[f < \frac{\alpha}{c}\right]$ when $c > 0$, $A[cf < \alpha] = A\left[f > \frac{\alpha}{c}\right]$ when $c < 0$, but be this as it may all displayed sets are measurable.

To show that $|f|$ is measurable when f is, we merely observe that, for an arbitrary α,

$$A[|f| < \alpha] = A[f < \alpha] \cap A[f > -\alpha].$$

In particular, we note that the function $-f = -1 \cdot f$ is always measurable when f is measurable.

40.7 / Theorem If f and g are measurable functions, then so are the functions $f + g$, $f \cdot g$, and f/g (provided $g(x) \neq 0$ everywhere).

Proof The first case is easily disposed of with the help of Theorem 40.5 since for each number α we have

$$A[f + g < \alpha] = A[f < \alpha - g],$$

where the function $\alpha - g(x)$ is measurable by Exercise 40-8.

To verify the second assertion we write

$$f \cdot g = \tfrac{1}{4}[f + g]^2 - \tfrac{1}{4}[f - g]^2.$$

From this it is seen that we really have only to demonstrate that the measurability of a function h implies that of the function h^2. But for each number α, $\frac{|\alpha| + \alpha}{2} \geq 0$ and clearly

$$A[f^2 > \alpha] = A\left[f^2 > \frac{|\alpha| + \alpha}{2}\right]$$

$$= A\left[f > \sqrt{\frac{|\alpha| + \alpha}{2}}\right] \cup A\left[f < -\sqrt{\frac{|\alpha| + \alpha}{2}}\right]$$

Since each set in the right side is measurable, so is the set $A[f^2 > \alpha]$. The last case is again left as an exercise.

The remainder of this section is devoted to the problem of convergence of sequences of measurable functions. An important feature of measurability is its preservation under pointwise convergence (compare this with the case of continuous functions). This will be obtained as a corollary of the following proposition (see Definition 12.8):

40.8 / Theorem Let $\{f_p\}$ $(p \in \mathfrak{N})$ be a pointwise bounded sequence of measurable functions $f_p : A \to \mathfrak{R}$; for each $x \in A$ put

$$f^*(x) = \limsup_{p \to \infty} f_p(x)$$

$$f_*(x) = \liminf_{p \to \infty} f_p(x).$$

Then the functions f^* and f_* are measurable.

When the sequence converges pointwise on A, then $f^* = f_*$ and this is formally stated as:

40.9 / Corollary If the sequence $\{f_p\}$ of measurable functions converges pointwise on A to the function f, then f is measurable.

The proof of the theorem is based on the following generalization of Lemma 35.13:

40.10 / Lemma Given measurable functions $\{f_p\}$, then the functions

$$a^*(x) = \underset{p \in \mathfrak{N}}{\text{lub}} \{f_p(x)\}$$

and

$$a_*(x) = \underset{p \in \mathfrak{N}}{\text{glb}} \{f_p(x)\},$$

which are defined pointwise for each $x \in A$, are measurable.

Proof Let $\alpha \in \mathcal{R}$ be given; let $x \in A$ be arbitrary. When $a^*(x) < \alpha$, then necessarily $f_p(x) < \alpha$ for all admitted values of p. This implies that

$$A[a^* < \alpha] = \bigcap_{p \in \mathfrak{N}} A[f_p < \alpha]$$

and hence the set $A[a^* < \alpha]$ is measurable.

On the other hand $a_* < \alpha$ implies the existence of an integer n such that $f_n(x) < \alpha$. This means that

$$A[a_* < \alpha] = \bigcup_{p \in \mathfrak{N}} A[f_p < \alpha]$$

and again the desired conclusion follows.

Proof of 40.8 We now proceed to establish Theorem 40.8. If we set for each x

$$a_n^*(x) = \operatorname*{lub}_{p>n} \{ f_p(x) \},$$

then the lemma tells us that the functions $a_n^*(x)$ are measurable,

$$f^*(x) = \lim_{n \to \infty} a_n^*(x),$$

and furthermore

$$a_1^*(x) \geq a_2^*(x) \geq a_3^*(x) \geq \cdots.$$

For an arbitrary number α, therefore,

$$A[f^* < \alpha] = \bigcup_{n \in \mathfrak{N}} A[a_n^* < \alpha],$$

and it follows that f^* is as claimed.

The measurability of f_* is established in a like manner and should be carried out by the student to test his grasp of the techniques involved here.

40.11 / ***Remark*** The class of measurable functions is distinguished from the class of continuous functions in another important respect. Namely, we recall that a continuous function $f: E \to \mathcal{R}$ is uniquely determined on a dense subset of E: if g coincides with f on, say, the rational points of E and if g is continuous, then $g = f$ everywhere in E. For measurable functions this is false: such a function may be arbitrarily redefined on a set of measure zero without effecting its measurability! This is the reason for introducing equality almost everywhere: $f = g$ *almost everywhere* if the equality holds everywhere except possibly on a set of measure zero. The functions f and g are said to be *equivalent*.

EXERCISES

40-1 Complete the proof of Theorem 40.5.

40-2 Prove: If the function f is measurable and $f \neq 0$, then also the function $1/f$ is measurable.

40-3 Prove: If $f:E \to \mathcal{R}$ is measurable and $g = f$ almost everywhere (on E), then also g is measurable.

40-4 Prove: If the sequence $\{f_p\}$ of measurable functions converges pointwise to the function f almost everywhere, then f is measurable.

40-5 Let f be differentiable on E. Is the function f' measurable?

40-6 If f is monotonic and g is measurable, then show that the composite function $f \circ g$ is measurable.

40-7 Prove: If f is measurable, then so is e^f.

40-8 Show that if f is measurable, then so is $\alpha - f$ for each $\alpha \in \mathcal{R}$.

41 / THE LEBESGUE INTEGRAL OF SIMPLE FUNCTIONS

The algebra of step functions (Example 35.10) is not dense in the space of measurable functions even under pointwise convergence. A case in point is the characteristic function of the irrational numbers. A simple generalization of the concept of a step function, however, relieves this difficulty. Namely, it is appropriate now to replace the class of functions which are constant on intervals with the class of functions which are constant on measurable sets:

41.1 / *Definition* The measurable function $\phi:A \to \mathcal{R}$ is *simple* if its set of values is finite or countable.

Remark: The term *simple function* is often reserved for measurable functions whose set of values is only finite. Under these circumstances, the functions in Definition 41.1 are called *μ-simple* or *countably valued*.

Thus, every step function is simple. In fact, a glance at Definition 35.6 and Example 35.10 shows the collection $\mathfrak{S}$ of simple functions to constitute an algebra; Definitions 35.8 and 35.9 are evidently complied with since $\mathfrak{S}$ contains the step functions as a subalgebra, and the uniform closure of the algebra $\mathfrak{S}$ will be shown to coincide with the set of measurable functions. Before doing this, however, we state the following characterization of simple functions:

41.2 / Theorem A function $\varphi:A \to \mathcal{R}$ which assumes the countably many values $a_1, a_2, a_3, \ldots$ is measurable if and only if the sets $A[\varphi = a_n]$ are measurable.

That this is, indeed, a faithful description of simple functions is evident from Example 40.4 where the theorem was shown to be false for arbitrary measurable functions. The verification of this proposition is left as an exercise. We prove instead:

41.3 / Theorem The following statements are equivalent:

(**1**) f is measurable;
(**2**) f is the uniform limit of a sequence of simple functions.

Proof Suppose (1) to be the case. For each fixed integer $n \in \mathfrak{N}$ and integer p, $p = 0, \pm 1, \pm 2, \ldots$, consider the sets $A\left[\frac{p}{n} \le f < \frac{p+1}{n}\right]$.

Each of these sets is measurable and, for fixed n,

$$A = \bigcup_p A\left[\frac{p}{n} \le f < \frac{p+1}{n}\right].$$

Setting

$$f_n(x) = \frac{p}{n}, \qquad \left(x \in A\left[\frac{p}{n} \le f < \frac{p+1}{n}\right]\right)$$

it follows that

$$\| f - f_n \| \le \frac{1}{n}.$$

This shows that (1) implies (2). The converse implication requires no further justification by virtue of Corollary 40.9.

A useful representation of simple functions is obtained as follows (see Example 35.10): suppose the function $\varphi: A \to \mathfrak{R}$ assumes the distinct values $a_1, a_2, a_3, \ldots$. Then there is a partition of A into pairwise disjoint measurable sets A_p such that $\varphi(x) = a_p$ for $x \in A_p$, and the formula

$$\varphi = \sum_{p=1}^{\infty} a_p \chi_{A_p}, \tag{i}$$

where χ_{A_p} is the characteristic function of A_p, completely specifies φ. Associated with φ is the numerical series

$$\sum_{p=1}^{\infty} a_p \mu(A_p), \tag{ii}$$

and, when φ is bounded, a constant $c > 0$ can be found such that $| a_p | < c$ for all $p \in \mathfrak{N}$. The resulting relations

$$\sum_{p=1}^{\infty} | a_p \mu(A_p) | = \sum_{p=1}^{\infty} | a_p | \mu(A_p) < c \cdot \sum_{p=1}^{\infty} \mu(A_p) = c \cdot \mu(A)$$

demonstrate the absolute convergence of the series (ii), and owing to Theorem 16.7 such a series is invariant under all rearrangements of its summands. In this light it is meaningful to define an integral for bounded simple functions thusly:

41.4 / *Definition* The *Lebesgue integral* of the bounded simple function φ in (i) over the set A is the number

$$\int_A \varphi \, d\mu = \sum_{n=1}^{\infty} a_n \mu(A_n).$$

The function φ is said to be *Lebesgue integrable* over A.

In particular, for an arbitrary measurable set $A \subset E$,

$$\int_A \chi_A \, d\mu = \mu(A);$$

the integral over a nonmeasurable set A is excluded *a priori.*

Reexamining the right side in (i) we notice that the function φ may have different representations when the stipulation that the a_n be distinct is replaced with the weaker requirement that the A_n are pairwise disjoint. It is rather important to know that the integral in Definition 41.4 is independent of the representation of φ. This is the subject of the next lemma:

41.5 / Lemma Let φ be specified by (i). If

$$\varphi = \sum_{n=1}^{\infty} b_n \chi_{B_n} \tag{iii}$$

is another representation of φ, where $A = \bigcup B_n$ and the B_n are mutually exclusive, then

$$\sum_{n=1}^{\infty} a_n \mu(A_n) = \sum_{n=1}^{\infty} b_n \mu(B_n).$$

Proof Holding n momentarily fixed, consider all values n_ι for which $b_{n_\iota} = a_n$. Evidently

$$A_n = \bigcup_\iota B_{n_\iota}$$

and as a result

$$\sum_{n=1}^{\infty} a_n \mu(A_n) = \sum_{n=1}^{\infty} a_n \sum_\iota \mu(B_{n_\iota}) = \sum_{n=1}^{\infty} \sum_\iota a_n \mu(B_{n_\iota}) = \sum_{n=1}^{\infty} \sum_\iota b_{n_\iota} \mu(B_{n_\iota}).$$

But the last series merely represents a rearrangement of the series in the right side of (iii) and an application of Theorem 16.7 shows the lemma to be true.

It is clear that the boundedness restriction present in Definition 41.4 is only incidental whereas the absolute convergence of the series is crucial. In order to extend the Lebesgue integral to unbounded simple functions we introduce:

41.6 / *Definition* The simple function $\varphi = \sum a_n \chi_{A_n}$ is *summable* when the series $\sum a_n \mu(A_n)$ converges absolutely.

41.7 / *Examples* Consider the situation $A = \bigcup A_n$ where

$$A_n = \left(\frac{n}{n+1}, \frac{n+1}{n+2}\right) \qquad (n \in \mathfrak{N}).$$

For each value of n,

$$\mu(A_n) = \frac{1}{(n+1)(n+2)}$$

and if we let $a_n = (-1)^n \sqrt{n}$, then

$$\sum_{n=1}^{\infty} |a_n| \mu(A_n) = \sum_{n=1}^{\infty} \frac{\sqrt{n}}{(n+1)(n+2)} < \sum_{n=1}^{\infty} n^{-3/2}.$$

The function

$$\varphi = \sum_{n=1}^{\infty} (-1)^n \sqrt{n} \chi_{A_n},$$

therefore, is summable.

The series

$$\sum_{n=1}^{\infty} \frac{(-1)^n n}{(n+1)(n+2)},$$

on the other hand, converges nonabsolutely, so that the associated simple function

$$\psi = \sum_{n=1}^{\infty} (-1)^n n \chi_{A_n}$$

is not summable.

As a final illustration consider the function

$$\eta = \sum_{n=1}^{\infty} (n+1)(n+2) \chi_{A_n}.$$

It is not summable and, in fact, the series

$$\sum_{n=1}^{\infty} (n+1)(n+2)\mu(A_n) = \sum_{n=1}^{\infty} 1$$

diverges to $+\infty$ irrespective of rearrangements.

A natural outcome of our reasoning is:

41.8 / ***Definition*** The Lebesgue integral of the summable simple function $\varphi = \sum a_n \chi_{A_n}$ over the set A is

$$\int_A \varphi \, d\mu = \sum_{n=1}^{\infty} a_n \mu(A_n).$$

Henceforth summable functions will be called *integrable* (in the Lebesgue sense).

Immediate consequences of the above definitions are these:

41.9 / Theorem If φ and ψ are integrable simple functions on A into $\mathfrak{R}$, then

(1) $\left| \int_A \varphi \, d\mu \right| \leq \alpha \mu(A)$ when $\| \varphi \| \leq \alpha$;

(2) $\int_A c \cdot \varphi \, d\mu = c \int_A \varphi \, d\mu$ for each constant c;

(3) $\int_A (\varphi + \psi) \, d\mu = \int_A \varphi \, d\mu + \int_A \psi \, d\mu.$

Proof No proof of the first two assertions is thought necessary. To establish the claim in (3), suppose

$$\varphi = \sum a_p \chi_{A_p}, \qquad \psi = \sum b_p \chi_{B_p}$$

where the constants $\{a_p\}$ as well as $\{b_p\}$ are all distinct, $A = \bigcup A_p = \bigcup B_p$. If we set

$$C_{pn} = A_p \cap B_n,$$

then

$$A = \bigcup_{p,n \in \mathfrak{N}} C_{pn}$$

and clearly

$$\mu(A_p) = \sum_{n=1}^{\infty} \mu(C_{pn}) \qquad (p \in \mathfrak{N}),$$

$$\mu(B_n) = \sum_{p=1}^{\infty} \mu(C_{pn}) \qquad (n \in \mathfrak{N}).$$

Owing to the fact that

$$(\varphi + \psi)(x) = a_p + b_n \qquad \text{for} \quad x \in C_{pn},$$

it follows from Lemma 41.5 that

$$\int_A (\varphi + \psi)\, d\mu = \sum_{p=1}^{\infty} \sum_{n=1}^{\infty} (a_p + b_n)\mu(C_{pn})$$

while at the same time the lemma and Theorem 16.7 tell us that

$$\int_A \varphi\, d\mu = \sum_{p=1}^{\infty} a_p\mu(A_p) = \sum_{p=1}^{\infty} a_p \sum_{n=1}^{\infty} \mu(C_{pn}) = \sum_{p=1}^{\infty} \sum_{n=1}^{\infty} a_p\mu(C_{pn})$$

and similarly

$$\int_A \psi\, d\mu = \sum_{n=1}^{\infty} b_n\mu(B_n) = \sum_{n=1}^{\infty} \sum_{p=1}^{\infty} b_n\mu(C_{pn}) = \sum_{p=1}^{\infty} \sum_{n=1}^{\infty} b_n\mu(C_{pn}).$$

The formula in (3) is now seen to be true.

41.10 / ***Remark*** When, as in Example 41.7, every series $\sum a_p(A_p)$ associated with a given simple function φ is strictly divergent to $+\infty$ or $-\infty$ (which is to say that every rearrangement diverges to $+\infty$ or $-\infty$), then we assign to the integral of φ the value $+\infty$ or $-\infty$, as the case may be.

EXERCISES

41-1 Prove Theorem 41.2

41-2 Show that the simple functions do, indeed, form an algebra.

41-3 Consider the Cantor set $\mathcal{P} \subset [0, 1]$ (refer to the construction in 20.2). Let the function f be defined as follows: $f(x) = 0$ for $x \in \mathcal{P}$, and on the 2^{n-1} open intervals discarded in the nth stage of the construction we set $f(x) = n$.

(a) Show that the function f is integrable and simple.

(b) Compute the integral

$$\int_{[0,1]} f\, d\mu.$$

41-4 If φ and ψ are equivalent integrable simple functions on A, then show that

$$\int_A \varphi \, d\mu = \int_A \psi \, d\mu.$$

41-5 If $\varphi \geq 0$ is an integrable simple function, then show that

$$\int_A \varphi \, d\mu = 0$$

if and only if φ vanishes almost everywhere.

41-6 The inequality $\varphi(x) \leq \psi(x)$ $(x \in A)$ implies that

$$\int_A \varphi \, d\mu \leq \int_A \psi \, d\mu.$$

for all admitted functions φ and ψ. Prove this fact.

41-7 Show that the inequality

$$\left| \int_A \varphi \, d\mu \right| \leq \int_A | \varphi | \, d\mu$$

holds for each integrable simple function φ.

41-8 An examination of the proof of Theorem 41.3 reveals that a bounded measurable function is the limit of a sequence of simple functions each of which assumes only finitely many values. Use Theorem 38.11 to prove that a bounded measurable function is the uniform limit almost everywhere of a sequence of step functions.

42 / THE LEBESGUE INTEGRAL

Consider an arbitrary uniformly convergent sequence $\{\varphi_n\}$ of integrable simple functions. This sequence gives rise to the numerical sequence of integrals

$$\int_A \varphi_n \, d\mu.$$

Regarding this sequence we have:

42.1 / Lemma The uniform convergence of the sequence of integrable simple functions $\varphi_n : A \to \mathfrak{R}$ carries with it the convergence of the sequence $\left\{ \int_A \varphi_n \, d\mu \right\}$.

Proof For each $\epsilon > 0$ there is an integer s such that $\|\varphi_n - \varphi_p\| < \epsilon$ whenever $n, p > s$. From Theorem 41.9 we now deduce that

$$\left| \int_A \varphi_n \, d\mu - \int_A \varphi_p \, d\mu \right| = \left| \int_A (\varphi_n - \varphi_p) \, d\mu \right| \leq \epsilon \cdot \mu(A)$$

for all integers $n, p > s$, and the lemma ensues.

We further establish:

42.2 / Lemma If $\{\varphi_n\}$ and $\{\psi_n\}$ are equivalent uniformly convergent sequences of simple integrable functions over A, then

$$\lim_{n \to \infty} \int_A \varphi_n \, d\mu = \lim_{n \to \infty} \int_A \psi_n \, d\mu.$$

Proof Arguing as in the preceding proof, we may replace φ_p by ψ_n to arrive at an inequality

$$\left| \int_A \varphi_n \, d\mu - \int_A \psi_n \, d\mu \right| < \epsilon \cdot \mu(A) \qquad (n > s),$$

from which the lemma is deduced.

These two lemmas in conjunction with Theorem 41.3 conveniently lead to the next definition:

42.3 / *Definition* The function $f: A \to \mathcal{R}$ is *Lebesgue integrable* over the set A if it is the uniform limit of a sequence $\{\varphi_n\}$ of integrable simple functions, and we set

$$\int_A f \, d\mu = \lim_{n \to \infty} \int_A \varphi_n \, d\mu.$$

The function f will also be said to be *integrable in the Lebesgue sense* or simply *integrable.* The class of integrable functions on A is designated with the symbol $\mathfrak{L}[A]$.

Looking back, it is clear that the new definition of integrability coincides with the old one as far as simple functions are concerned, and the fundamental properties of the Lebesgue integral for simple functions remain intact also in the general case. The most significant outcome of our procedure is the automatic inclusion of all bounded measurable functions in the class of Lebesgue integrable functions:

42.4 / Theorem Every bounded measurable function on A is integrable over A.

A casual inspection of the proof of Theorem 41.3 shows that a bounded measurable function is the uniform limit of a uniformly bounded sequence of simple functions. But a bounded simple function is integrable and the theorem follows.

We formally summarize the basic properties of the Lebesgue integral in the following theorem:

42.5 / Theorem Let f and g be in $\mathfrak{L}[A]$. Then

(1) $\int_A c\cdot f\,d\mu = c\cdot\int_A f\,d\mu$ for each constant c;

(2) $\int_A (f+g)\,d\mu = \int_A f\,d\mu + \int_A g\,d\mu$;

(3) $\int_A f\,d\mu \geq 0$ when $f \geq 0$;

(4) $\int_A f\,d\mu \geq \int_A g\,d\mu$ when $f \geq g$;

(5) if $m_1 \leq f \leq m_2$, then $m_1\mu(A) \leq \int_A f\,d\mu \leq m_2\mu(A)$.

We merely indicate the proof of the theorem: The assertion made in (1) is verified as follows: Let $\{\varphi_n\}$ be a sequence of integrable simple functions which converges uniformly to f. Then the sequence $\{c\cdot\varphi_n\}$ converges uniformly to $c\cdot f$ and from Theorem 41.9(2) we conclude that

$$\int_A c\cdot f\,d\mu = \lim_{n\to\infty}\int_A c\cdot\varphi_n\,d\mu = c\cdot\lim_{n\to\infty}\int_A \varphi_n\,d\mu = c\cdot\int_A f\,d\mu.$$

The equality in (2) is verified in the same way. The inequality in (3) is clearly true for nonnegative simple functions, and the general case is deduced from Theorem 41.3, for it follows from the construction there that a nonnegative measurable function can be uniformly approximated with nonnegative simple functions. The claim made in (4) is an immediate corollary of (2) and (3), whereas the inequalities in (5) are obtained from a simple application of (1) and (4).

The Lebesgue integral has a very useful interpretation as a set function: An example of a set function is furnished by the Lebesgue measure which is defined on the bounded subsets of $\mathfrak{R}$. In general, given a family $\mathfrak{F}$ of sets, then a *set function* σ is simply a mapping $\sigma:\mathfrak{F}\to\mathfrak{R}$. If, for a fixed function $f \in \mathfrak{L}[A]$ we set

$$\sigma_f(B) = \int_B f \, d\mu,$$

where $B \subset A$ is measurable, then σ_f is a set function on the measurable subsets of A. The verification of this statement hinges on the simple observation that if we set

$$g(x) = \begin{cases} f(x) & (x \in B) \\ 0 & (x \in A \setminus B), \end{cases}$$

then g is integrable over A, and moreover,

$$\int_A g \, d\mu = \int_B g \, d\mu = \int_B f \, d\mu.$$

That is, we have:

42.6 / Theorem If f is integrable over the set A, then f is integrable over every measurable subset $B \subset A$.

Pursuing this line of thought we observe that if f is the characteristic function of the set A, then $\sigma_f(B) = \mu(B)$ for each admitted $B \subset A$, so that σ_f is completely additive. An intriguing question is, therefore, if σ_f is a measure for an arbitrary positive function $f \in \mathfrak{L}[A]$. When cast against this background, the ensuing proposition is seen to be very fundamental and far reaching. To wit, the complete additivity, which was among the outstanding properties of the Lebesgue measure, carries over to σ_f for each $f \in \mathfrak{L}[A]$.

42.7 / Theorem The Lebesgue integral is completely additive: Let $f \in \mathfrak{L}[A]$ and suppose

$$A = \bigcup_{n \in \mathfrak{N}} A_n$$

where each set A_n is measurable and

$$A_n \cap A_p = \varnothing \qquad (n \neq p).$$

Then

$$\int_A f \, d\mu = \sum_{n=1}^{\infty} \int_{A_n} f \, d\mu. \tag{i}$$

It should be pointed out that the conclusion of the theorem implies the absolute convergence of the series in the right side of (i) to the proper value.

Proof Let the theorem be first verified for the case of a simple function f, say, $f = \sum b_t \chi_{B_t}$, and we might as well assume that the numbers b_t are all distinct. The argument is modeled after that employed in the proof of Lemma 41.5: Setting $C_{nt} = A_n \cap B_t$, we find that

$$\int_A f\,d\mu = \sum_{t=1}^{\infty} b_t\mu(B_t) = \sum_{t=1}^{\infty} b_t \sum_{n=1}^{\infty} \mu(C_{nt}) = \sum_{n=1}^{\infty}\sum_{t=1}^{\infty} b_t\mu(C_{nt}), \tag{ii}$$

where each of the series is absolutely convergent. For each value of n, however,

$$A_n = \bigcup_t C_{nt},$$

and the formula

$$\int_{A_n} f\,d\mu = \sum_{t=1}^{\infty} b_t\mu(C_{nt})$$

is provided by Lemma 41.5. A substitution in the right side of (ii) yields the relation in (i) for the stipulated function f.

Regretably, the general case is more difficult to establish because the absolute convergence of the series in (i) is not as apparent as in the instance above. Now, since $f \in \mathfrak{L}[A]$, there is a sequence of simple functions $\varphi_t \in \mathfrak{L}[A]$ whose uniform limit is f. Owing to the preceding step

$$\int_A \varphi_t\,d\mu = \sum_{n=1}^{\infty} \int_{A_n} \varphi_t\,d\mu \tag{iii}$$

for each $t \in \mathfrak{N}$: Theorem 42.6 and Definition 42.3, when applied to the integral over a fixed set A_n, show that

$$\int_{A_n} f\,d\mu = \lim_{t\to\infty} \int_{A_n} \varphi_t\,d\mu.$$

Given an arbitrary $\epsilon > 0$, then owing to the uniform convergence of the sequence φ_t to f there is an integer k such that

$$\|f - \varphi_k\| < \epsilon$$

and accordingly

$$\left| \int_{A_n} f\,d\mu - \int_{A_n} \varphi_k\,d\mu \right| < \epsilon\mu(A_n).$$

The estimate

$$\sum_{n=1}^{\infty}\left|\int_{A_n} f\,d\mu\right| \le \sum_{n=1}^{\infty}\left|\int_{A_n} f\,d\mu - \int_{A_n}\varphi_k\,d\mu\right| + \sum_{n=1}^{\infty}\left|\int_{A_n}\varphi_k\,d\mu\right|$$

$$\le \sum_{n=1}^{\infty}\epsilon\mu(A_n) + \sum_{n=1}^{\infty}\left|\int_{A_n}\varphi_k\,d\mu\right| = \epsilon\mu(A) + \sum_{n=1}^{\infty}\left|\int_{A_n}\varphi_k\,d\mu\right|$$

shows that the series

$$\sum_{n=1}^{\infty}\int_{A_n} f\,d\mu$$

converges absolutely. Concurrently we also have the inequalities

$$\left|\sum_{n=1}^{\infty}\int_{A_n} f\,d\mu - \int_A f\,d\mu\right|$$

$$\le \left|\sum_{n=1}^{\infty}\int_{A_n} f\,d\mu - \sum_{n=1}^{\infty}\int_{A_n}\varphi_k\,d\mu\right| + \left|\int_A \varphi_k\,d\mu - \int_A f\,d\mu\right|$$

$$\le \left|\sum_{n=1}^{\infty}\int_{A_n}(f-\varphi_k)\,d\mu\right| + \left|\int_A(\varphi_k - f)\,d\mu\right|$$

$$\le \sum_{n=1}^{\infty}\epsilon\mu(A_n) + \epsilon\mu(A) = 2\epsilon\mu(A),$$

the theorem now following from the fact that $\epsilon > 0$ was arbitrary.

The following statement needs no further clarification:

42.8 / Corollary If f is integrable over the set A, and if the set $B \subset A$ is such that $\mu(A \setminus B) = 0$, then

$$\int_A f\,d\mu = \int_B f\,d\mu.$$

Another important property of the Lebesgue integral is formulated in the following theorem:

42.9 / Theorem Let f be measurable over A. Then $f \in \mathfrak{L}[A]$ if and only if $|f| \in \mathfrak{L}[A]$.

Proof Suppose $f \in \mathfrak{L}[A]$. Let $\{\varphi_n\}$ be a sequence of simple functions in $\mathfrak{L}[A]$ which converges uniformly to f. By virtue of the inequality

$$\big||\varphi_n(x)| - |f(x)|\big| \le |\varphi_n(x) - f(x)|,$$

which holds for each $x \in A$, it follows that $|f|$ is the uniform limit of the sequence $\{|\varphi_n|\}$. Since the functions $|\varphi_n|$ are in $\mathfrak{L}[A]$, so is $|f|$.

Conversely, let $|f| \in \mathfrak{L}[A]$. Consider the sets $A_1 = A[f < 0]$ and $A_2 = A[f \geq 0]$. Then $A = A_1 \cup A_2$, $A_1 \cap A_2 = \emptyset$ and owing to Theorem 42.7,

$$\int_A |f| \, d\mu = \int_{A_1} |f| \, d\mu + \int_{A_2} |f| \, d\mu = \int_{A_1} -f \, d\mu + \int_{A_2} f \, d\mu.$$

According to Theorem 42.5(1),

$$\int_{A_1} -f \, d\mu = -\int_{A_1} f \, d\mu,$$

and since f itself is thus seen to be integrable over the sets A_1 and A_2, it follows that $f \in \mathfrak{L}[A]$, as was to be shown.

Theorem 42.5(4) can now be used in conjunction with the inequalities $f \leq |f|$ and $-f \leq |f|$ to yield:

42.10 / Corollary For an arbitrary function $f \in \mathfrak{L}[A]$,

$$\left| \int_A f \, d\mu \right| \leq \int_A |f| \, d\mu.$$

EXERCISES

42-1 Let $g \in \mathfrak{L}[A]$. If f is measurable over A and $|f| \leq g$, then show that $f \in \mathfrak{L}[A]$, also.

42-2 Show that if $f \in \mathfrak{L}[A]$ and $g \in \mathfrak{L}[A]$, then also $f \cdot g \in \mathfrak{L}[A]$.

42-3 If $g: A \to A$ is continuous and $f \in \mathfrak{L}[A]$, does it follow that $f \circ g \in \mathfrak{L}[A]$?

42-4 Prove the following *Chebyshev inequality*: If $g \in \mathfrak{L}[A]$ and $g \geq 0$, then for each number α,

$$\mu(A[g \geq \alpha]) \leq \frac{1}{\alpha} \int_A g \, d\mu.$$

42-5 Prove or disprove: If $f \in \mathfrak{L}[A]$, then

$$\int_A |f| \, d\mu = 0$$

if and only if $f = 0$ almost everywhere.

42-6 Is the function $f(x) = 1/x$ Lebesgue integrable on $(0, 1)$?

42-7 Prove or disprove: If $f \in \mathfrak{L}[A]$ and $A = A_1 \cup A_2$, where A_1 and A_2 are measurable, then

$$\int_{A_1} f\,d\mu + \int_{A_2} f\,d\mu = \int_A f\,d\mu + \int_{A_1 \cap A_2} f\,d\mu.$$

42-8 A set function $\varphi(B)$ defined on a class of measurable sets $B \subset A$ is *absolutely continuous* if for each $\epsilon > 0$ there is a $\delta > 0$ with the following property: If $A_1, \ldots, A_n$ is an arbitrary family of mutually exclusive subsets of A such that

$$\sum_{n=1}^{n} \mu(A_n) < \delta,$$

then

$$\sum_{n=1}^{n} | \varphi(A_n) | < \epsilon.$$

(**1**) Prove or disprove the absolute continuity of the set function σ_f, where $f \in \mathfrak{L}[A]$ is arbitrary.

(**2**) Can you characterize the family of set functions $\sigma_{|f|}$ $(f \in \mathfrak{L}[A])$?

42-9 Verify the equivalence of the following statements:

(a) The set function $\varphi(B)$ is absolutely continuous;

(b) For each $\epsilon > 0$ a $\delta > 0$ can be found such that if $\{A_n\}$ $(n \in \mathfrak{N})$ is a family of pairwise disjoint sets in A for which

$$\sum_{n=1}^{\infty} \mu(A_n) < \delta,$$

then

$$\sum_{n=1}^{\infty} | \varphi(A_n) | < \epsilon.$$

42-10 Given a function $f: A \to \mathfrak{R}$, associate with it the functions

$$f^+(x) = \max(f(x), 0) \qquad \text{and} \qquad f^-(x) = -\min(f(x), 0).$$

Prove that $f \in \mathfrak{L}[A]$ implies that $f^+ \in \mathfrak{L}[A]$ and $f^- \in \mathfrak{L}[A]$. Is the converse true?

42-11 Let us agree to call the measurable function *f-summable* when $f^+ \in \mathfrak{L}[A]$ and $f^- \in \mathfrak{L}[A]$; let us consent further to define the integral of f by the formula

$$\int_A f\,d\mu = \int_A f^+\,d\mu - \int_A f^-\,d\mu.$$

How does this definition compare with Definition 42.3? In particular, are the integrals equivalent?

42-12 If we call f integrable when at least *one* of the integrals $\int_A f^+ \, d\mu$ or $\int_A f^- \, d\mu$ is finite, and if we define the integral of f with the formula

$$\int_A f \, d\mu = \int_A f^+ \, d\mu - \int_A f^- \, d\mu,$$

does it then follow that every function which is integrable in the sense of Definition 42.3 or Exercise 42-11 is integrable in the new sense? What about the converse?

42-13 Let $A = [a, b]$. The function $f: A \to \mathcal{R}$ is of *bounded variation* if there is a constant α such that for an arbitrary partition p: $a = a_0 < a_1 < a_2 < \cdots < a_n = b$ we have the inequality

$$v(p) = \sum_{k=1}^{n} | f(a_k) - f(a_{k-1}) | \leq \alpha.$$

The *total variation* of f is the number

$$T(a, b) = \operatorname*{lub}_{p} \{v(p)\},$$

where the least upper bound is taken over all partitions of A. Show that if f is of bounded variation, then f is Lebesgue integrable. (*Hint*: For each $x \in A$ let the variation of f on the interval $[a, x]$ be $T(a, x)$. Write

$$f(x) = T(a, x) + [f(x) - T(a, x)],$$

and show that each of the functions $T(a, x)$ and $f(x) - T(a, x)$ is nondecreasing.)

43 / THEOREMS ON LIMITS UNDER THE INTEGRAL SIGN

Many problems involving integration require in the course of their solution the evaluation of a limit under the integral sign. Already the definition of the integral of simple functions entails a limit, namely, that of an absolutely convergent series. The formula in Definition 42.3, therefore, intrinsically implies the interchangeability of limits in the case of a uniformly convergent sequence of integrable simple functions:

$$\int_A \lim_{n\to\infty} \varphi_n \, d\mu = \lim_{n\to\infty} \int_A \varphi_n \, d\mu.$$

The status of this recipe in the general case, when the φ_n are no longer

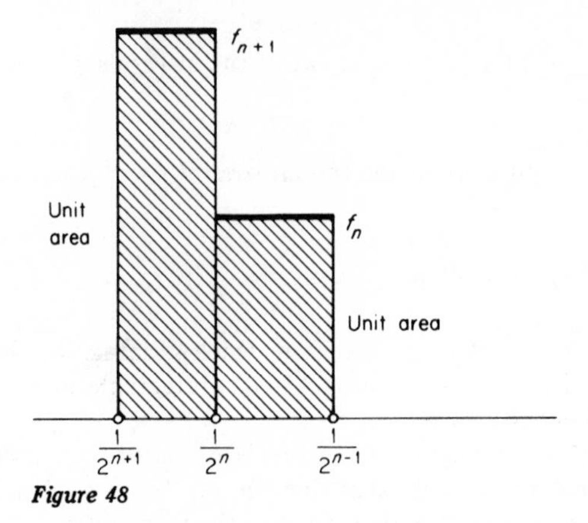

Figure 48

simple and when the limit is not uniform, is going to be the dominant theme in this section. That we do have, indeed, a problem on our hands follows from the examples:

43.1 / ***Examples***

(**1**) In the interval $E = (0, 1)$ let there be given the subintervals

$E_n = \left(\frac{1}{2^n}, \frac{1}{2^{n-1}}\right)$ for $n \in \mathfrak{N}$. Consider the functions

$$f_n(x) = \begin{cases} 2^n & (x \in E_n) \\ 0 & (x \in E \setminus E_n) \end{cases}$$

(see Figure 48) : each function f_n is integrable and simple. In fact,

$$\int_E f_n \, d\mu = 2^n \int_{E_n} 1 \cdot d\mu = 2^n \frac{1}{2^n} = 1$$

for each admitted value of n, so that

$$\lim_{n \to \infty} \int_E f_n \, d\mu = 1.$$

At the same time, however, the sequence $\{ f_n\}$ converges pointwise on A to the limit function $f = 0$, and accordingly

$$\int_E \lim_{n \to \infty} f_n \, d\mu = \int_E 0 \cdot d\mu = 0.$$

(**2**) Consider the function $g(x) = x$ on $A = [-1, 1]$. Setting

$g_{2n-1} = g$, $g_{2n} = -g$ $(n \in \mathfrak{N})$, it follows that

$$\int_A g_t \, d\mu = 0 \qquad (t \in \mathfrak{N})$$

and hence also

$$\lim_{t\to\infty} \int_A g_t \, d\mu = 0.$$

Yet the sequence $\{g_t\}$ has no limit!

We now address ourselves to proving one of the most significant results in this integration theory:

43.2 / Lebesgue's Dominated Convergence Theorem Consider a sequence of functions $f_n \in \mathfrak{L}[A]$ which converges pointwise on A to the function f. If there exists a function $g \in \mathfrak{L}[A]$ such that

$$|f_n(x)| \leq g(x) \qquad (x \in A; \;\; n \in \mathfrak{N}), \tag{i}$$

then also $f \in \mathfrak{L}[A]$ and furthermore

$$\lim_{n\to\infty} \int_A f_n \, d\mu = \int_A f \, d\mu.$$

Proof Two properties of f stand out from the beginning: Owing to Corollary 40.9 the function is measurable, and by virtue of (i),

$$|f(x)| \leq g(x) \qquad (x \in A): \tag{ii}$$

Exercise 42-1 therefore tells us that $f \in \mathfrak{L}[A]$. The second assertion in the theorem is not come by as easily.

Let g be as demanded. Regarding the sets

$$A_t = A[t - 1 \leq g < t] \qquad (t \in \mathfrak{N}), \tag{iii}$$

Theorem 42.7 can be called on to obtain the representation

$$\int_A g \, d\mu = \sum_{t=1}^{\infty} \int_{A_t} g \, d\mu,$$

where the series in the right side converges absolutely. Let an arbitrary number $\epsilon > 0$ be fixed. Then an integer k can be found such that

$$\sum_{t=k+1}^{\infty} \int_{A_t} g \, d\mu < \frac{\epsilon}{5}.$$

With the notation

$$A^k = \bigcup_{t=k+1}^{\infty} A_t \tag{iv}$$

this estimate is expressible as

$$\int_{A^k} g \, d\mu < \frac{\epsilon}{5}.$$

The estimates

$$\int_{A^k} |f_n| \, d\mu < \frac{\epsilon}{5} \qquad (n \in \mathfrak{N}) \tag{v}$$

and

$$\int_{A^k} |f| \, d\mu < \frac{\epsilon}{5} \tag{vi}$$

are now obtained with the help of (i), (ii), and Theorem 42.5(4). These inequalities will be used later.

With $\epsilon > 0$ and k still fixed, let us proceed now to obtain an estimate on the differences $|f_n(x) - f(x)|$. For this purpose we introduce the sets

$$A'_m = A\left[|f_n - f| < \frac{\epsilon}{5\mu(A)}\right] \qquad (n > m). \tag{vii}$$

These sets are nested,

$$A'_1 \subset A'_2 \subset A'_3 \subset \cdots$$

and are such that

$$A = \bigcup_{m \in \mathfrak{N}} A'_m.$$

But these are just the conditions needed in order to apply Exercise 39-5 to the situation at hand. The resulting relation,

$$\lim_{m \to \infty} \mu(A'_m) = \mu(A),$$

tells us that an integer m exists such that

$$\mu(A \setminus A'_m) = \mu(A) - \mu(A'_m) < \frac{\epsilon}{5k}. \tag{viii}$$

The machinery for the final argument is now available. On the set

$A \setminus A^k$ the relation $g(x) < k$ is seen to hold (see (iii) and (iv)); by (i) and (ii), therefore,

$$| f(x) | < k \quad \text{and} \quad | f_m(x) | < k \quad (x \in A \setminus A^k). \tag{ix}$$

Because of (vii) and (viii), the set $A \setminus A^k$ can be expressed in the form $A \setminus A^k = B_1 \cup B_2$, where $B_1 \cap B_2 = \varnothing$,

$$\mu(B_1) < \frac{\epsilon}{5k}, \tag{x}$$

and

$$| f_m(x) - f(x) | < \frac{\epsilon}{5\mu(A)} \quad (x \in B_2). \tag{xi}$$

Writing

$$A = A^k \cup (A \setminus A^k) = A^k \cup B_1 \cup B_2,$$

it is evident that

$$\left| \int_A f_m \, d\mu - \int_A f \, d\mu \right| \leq \int_{A^k} | f_m | \, d\mu + \int_{A^k} | f | \, d\mu + \int_{B_1} | f_m | \, d\mu$$
$$+ \int_{B_1} | f | \, d\mu + \int_{B2} | f_m - f | \, d\mu \leq 5 \cdot \frac{\epsilon}{5} = \epsilon,$$

where we have used, in turn, (v), (vi), and the relations (ix)–(xi) in conjunction with Theorem 42.5(5). The theorem is now seen to be true.

The preceding theorem can be sharpened in two ways. The first involves an application of Corollary 42.8.

43.3 / Corollary The conclusion of Theorem 43.2 remains valid when $f_n(x) \to f(x)$ and $| f_n(x) | \leq g(x)$ hold almost everywhere on A.

The second generalization is this:

43.4 / Corollary Under the conditions of Theorem 43.2,

$$\lim_{n \to \infty} \int_A | f_n - f | \, d\mu = 0.$$

We now prove:

43.5 / Lebesgue's Monotone Convergence Theorem Imagine a sequence of functions $f_n \in \mathfrak{L}[A]$ such that

(a) $f_1(x) \leq f_2(x) \leq f_3(x) \leq \cdots \qquad (x \in A)$;
(b) there is a constant c with the property that

$$\int_A f_n \, d\mu \leq c \qquad (n \in \mathfrak{N}).$$

Then the pointwise limit

$$f(x) = \lim_{n\to\infty} f_n(x)$$

exists almost everywhere, $f \in \mathfrak{L}[A]$, and furthermore

$$\lim_{n\to\infty} \int_A f_n \, d\mu = \int_A f \, d\mu.$$

Proof First, it may be assumed without loss of generality that $f_1 \geq 0$; otherwise we could introduce the functions $g_n = f_n - f_1$ which have this property, and it is clear that the validity of the theorem for the functions g_n implies the original theorem.

Next, let it be shown that the function f is finite almost everywhere. For this, consider for each pair n, t of positive integers the set

$$A_n^t = A[f_n > t].$$

Owing to (a),

$$A_1^t \subset A_2^t \subset A_3^t \subset \cdots \qquad (t \in \mathfrak{N}), \tag{xii}$$

and by virtue of (b) and Exercise 42-4,

$$\mu(A_n^t) \leq \frac{c}{t} \qquad (n, t \in \mathfrak{N}). \tag{xiii}$$

Since the estimate in (xiii) is independent of n, it follows from (xii) that

$$\mu(\bigcup_{n\in\mathfrak{N}} A_n^t) \leq \frac{c}{t}. \tag{xiv}$$

Now let $x \in A$ be such that $f(x) = +\infty$. Let $t \in \mathfrak{N}$ be arbitrary. Then there is an integer k such that $f_k(x) > t$. Accordingly, $x \in \bigcup_{n\in\mathfrak{N}} A_n^t$ and hence

$$A[f = +\infty] \subset \bigcup_{n\in\mathfrak{N}} A_n^t.$$

In view of (xiv),

$$\mu(A[f = +\infty]) \leq \frac{c}{t}$$

and since t was arbitrary,

$$\mu(A[f = +\infty]) = 0.$$

To show that $f \in \mathfrak{L}[A]$, consider the sets

$$A_t = A[t - 1 \leq f < t] \qquad (t \in \mathfrak{N}).$$

Let the function g be defined as follows:

$$g(x) = t \qquad \text{for} \quad x \in A_t.$$

Since each set A_t is measurable, we conclude that g is simple, and moreover, g dominates f:

$$|f(x)| \leq g(x) \qquad (x \in A).$$

In view of Exercise 42-1 it suffices to show that $g \in \mathfrak{L}[A]$. To accomplish this it is being observed that the functions f_n and f are bounded on each set

$$B_m = \bigcup_{t=1}^{m} A_t.$$

According to the definition of g,

$$g(x) \leq f(x) + 1 \qquad (x \in B_m)$$

and hence

$$\sum_{t=1}^{m} t \cdot \mu(A_t) = \int_{B_m} g \, d\mu \leq \int_{B_m} (f + 1) \, d\mu$$

$$\leq \lim_{m \to \infty} \int_{B_m} f \, d\mu + \mu(A) \leq c + \mu(A),$$

but this implies that the series

$$\sum_{t=1}^{\infty} t \cdot \mu(A_t) = \int_A g \, d\mu$$

converges (absolutely). Hence f is integrable, and since

$$f_n(x) \leq f(x) \qquad (n \in \mathfrak{N})$$

we apply Theorem 43.2 to complete the proof.

From this theorem follows:

43.6 / Beppo Levi's Theorem If the sequence $\{f_n\}$ of nonnegative functions in $\mathfrak{L}[A]$ is such that

$$\sum_{n=1}^{\infty} \int_A f_n \, d\mu \leq c$$

for some constant c, then the series $\sum f_n$ converges almost everywhere on A, and moreover

$$\sum_{n=1}^{\infty} \int_A f_n \, d\mu = \int_A \left(\sum_{n=1}^{\infty} f_n \right) d\mu.$$

An equivalent version of this theorem is stated in terms of nondecreasing sequences of integrable functions whose integrals have a common bound.

EXERCISES

43-1 Carry out the proof of Theorem 43.6.

43-2 Derive the following theorem of Fatou from Theorem 43.5: If the sequence $\{f_n\}$ of nonnegative integrable functions converges almost everywhere on A to the function f, and if there is a constant c such that

$$\int_A f_n \, d\mu \leq c,$$

then $f \in \mathfrak{L}[A]$, and

$$\int_A f \, d\mu \leq c.$$

(*Hint*: For each $n \in \mathfrak{N}$ consider the functions $g_n = \text{glb}\{f_k \mid k \geq n\}$).

44 / THE RIEMANN INTEGRAL

Because of the prevailing status of the Riemann integral in the literature it is appropriate to define here this integral at least for the purpose of its comparison with the Lebesgue integral, but we have another motive. Namely, we shall show that when the Riemann integral of a bounded function f exists, it coincides with the Lebesgue integral of f, thereby demonstrating that all the basic theorems of the integral calculus remain valid in the realm of the Lebesgue integral. The reader may be surprised to learn that a bounded function is Riemann integrable if and only if it is continuous almost everywhere (Theorem 44.5). Under the assumption that the reader has more than a nodding acquaintance with the Riemann integral we dispense here with both motivation and detailed discussion. As before, the arguments are restricted to functions defined on a fixed compact interval $\mathcal{I} = [a, b]$. The familiar approximations which are the trademark of the Riemann integral are launched with the following definition:

44.1 / *Definition* A *partition* σ of $[a, b]$ is an arbitrary finite set of numbers

$$a = a_0 < a_1 < a_2 < \cdots < a_n = b$$

given in their natural order. The intervals

$$I_k = [a_k, a_{k+1}] \qquad (0 \leq k \leq n - 1)$$

are the *component intervals* of σ.

Let $f: \mathcal{I} \to \mathcal{R}$ be a fixed bounded function and let σ be an arbitrary partition of $\mathcal{I}$. Putting

$$M(f; I_k) = \operatorname*{lub}_{x \in I_k} \{f(x)\} \tag{i}$$

and

$$m(f; I_k) = \operatorname*{glb}_{x \in I_k} \{f(x)\} \tag{ii}$$

we associate with f and σ the step functions

$$M(f; \sigma) = \sum_{k=0}^{n-1} M(f; I_k)\chi_{I_k} \tag{iii}$$

and

$$m(f; \sigma) = \sum_{k=0}^{n-1} m(f; I_k)\chi_{I_k} \tag{iv}$$

with corresponding numbers

$$U(f; \sigma) = \sum_{k=0}^{n-1} M(f; I_k)\mu(I_k) \tag{v}$$

and

$$L(f; \sigma) = \sum_{k=0}^{n-1} m(f; I_k)\mu(I_k) \tag{vi}$$

(see Figure 49). While the functions in (iii) and (iv) approximate f, the numbers in (v) and (vi) approximate the area as indicated in the figure.

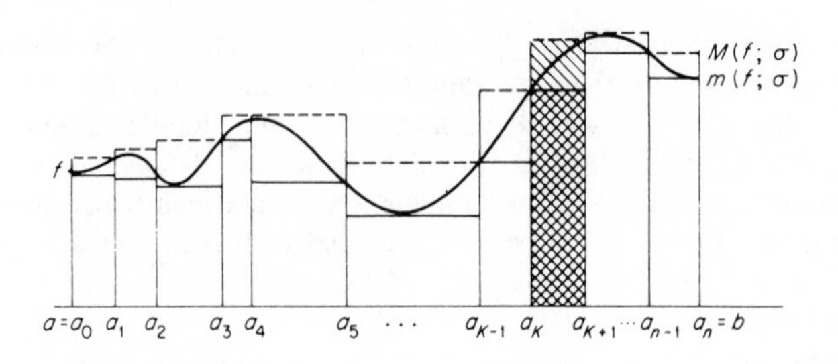

Figure 49

We say that the partition τ is a *refinement* of σ when $\sigma \subset \tau$ (that is, each point in σ is a member of τ). It is quite clear that if σ and τ are thusly related, then

$$U(f;\sigma) \geq U(f;\tau) \tag{vii}$$

whereas

$$L(f;\sigma) \leq L(f;\tau). \tag{viii}$$

These inequalities yield at once:

44.2 / Lemma If $f\colon \mathcal{I} \to \mathcal{R}$ is bounded and if τ and σ are arbitrary partitions of $\mathcal{I}$, then

$$U(f;\sigma) \geq L(f;\tau).$$

In particular, the lemma is true when τ is a refinement of σ.

The Riemann integral is conveniently introduced with the help of the next definition:

44.3 / *Definition* Let $f\colon \mathcal{I} \to \mathcal{R}$ be bounded. The *upper Riemann integral* of f over $\mathcal{I}$ is the number

$$R\,\overline{\int_a^b} f(x)\,dx = \operatorname*{glb}_{\sigma} U(f;\sigma); \tag{ix}$$

the number

$$R\,\underline{\int_a^b} f(x)\,dx = \operatorname*{lub}_{\sigma} L(f;\sigma) \tag{x}$$

is the *lower Riemann integral* of f over $\mathcal{I}$, where, of course, the greatest lower bound and least upper bound are taken over all partitions σ of $\mathcal{I}$.

44.4 / *Definition* The function f is *Riemann integrable* over $\mathcal{I}$, written

$$f \in R[\mathcal{I}],$$

when

$$R\overline{\int_a^b} f(x)\,dx = R\underline{\int_a^b} f(x)\,dx.$$

The common value of the upper and lower integrals is written as

$$R\int_a^b f(x)\,dx$$

and called the *Riemann integral* of f (over $\mathcal{I}$).

This definition should be carefully compared with Definition 42.3. A complete description of the class of Riemann integrable functions is contained in the following theorem:

44.5 / Theorem The bounded function $f: \mathcal{I} \to \mathcal{R}$ is Riemann integrable over $\mathcal{I}$ if and only if its set of discontinuities has measure zero.

The proof of this theorem is based on the following:

44.6 / Lemma If f is bounded, then $f \in R[\mathcal{I}]$ if and only if for each $\epsilon > 0$ there is a partition σ_ϵ of $\mathcal{I}$ such that

$$U(f; \sigma_\epsilon) - L(f; \sigma_\epsilon) < \epsilon. \tag{xi}$$

Proof Suppose $f \in R[\mathcal{I}]$ and let a number $\epsilon > 0$ be given. From Exercise 9-4 and the formulas in (ix) and (x) we conclude that there are partitions σ_1 and σ_2 of $\mathcal{I}$ such that

$$\overline{\int_a^b} f(x)\,dx > U(f; \sigma_1) - \frac{\epsilon}{2}$$

whereas

$$\underline{\int_a^b} f(x)\,dx < L(f; \sigma_2) + \frac{\epsilon}{2}.$$

By virtue of the hypothesis $f \in R[\mathcal{I}]$, therefore,

$$U(f; \sigma_1) - L(f; \sigma_2) < \epsilon.$$

Owing to the inequalities in (vii) and (viii), however,

$$U(f; \sigma_1 \cup \sigma_2) \le U(f; \sigma_1) < L(f; \sigma_2) + \epsilon \le L(f; \sigma_1 \cup \sigma_2) + \epsilon,$$

since $\sigma_1 \cup \sigma_2$ refines simultaneously σ_1 and σ_2. Setting $\sigma_\epsilon = \sigma_1 \cup \sigma_2$ we have the desired result.

Conversely, let $\epsilon > 0$ be given and let σ_ϵ be a partition of $\mathcal{I}$ for which (xi) is true. The inequalities

$$\overline{\int_a^b} f(x)\,dx \leq U(f;\sigma_\epsilon)$$

and

$$\underline{\int_a^b} f(x)\,dx \geq L(f;\sigma_\epsilon),$$

which are a consequence of Definition 44.3, show that

$$\overline{\int_a^b} f(x)\,dx < \underline{\int_a^b} f(x)\,dx + \epsilon.$$

The number $\epsilon > 0$ being arbitrary, it follows that

$$\overline{\int_a^b} f(x)\,dx \leq \underline{\int_a^b} f(x)\,dx.$$

Finally, the converse inequality is a direct consequence of Lemma 44.2 and Definition 44.3 and the lemma follows.

Proof of 44.5 Suppose $f \in R[\mathcal{I}]$. Referring to Exercise 25-6, let

$$D_n = \left\{x \in \mathcal{I} \mid \omega(f;x) \geq \frac{1}{n}\right\} \qquad (n \in \mathfrak{N}).$$

Then the set D of discontinuities of f can be expressed as

$$D = \bigcup_{n \in \mathfrak{N}} D_n,$$

and to show that $\mu(D) = 0$ we shall show that $\mu(D_n) = 0$ for each $n \in \mathfrak{N}$. Holding n fixed, let $\epsilon > 0$ be an arbitrary given number. Owing to Lemma 44.6 there is a partition σ_ϵ of $\mathcal{I}$ such that

$$U(f;\sigma_\epsilon) - L(f;\sigma_\epsilon) < \frac{\epsilon}{2n}.$$

If $\sigma_\epsilon = \{a_0, a_1, \ldots, a_m\}$, $I_j = [a_j, a_{j+1}]$ $(0 \leq j \leq m-1)$, then clearly

$$\begin{aligned} U(f;\sigma_\epsilon) - L(f;\sigma_\epsilon) &= \sum_{j=0}^{m-1} M(f;I_j)\mu(I_j) - \sum_{j=0}^{m-1} m(f;I_j)\mu(I_j) \\ &= \sum_{j=0}^{m-1} [M(f;I_j) - m(f;I_j)]\mu(I_j) \\ &= \sum_{j=0}^{m-1} \omega(f;I_j)\mu(I_j) < \frac{\epsilon}{2n}. \end{aligned} \qquad \text{(xii)}$$

Now let

$$E_n = D_n \cap \sigma_\epsilon \qquad \text{and} \qquad F_n = D_n \setminus E_n,$$

then

$$D_n = E_n \cup F_n.$$

The set E_n is finite since σ_ϵ is; in particular

$$E_n \subset \bigcup_{j=0}^{m} N\left(a_j; \frac{\epsilon}{4(m+1)}\right)$$

and

$$\sum_{j=0}^{m} \mu\left[N\left(a_j; \frac{\epsilon}{4(m+1)}\right)\right] \leq \frac{\epsilon}{2}. \tag{xiii}$$

When $x \in F_n$, then there is an integer j such that $x \in \mathring{I}_j$, so that

$$\omega(f; I_j) \geq \omega(f; x) \geq \frac{1}{n}. \tag{xiv}$$

Let

$$I_{j_1}, \ldots, I_{j_p}$$

be the component intervals of σ_ϵ which intersect F_n. Then owing to (xiv) and (xii)

$$\frac{\epsilon}{2n} > \sum_{k=1}^{p} \omega(f; I_{j_k})\mu(I_{j_k}) \geq \sum_{k=1}^{p} \frac{1}{n} \mu(I_{j_k})$$

and consequently

$$\sum_{k=1}^{p} \mu(I_{j_k}) < \frac{\epsilon}{2}.$$

Combining this estimate with that in (xiii) we see that for each $\epsilon > 0$ the set D_n has an open covering of measure ϵ. Hence $\mu(D_n) = 0$, as was to be shown.

To prove the converse let us suppose that the set D of discontinuities of f has measure zero. By virtue of Lemma 44.6 it suffices to show that for each $\epsilon > 0$ there is a partition σ_ϵ for which formula (xi) holds.

As in the first part of the proof let D be expressed in terms of the sets D_n defined there. Then each set D_n is closed, has measure zero, and as such is nowhere dense (verify these facts). Let $\epsilon > 0$ be an arbitrary given number and let an integer n be fixed subject to the requirement

$$\frac{1}{n} < \frac{\epsilon}{2\mu(\mathcal{J})}.$$

Then there are mutually exclusive closed intervals $I_1, \ldots, I_p$ such that

$$I_j \subset \mathcal{J} \setminus D_n \qquad (1 \leq j \leq p)$$

and

$$\sum_{j=1}^{p} \mu(I_j) > \mu(\mathcal{J}) - \frac{\epsilon}{4\,\|f\|}$$

(clearly if $\|f\|$ were zero there would be nothing to prove). Regarding the intervals I_j we find that

$$\omega(f; x) \leq \frac{1}{n} < \frac{\epsilon}{2\mu(\mathcal{J})} \qquad (x \in I_j).$$

Let j be momentarily fixed and let $I_j = [a_j, b_j]$. We claim that there is a partition $a_j = c_1 < c_2 < \cdots < c_k = b_j$ such that

$$\omega(f; [c_j, c_{j+1}]) < \frac{\epsilon}{2\mu(\mathcal{J})} \qquad (1 \leq j \leq k - 1). \tag{xv}$$

To verify this, we note that for each $x \in I_j$ there is a neighborhood of radius $\delta(x)$ such that

$$\omega(f; N(x; \delta(x)) \cap \mathcal{J}) < \frac{\epsilon}{2\mu(\mathcal{J})}.$$

Because of the compactness of I_j there is a finite number of sets $N(x_1; \delta(x_1)), \ldots, N(x_k, \delta(x_k))$ which cover I_j. If we list the endpoints of the sets $N(x_j; \delta(x_j)) \cap \mathcal{J}$ $(1 \leq j \leq k)$ in increasing order, then these points are as demanded in (xv). Hence, *each* interval I_j can be expressed as a finite union of closed intervals J_{ij} which share only endpoints with each other and such that

$$\omega(f; J_{ij}) < \frac{\epsilon}{2\mu(\mathcal{J})} \qquad (1 \leq i \leq m_j).$$

Now put $H_j = [b_j, a_{j+1}]$ $(0 \leq j \leq p)$, where $b_0 = a$ and $a_{p+1} = b$. Then the following relations are easily verified:

If $\sigma_\epsilon = \{a_0, a_1, b_1, \ldots, a_p, b_p, b\}$, then

$$U(f;\sigma_\epsilon) - L(f;\sigma_\epsilon) = \sum_{j=1}^{p}\sum_{i=1}^{m_j} M(f;J_{ij})\mu(J_{ij}) + \sum_{j=0}^{p} M(f;H_j)\mu(H_j)$$

$$- \sum_{j=1}^{p}\sum_{i=1}^{m_j} m(f;J_{ij})\mu(J_{ij}) - \sum_{j=0}^{p} m(f;H_j)\mu(H_j)$$

$$= \sum_{j=1}^{p}\sum_{i=1}^{m_j} \omega(f;J_{ij})\mu(J_{ij}) + \sum_{j=0}^{p} \omega(f;H_j)\mu(H_j)$$

$$< \frac{\epsilon}{2\mu(\mathcal{I})}\sum_{j=1}^{p}\sum_{i=1}^{m_j} \mu(J_{ij}) + 2\,\|f\|\sum_{j=0}^{p}\mu(H_j)$$

$$< \frac{\epsilon}{2\mu(\mathcal{I})}\mu(\mathcal{I}) + 2\,\|f\|\frac{\epsilon}{4\,\|f\|} = \epsilon.$$

According to Lemma 44.6 f is Riemann integrable over $\mathcal{I}$.

With the help of Theorem 44.5 we prove

44.7 / Theorem If $f \in R[\mathcal{I}]$, then $f \in \mathfrak{L}[\mathcal{I}]$. Moreover,

$$R\int_a^b f(x)\,dx = \int_{\mathcal{I}} f\,d\mu.$$

From the introductory remarks of this chapter we know that this theorem has no true converse.

Proof Consider an arbitrary partition σ of $\mathcal{I}$ with components $I_k = [a_k, a_{k+1}]$, $a_0 = a$, $a_n = b$, and $0 \le k \le n-1$. Let us associate with this partition a simple function $\varphi(\sigma; x)$ as follows: If A stands for the set consisting of the points of discontinuity of f and the points a_k of σ, then we define pointwise the function:

$$\varphi(\sigma;x) = \begin{cases} 0 & (x \in A) \\ m(f;I_k) & (x \in I_k \setminus A). \end{cases}$$

Let an arbitrary number $\epsilon > 0$ be given. Then a partition σ can be fixed such that the two inequalities,

$$|f(x) - \varphi(\sigma;x)| < \frac{\epsilon}{2\mu(\mathcal{I})} \qquad (x \in \mathcal{I} \setminus A)$$

and

$$\left| R\int_a^b f(x)\,dx - L(f;\sigma)\right| < \frac{\epsilon}{2}.$$

are met. From Theorem 44.5 we conclude that the function f is measurable. Noting that a Riemann integrable function on $\mathcal{I}$ is of necessity bounded, we see that, in fact, $f \in \mathfrak{L}[\mathcal{I}]$. In particular, the function φ is seen to be bounded, and so also $\varphi \in \mathfrak{L}[\mathcal{I}]$. Clearly $\mu(A) = 0$ and we see at once that

$$\int_{\mathcal{I}} \varphi \, d\mu = \int_{\mathcal{I} \setminus A} \varphi \, d\mu = \sum_{k=0}^{n-1} m(f; I_k) \cdot \mu(I_k) = L(f; \sigma).$$

We now find that

$$\left| \int_{\mathcal{I}} f \, d\mu - R \int_a^b f(x) \, dx \right|$$

$$\leq \left| \int_{\mathcal{I}} f \, d\mu - L(f; \sigma) \right| + \left| L(f; \sigma) - R \int_a^b f(x) \, dx \right| = \alpha + \beta.$$

Regarding α, we have

$$\alpha = \left| \int_{\mathcal{I}} f \, d\mu - \int_{\mathcal{I}} \varphi \, d\mu \right| \leq \int_{\mathcal{I}} | f - \varphi | \, d\mu \leq \frac{\epsilon}{2\mu(\mathcal{I})} \cdot \mu(\mathcal{I}) = \frac{\epsilon}{2}.$$

Owing to the above, $\beta < \epsilon/2$, and the theorem follows from the fact that $\epsilon > 0$ was arbitrary.

EXERCISES

44-1 Referring to Example 35.10, is every member in the uniform closure of the algebra of step functions Riemann integrable?

44-2 Prove or disprove the following statement: The bounded function $f: \mathcal{I} \to \mathcal{R}$ is Riemann integrable if and only if f is the uniform limit of a sequence of step functions.

44-3 Review the basic theorems of the integral calculus.

TEN

FOURIER SERIES

By way of motivation we may do well to consider a concrete instance in which trigonometric series arise. Like Fourier, let us investigate a heat conduction problem. To simplify matters, we imagine an infinitely thin rod, of length l, and we assume that there is no heat exchange between the rod and its surrounding environment. We wish to determine the temperature distribution along the rod at time $t > 0$ when the temperature distribution f at time $t = 0$ is given and when the endpoints of the rod are held at a constant temperature of 0 degrees.

For convenience, let the rod be laid along the positive x-axis in the xt-plane, with one end at the origin. If the function $u(x, t)$ designates the temperature of the rod at the point x and time t, then we can express the problem at hand as the following *initial value problem*:

$$\begin{aligned} &\frac{\partial u}{\partial t} = c^2 \frac{\partial^2 u}{\partial x^2} \\ &u(0, t) = u(l, t) = 0 \qquad (t > 0) \\ &u(x, 0) = f(x), \end{aligned} \tag{i}$$

where c is an appropriate constant. Solutions in which the variables are separated are easy to handle. Assuming such a solution, $u(x, t) = v(x)w(t)$, to exist, we proceed quite formally to transform the above differential equation into

$$\frac{1}{c^2 w}\frac{dw}{dt} = \frac{1}{v}\frac{d^2v}{dx^2}.$$

Since the left side of this equation is independent of x and the right side is independent of t, it follows that both sides of the differential equation must equal a constant, say, α. This being so, we arrive at the two differential equations

$$\frac{dw}{dt} = \alpha c^2 w$$

$$\frac{d^2v}{dx^2} = \alpha v.$$

The first equation is seen to have a solution

$$w(t) = \exp(\alpha c^2 t);$$

for the second equation we see that the function $v(x) = \sin \beta x$ is a solution when $\alpha = -\beta^2$. The condition $u(0, t) = u(l, t) = 0$, which now assumes the form $v(0) = v(l) = 0$, will not be satisfied, however, unless β has the form $k\pi/l$, where k is an integer. Putting

$$w_k(t) = \exp\left(-\frac{k^2\pi^2c^2}{l^2}t\right) \qquad \text{and} \qquad v_k(x) = \sin\frac{k\pi}{l}x$$

for $k \in \mathfrak{N}$, we conclude that each of the functions $u_k(x, t) = v_k(x)w_k(t)$ is a solution of the initial value problem

$$\frac{\partial u}{\partial t} = c^2\frac{\partial^2 u}{\partial x^2},$$

$$u(0, t) = u(l, t) = 0 \qquad (t > 0).$$

Clearly, the series

$$u(x, t) = \sum_{k=1}^{\infty} a_k v_k(x) w_k(t)$$

with constants a_k, is a formal solution of the same problem. To comply with the last condition in (i) we must be able to find coefficients a_k such that

$$f(x) = \sum_{k=1}^{\infty} a_k \sin\frac{k\pi}{l}x.$$

It is necessary, in other words, to decide if f can be expressed as a trigonometric series.

In the coming discussion it will be convenient to adopt the notation

$$\int_a^b f(x)\, d\mu$$

for the Lebesgue integral of f.

45 / BASIC FACTS

The trigonometric functions

$$1, \quad \cos x, \quad \sin x, \quad \cos 2x, \quad \sin 2x, \ldots, \cos nx, \quad \sin nx, \ldots$$

satisfy the following orthogonality relations:

45.1 / Theorem

$$\int_{-\pi}^{\pi} \cos kx \cos nx\, d\mu = \begin{cases} 0 & k \neq n, \\ 2\pi & k = n = 0 \\ \pi & k = n \neq 0 \end{cases}$$

$$\int_{-\pi}^{\pi} \sin kx \sin nx\, d\mu = \begin{cases} 0 & k \neq n \\ 0 & k = n = 0 \\ \pi & k = n \neq 0 \end{cases}$$

$$\int_{-\pi}^{\pi} \sin kx \cos nx\, d\mu = 0.$$

Proof These relations follow at once from the formulas

$$\begin{aligned} \cos kx \cos nx &= \tfrac{1}{2}[\cos(k+n)x + \cos(k-n)x], \\ \sin kx \sin nx &= \tfrac{1}{2}[\cos(k-n)x - \cos(k+n)x], \qquad \text{(ii)} \\ \cos kx \sin nx &= \tfrac{1}{2}[\sin(k+n)x + \sin(k-n)x]. \end{aligned}$$

From the periodicity of the trigonometric functions we easily deduce that similar orthogonality relations hold on an arbitrary interval of length 2π. Formally we state the following definition:

45.2 / *Definition* The functions $f_1, f_2, f_3, \ldots$ in $\mathfrak{L}[-\pi, \pi]$ are *orthogonal* with respect to the interval $[-\pi, \pi]$ if

$$\int_{-\pi}^{\pi} f_k(x) f_n(x)\, d\mu = 0 \qquad (k \neq n).$$

The functions are *orthonormal* when

$$\int_{-\pi}^{\pi} f_k(x) f_n(x)\, d\mu = \begin{cases} 0 & (k \neq n) \\ 1 & (k = n). \end{cases}$$

The set of functions

$$\frac{1}{\sqrt{2\pi}}, \quad \frac{\cos x}{\sqrt{\pi}}, \quad \frac{\sin x}{\sqrt{\pi}}, \ldots, \frac{\cos nx}{\sqrt{\pi}}, \quad \frac{\sin nx}{\sqrt{\pi}}, \ldots$$

is readily seen to be orthonormal.

Consider an arbitrary trigonometric polynomial

$$Q_p(x) = \sum_{k=0}^{p} c_k \cos^{\alpha_k} x \, \sin^{\beta_k} x.$$

With the help of the identities in (ii) each such polynomial can be reduced to the form

$$T_n(x) = \sum_{k=0}^{n} a_k \cos kx + b_k \sin kx;$$

by a *trigonometric polynomial* we will henceforth mean a function of this form. The set of trigonometric polynomials on $[-\pi, \pi]$ clearly constitutes an algebra which vanishes at no point of the interval (see Definitions 35.6 and 35.9). Definition 35.8, however, is not complied with in the closed interval, since for an arbitrary trigonometric polynomial T_n, $T_n(-\pi) = T_n(\pi)$. With this in mind we state the following:

45.3 / *Definition* The function $f \in \mathfrak{L}[-\pi, \pi]$ is *periodic* when

$$f(-\pi) = f(\pi).$$

Clearly, the period of f must have length $2\pi/k$ for some positive integer k. Such a function f has a periodic extension on $\mathcal{R}$, again denoted by f, which is specified as

$$f(x + 2k\pi) = f(x) \qquad \text{for} \quad k = 0, \pm 1, \pm 2, \ldots.$$

The following result is seen to be true by virtue of Theorem 35.14:

45.4 / Weierstrass Second Approximation Theorem Let f be an arbitrary continuous periodic function on $[-\pi, \pi]$. Then for each number $\epsilon > 0$ there is a trigonometric polynomial T_n such that

$$\| f - T_n \| < \epsilon.$$

Clearly, the interval in the theorem can be replaced by an arbitrary interval of length 2π.

Now consider a *trigonometric series*

$$f(x) = \sum_{k=0}^{\infty} a_k \cos kx + b_k \sin kx.$$

When we formally integrate the series term by term we see that

$$\int_{-\pi}^{\pi} f(x)\, d\mu = \sum_{k=0}^{\infty} \left[a_k \int_{-\pi}^{\pi} \cos kx\, d\mu + b_k \int_{-\pi}^{\pi} \sin kx\, d\mu \right] = 2\pi a_0. \tag{iii}$$

For a fixed integer $n \in \mathfrak{N}$ we multiply both sides of the series by $\cos nx$, and again integrate the resulting series term by term. This yields the integrals

$$\int_{-\pi}^{\pi} f(x) \cos nx\, d\mu$$

$$= \sum_{k=0}^{\infty} \left[a_k \int_{-\pi}^{\pi} \cos kx \cos nx\, d\mu + b_k \int_{-\pi}^{\pi} \sin kx \cos nx\, d\mu \right].$$

According to the orthogonality relations in Theorem 45.1, all these integrals vanish except the one associated with a_n. In particular,

$$\int_{-\pi}^{\pi} f(x) \cos nx\, d\mu = a_n \int_{-\pi}^{\pi} \cos^2 nx\, d\mu = \pi a_n,$$

and since n was arbitrary, we find that

$$a_n = \frac{1}{\pi} \int_{-\pi}^{\pi} f(x) \cos nx\, d\mu \qquad (n \in \mathfrak{N}). \tag{iv}$$

Similarly, when we repeat the above procedure with $\sin nx$ rather than $\cos nx$ we obtain the expressions

$$b_n = \frac{1}{\pi} \int_{-\pi}^{\pi} f(x) \sin nx\, d\mu \qquad (n \in \mathfrak{N}). \tag{v}$$

Thus, if f has a representation as a trigonometric series, then the coefficients a_n and b_n can be formally calculated from the formulas in (iii) to (v). On account of (iii) we consider trigonometric series in which a_0 is replaced by $\frac{1}{2}a_0$. Noting that the above integrals exist whenever f is Lebesgue integrable (could you prove this?), we state the next definition:

45.5 / ***Definition*** Let $f \in \mathfrak{L}[-\pi, \pi]$. The *Fourier series* of f is the trigonometric series

$$\tfrac{1}{2}a_0 + \sum_{k=1}^{\infty} (a_k \cos kx + b_k \sin kx),$$

in which the coefficients are specified by the formulas

$$a_k = \frac{1}{\pi}\int_{-\pi}^{\pi} f(x)\cos kx\,d\mu \qquad (k = 0, 1, 2, \ldots),$$

$$b_k = \frac{1}{\pi}\int_{-\pi}^{\pi} f(x)\sin kx\,d\mu \qquad (k = 1, 2, 3, \ldots).$$

The numbers a_k and b_k are the *Fourier coefficients* of f. The association between f and its Fourier series is expressed by writing

$$f(x) \sim \tfrac{1}{2}a_0 + \sum_{k=1}^{\infty} (a_k \cos kx + b_k \sin kx).$$

In this chapter we shall be concerned with the convergence to a given integrable function f of its Fourier series. Before formulating the problem more precisely we shall prove some basic results. First, however, we consider some examples:

45.6 / *Examples*

(**1**) Let us regard the function $f(x) = x$. This function is odd and hence

$$a_k = \frac{1}{\pi}\int_{-\pi}^{\pi} x\cos kx\,d\mu = -\frac{1}{\pi}\int_{0}^{-\pi} x\cos kx\,d\mu + \frac{1}{\pi}\int_{0}^{\pi} x\cos kx\,d\mu$$

$$= -\frac{1}{\pi}\int_{0}^{\pi} x\cos kx\,d\mu + \frac{1}{\pi}\int_{0}^{\pi} x\cos kx\,d\mu = 0$$

for $k = 0, 1, 2, 3, \ldots$, whereas a simple transformation shows that

$$b_k = \frac{2}{\pi}\int_{0}^{\pi} x\sin kx\,d\mu = -\frac{2}{\pi k}\,x\cos kx\,\Big|_0^{\pi} + \frac{2}{\pi k}\int_{0}^{\pi}\cos kx\,d\mu$$

$$= -\frac{2}{k}\cos k\pi = \frac{2}{k}(-1)^{k+1}.$$

It follows that

$$x \sim 2(\sin x - \tfrac{1}{2}\sin 2x + \tfrac{1}{3}\sin 3x - \cdots).$$

(**2**) Now take the even function $g(x) = |x|$. Here a simple calculation shows that $b_k = 0$ for each admitted value of k. At the same time

$$a_0 = \frac{1}{\pi}\int_{-\pi}^{\pi} |x|\,d\mu = \frac{2}{\pi}\int_{0}^{\pi} x\,d\mu = \frac{2}{\pi}\frac{x^2}{2}\,\Big|_0^{\pi} = \pi$$

and, for $k > 0$,

$$a_k = \frac{2}{\pi}\int_0^{\pi} x \cos kx \, d\mu = -\frac{2}{\pi k}\int_0^{\pi} \sin kx \, d\mu$$

$$= \frac{2}{\pi k^2} \cos kx \Big|_0^{\pi} = \frac{2}{\pi k^2}[\cos k\pi - 1] = \frac{2}{\pi k^2}[(-1)^k - 1].$$

Accordingly

$$a_{2k} = 0 \qquad \text{and} \qquad a_{2k+1} = \frac{-4}{\pi(2k+1)^2},$$

and so

$$|x| \sim \frac{\pi}{2} - \frac{4}{\pi}\left(\cos x + \frac{1}{3^2}\cos 3x + \frac{1}{5^2}\cos 5x + \cdots\right).$$

(**3**) Let us obtain the Fourier series of the function $h(x) = (x + \pi)^2$ which is neither even nor odd. We find that

$$a_0 = \frac{1}{\pi}\int_{-\pi}^{\pi} (x + \pi)^2 \, d\mu = \frac{1}{\pi}\int_0^{2\pi} x^2 \, d\mu = \frac{8\pi^2}{3},$$

$$a_k = \frac{1}{\pi}\int_{-\pi}^{\pi} (x + \pi)^2 \cos kx \, d\mu = \frac{(-1)^k}{\pi}\int_0^{2\pi} x^2 \cos kx \, d\mu$$

$$= -\frac{2(-1)^k}{\pi k}\int_0^{2\pi} x \sin kx \, d\mu$$

$$= \frac{2(-1)^k}{\pi k^2} x \cos kx \Big|_0^{2\pi} - \frac{2(-1)^k}{\pi k^2}\int_0^{2\pi} \cos kx \, d\mu = \frac{4(-1)^k}{k^2},$$

$$b_k = \frac{1}{\pi}\int_{-\pi}^{\pi} (x + \pi)^2 \sin kx \, d\mu = \frac{(-1)^k}{\pi}\int_0^{2\pi} x^2 \sin kx \, d\mu$$

$$= -\frac{(-1)^k}{\pi k} x^2 \cos kx \Big|_0^{2\pi} + \frac{2(-1)^k}{\pi k}\int_0^{2\pi} x \cos kx \, d\mu$$

$$= \frac{4(-1)^{k+1}\pi}{k} - \frac{2(-1)^k}{\pi k^2}\int_0^{2\pi} \sin kx \, d\mu = -\frac{4(-1)^k\pi}{k}.$$

Accordingly,

$$(x + \pi)^2 \sim \frac{4\pi^2}{3} + 4\sum_{k=1}^{\infty} (-1)^k \frac{\cos kx}{k^2} - 4\pi \sum_{k=1}^{\infty} (-1)^k \frac{\sin kx}{k}.$$

Convergence of Fourier series is guaranteed when we deal with uniformly convergent trigonometric series:

45.7 / Theorem If the trigonometric series

$$\tfrac{1}{2}a_0 + \sum_{k=1}^{\infty} (a_k \cos kx + b_k \sin kx)$$

converges uniformly on $[-\pi, \pi]$ to the integrable function f, then it is the Fourier series of f.

Proof This theorem states, of course, that the coefficients a_k and b_k are those computed through the formulas in Definition 45.5. To show this, let us first demonstrate that the series

$$\tfrac{1}{2}a_0 \cos nx + \sum_{k=1}^{\infty} (a_k \cos kx \cos nx + b_k \sin bx \cos nx)$$

converges uniformly to the function $f(x) \cos nx$ for each fixed integer $n \in \mathfrak{N}$. This is easy, for let an arbitrary number $\epsilon > 0$ be given. Then there is an integer r_ϵ such that

$$\| f(x) - \frac{1}{2} a_0 - \sum_{k=1}^{r} (a_k \cos kx + b_k \sin kx) \| < \frac{\epsilon}{2\pi}$$

for all integers $r > r_\epsilon$. Hence

$$\left\| f(x) \cos nx - \frac{1}{2} a_0 \cos nx - \sum_{k=1}^{r} (a_k \cos kx \cos nx + b_k \sin kx \sin nx) \right\|$$

$$\leq \|\cos nx\| \cdot \left\| f(x) - \frac{1}{2} a_0 - \sum_{k=1}^{r} (a_k \cos kx + b_k \sin kx) \right\| < \frac{\epsilon}{2\pi}$$

since $\| \cos nx \| \leq 1$.

Now let n be given and let $r > \max\{n, r_\epsilon\}$. Then the following

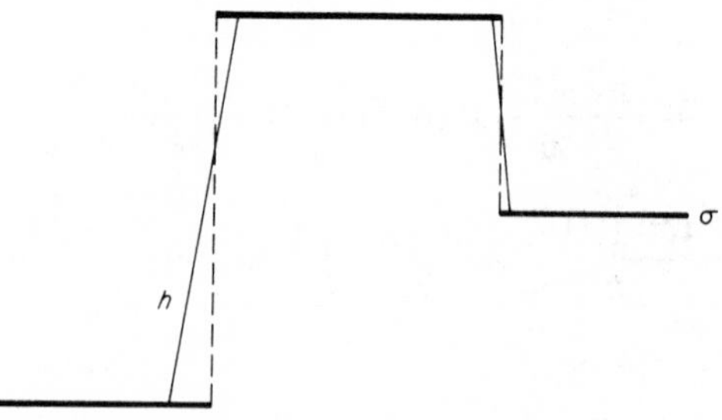

Figure 50

steps are justified:

$$0 \le \left| \int_{-\pi}^{\pi} f(x) \cos nx \, d\mu - \pi a_n \right|$$

$$= \left| \int_{-\pi}^{\pi} f(x) \cos nx \, d\mu - \int_{-\pi}^{\pi} \tfrac{1}{2} a_0 \cos nx \, d\mu \right.$$

$$\left. - \sum_{k=1}^{r} \int_{-\pi}^{\pi} (a_k \cos kx \cos nx + b_k \sin kx \cos nx) \, d\mu \right|$$

$$= \left| \int_{-\pi}^{\pi} f(x) \cos nx \, d\mu - \int_{-\pi}^{\pi} [\tfrac{1}{2} a_0 \cos nx \right.$$

$$\left. + \sum_{k=1}^{r} (a_k \cos kx \cos nx + b_k \sin kx \cos nx)] \, d\mu \right|$$

$$\le \int_{-\pi}^{\pi} \left| f(x) \cos nx - \tfrac{1}{2} a_0 \cos nx \right.$$

$$\left. - \sum_{k=1}^{r} (a_k \cos kx \cos nx + b_k \sin kx \cos nx) \right| d\mu$$

$$< 2\pi \frac{\epsilon}{2\pi} = \epsilon,$$

and the theorem follows.

We now prove the following fundamental result:

45.8 / Theorem Let $f \in \mathfrak{L}[-\pi, \pi]$ be a periodic function. If

$$\int_{-\pi}^{\pi} f(x) \, d\mu = 0, \int_{-\pi}^{\pi} f(x) \cos kx \, d\mu = 0, \int_{-\pi}^{\pi} f(x) \sin kx \, d\mu = 0 \qquad (k \in \mathfrak{N}),$$

then $f(x) = 0$ almost everywhere.

Proof Let g be an arbitrary bounded measurable function. We shall show that if f is as demanded in the theorem, then

$$\int_{-\pi}^{\pi} f(x) g(x) \, d\mu = 0.$$

An inspection of the proof of Theorem 41.3 shows that g is the limit of a sequence of simple functions each of which assumes only finitely many values. With the help of Theorem 38.11 it can be shown that each such simple function is the limit almost everywhere of a sequence of step functions (see also Exercise 41-4). We can therefore claim that if arbitrary numbers $\epsilon > 0$ and $\delta > 0$ are given, then there is a step function σ such that

$$| g(x) - \sigma(x) | < \frac{\epsilon}{3}$$

except for a set of at most measure δ. For this step function there is a continuous function h, such that $h(-\pi) = h(\pi)$ and

$$| \sigma(x) - h(x) | < \frac{\epsilon}{3}$$

except possibly for a set of at most measure δ (this set contains the points $\pm\pi$ and the points of discontinuity of σ (see Figure 50)). Finally, Theorem 45.4 tells us that there is a trigonometric polynomial T_n such that

$$|| h - T_n || < \frac{\epsilon}{3}.$$

Now, the conditions in the theorem guarantee that

$$\int_{-\pi}^{\pi} f(x) T_n(x) \, d\mu = 0.$$

Consequently

$$\int_{-\pi}^{\pi} f(x) \cdot g(x) \, d\mu = \int_{-\pi}^{\pi} f(x) [g(x) - \sigma(x)] \, d\mu$$

$$+ \int_{-\pi}^{\pi} f(x) [\sigma(x) - h(x)] \, d\mu + \int_{-\pi}^{\pi} f(x) [h(x) - T_n(x)] \, d\mu,$$

and hence

$$0 \leq \left| \int_{-\pi}^{\pi} f(x) g(x) \, d\mu \right|$$

$$\leq \int_{-\pi}^{\pi} | f(x) | \cdot | g(x) - \sigma(x) | \, d\mu + \int_{-\pi}^{\pi} | f(x) | \cdot | \sigma(x) - h(x) | \, d\mu$$

$$+ \int_{-\pi}^{\pi} | f(x) | \cdot | h(x) - T_n(x) | \, d\mu.$$

But each of the integrals in the right side is bounded above outside of a set of at most measure δ by

$$\frac{\epsilon}{3}\int_{-\pi}^{\pi} |f(x)|\, d\mu = c\cdot\frac{\epsilon}{3}.$$

Thus

$$0 \le \left|\int_{-\pi}^{\pi} f(x)g(x)\, d\mu\right| \le c\cdot\left|\frac{\epsilon}{3}+\frac{\epsilon}{3}+\frac{\epsilon}{3}\right| = c\cdot\epsilon,$$

except for a set of at most measure δ. We can now let $\delta \to 0$, and the assertion that the integral of $f\cdot g$ vanishes follows from the fact that $\epsilon > 0$ was arbitrary.

Since g was arbitrary, let us select it to be

$$g(x) = \begin{cases} 0 & \text{when} \quad f(x) = 0 \\ \operatorname{sgn} f(x) & \text{when} \quad f(x) \neq 0. \end{cases}$$

Then $f(x)g(x) = |f(x)|$ and consequently

$$0 = \int_{-\pi}^{\pi} f(x)g(x)\, d\mu = \int_{-\pi}^{\pi} |f(x)|\, d\mu.$$

But this implies that $f = 0$ almost everywhere, as was to be shown.

From the above theorem we obtain the following:

45.9 / Uniqueness Theorem Let f and g be periodic functions in $\mathfrak{L}[-\pi,\pi]$. If the Fourier coefficients of the two functions coincide, then $f = g$ almost everywhere.

EXERCISES

45-1 We recall that the function $f \in \mathfrak{L}[-\pi, \pi]$ is *even* when $f(x) = f(-x)$, *odd* when $f(x) = -f(-x)$ for all $x \in [-\pi, \pi]$. Show that:

(a) If f is even, then $b_k = 0$ for $k = 1, 2, 3, \ldots$;

(b) If f is odd, then $a_k = 0$ for $k = 0, 1, 2, 3, \ldots$;

(c) If f is even and g is odd, then the function $f\cdot g$ is odd;

(d) When f and g are both even or both odd, then the function $f\cdot g$ is even.

45-2 Determine the Fourier series of the following functions:

(a) $f(x) = (\sin x)^2$ on $[-\pi, \pi]$,

(b) $f(x) = x^2$ on $[0, 2\pi]$,

(c) $f(x) = x^2$ on $[-\pi, \pi]$,

(d) $$h(x) = \begin{cases} 1 & (0 \le x \le \pi) \\ -1 & (-\pi \le x < 0). \end{cases}$$

45-3 Let

$$f(x) \sim \tfrac{1}{2}a_0 + \sum_{k=1}^{\infty} (a_k \cos kx + b_k \sin kx),$$

$$g(x) \sim \tfrac{1}{2}a_0' + \sum_{k=1}^{\infty} (a_k' \cos kx + b_k' \sin kx).$$

Show that

(a) If

$$(f + g)(x) \sim \tfrac{1}{2}c_0 + \sum_{k=1}^{\infty} (c_k \cos kx + d_k \sin kx),$$

then $c_k = a_k + a_k'$ and $d_k = b_k + b_k'$ for all admitted values of k;

(b) For any arbitrary constant c,

$$cf(x) \sim \tfrac{1}{2}(a_0c) + \sum_{k=1}^{\infty} ((a_kc) \cos kx + (b_kc) \sin kx).$$

46 / THE SPACE $\mathfrak{L}^2$

Several important results in the theory of Fourier series involve measurable functions whose square is integrable.

46.1 / *Definition* The function f is *square integrable* over the interval $[-\pi, \pi]$ when f is measurable and $f^2 \in \mathfrak{L}[-\pi, \pi]$. The space of square integrable functions is designated with the symbol $\mathfrak{L}^2 = \mathfrak{L}^2[-\pi, \pi]$.

The space $\mathfrak{L}^2$ is endowed with numerous interesting properties, and it is of great importance in various branches of analysis. For these reasons we digress momentarily from our study of Fourier series to examine some of the salient features of this space. In the sequel we shall have ample occasion to call on the inequality

$$|fg| \le \tfrac{1}{2}(f^2 + g^2), \tag{i}$$

where f and g are arbitrary functions: it is a consequence of the inequality $(|f| - |g|)^2 \geq 0$. As a first application of (i) we note that

$$|f| \leq \tfrac{1}{2}(f^2 + 1).$$

This tells us, of course, that if $f \in \mathfrak{L}^2$, then $f \in \mathfrak{L}$. On the other hand, consider the function $f(x) = x^{-1/2}$. This time $f \in \mathfrak{L}$, yet $f \notin \mathfrak{L}^2$, and we have thus shown that $\mathfrak{L}^2$ is a proper subspace of $\mathfrak{L}$.

46.2 / Theorem If f and g are functions in $\mathfrak{L}^2$, then so are the functions $f + g$ and $a \cdot f$ for each $a \in \mathfrak{R}$. The function $f \cdot g$ is in $\mathfrak{L}$.

Proof The last assertion is another consequence of the inequality in (i); the first assertion follows from the inequalities

$$(f + g)^2 = f^2 + 2fg + g^2 \leq f^2 + 2|fg| + g^2 \leq 2(f^2 + g^2).$$

The second claim is clear.

Of fundamental importance is:

46.3 / The Bunyakovsky–Cauchy–Schwarz Inequality If f and g are functions in $\mathfrak{L}^2$, then

$$\left| \int_{-\pi}^{\pi} f(x)g(x)\, d\mu \right| \leq \left(\int_{-\pi}^{\pi} f^2(x)\, d\mu \right)^{1/2} \left(\int_{-\pi}^{\pi} g^2(x)\, d\mu \right)^{1/2}.$$

Proof Let f and g be square integrable over the interval $[-\pi, \pi]$. We leave it to the reader to reason why this inequality is true when either integral in the right side vanishes. When this is not the case, we again utilize the inequality in (i). Namely, when we replace f by αf, where α is an arbitrary positive constant, then we arrive at the inequality

$$|fg| \leq \frac{1}{2} \left| \alpha f^2 + \frac{1}{\alpha} g^2 \right|.$$

Therefrom we deduce that

$$\int_{-\pi}^{\pi} |f(x)g(x)|\, d\mu \leq \frac{1}{2}\left(\alpha \int_{-\pi}^{\pi} f^2(x)\, d\mu + \frac{1}{\alpha} \int_{-\pi}^{\pi} g^2(x)\, d\mu \right).$$

Since α is arbitrary, we select it as the positive solution of the equation

$$\alpha \int_{-\pi}^{\pi} f^2(x)\, d\mu = \frac{1}{\alpha} \int_{-\pi}^{\pi} g^2(x)\, d\mu.$$

A substitution for the resulting value of α in the last inequality yields the theorem.

For ease of reference we shall call the above inequality the *Schwarz Inequality*. From it we deduce:

46.4 / The Minkowski Inequality If f and g are functions in $\mathfrak{L}^2$, then

$$\left(\int_{-\pi}^{\pi} [f(x) + g(x)]^2 \, d\mu\right)^{1/2} \leq \left(\int_{-\pi}^{\pi} f^2(x) \, d\mu\right)^{1/2} + \left(\int_{-\pi}^{\pi} g^2(x) \, d\mu\right)^{1/2}.$$

Proof Indeed,

$$\int_{-\pi}^{\pi} (f(x) + g(x))^2 \, d\mu = \int_{-\pi}^{\pi} f(x)[f(x) + g(x)] \, d\mu$$
$$+ \int_{-\pi}^{\pi} g(x)[f(x) + g(x)] \, d\mu \leq \left|\int_{-\pi}^{\pi} f(x)[f(x) + g(x)] \, d\mu\right|$$
$$+ \left|\int_{-\pi}^{\pi} g(x)[f(x) + g(x)] \, d\mu\right|.$$

According to the Schwarz Inequality 46.3,

$$\left|\int_{-\pi}^{\pi} f(x)[f(x) + g(x)] \, d\mu\right|$$
$$\leq \left(\int_{-\pi}^{\pi} f^2(x) \, d\mu\right)^{1/2} \left(\int_{-\pi}^{\pi} [f(x) + g(x)]^2 \, d\mu\right)^{1/2}$$

and

$$\left|\int_{-\pi}^{\pi} g(x)[f(x) + g(x)] \, d\mu\right|$$
$$\leq \left(\int_{-\pi}^{\pi} g^2(x) \, d\mu\right)^{1/2} \left(\int_{-\pi}^{\pi} [f(x) + g(x)]^2 \, d\mu\right)^{1/2},$$

and the desired inequality is now in sight.

Let us now introduce some convenient nomenclature:

46.5 / *Definition* Let f and g belong to $\mathfrak{L}^2$. The *inner product* of f and g is the number

$$(f, g) = \frac{1}{\pi} \int_{-\pi}^{\pi} f(x)g(x) \, d\mu.$$

The number

$$\|f\| = (f, f)^{1/2}$$

is the *norm* of f.

The reader should have no difficulties in establishing:

46.6 / Theorem If f, g, and h are arbitrary functions in $\mathfrak{L}^2$, then

(1) $(f, g) = (g, f)$,

(2) $(f + h, g) = (f, g) + (h, g)$,

(3) $(af, g) = a(f, g)$ for each $a \in \mathfrak{R}$,

(4) $(f, f) > 0$ if f does not vanish almost everywhere.

With the above notation, the Schwarz Inequality in 46.3 takes the form

$$|(f, g)| \leq \|f\| \cdot \|g\|, \tag{ii}$$

and Minkowski's Inequality can be written as

$$\|f + g\| \leq \|f\| + \|g\|. \tag{iii}$$

We are led to the next theorem:

46.7 / Theorem If f, g, and h are arbitrary functions in $\mathfrak{L}^2$, then

(1) $\|f - g\| \geq 0$, equality holding if and only if $f = g$ almost everywhere,

(2) $\|f - g\| = \|g - f\|$,

(3) $\|f - g\| \leq \|f - h\| + \|h - g\|$.

It is thus seen that the norm $\| \ \|$, called the $\mathfrak{L}^2$-*norm*, meets all the qualifications demanded of a metric function, except that equality has to be replaced with equality almost everywhere. Consequently, the space $\mathfrak{L}^2$, when regarded as the space of equivalence classes of square integrable functions which are equal almost everywhere, is a *metric space* with metric $\| \ \|$. Moreover, this space is a linear space with an inner product, and as such it shares many of the properties of the (finite dimensional) Euclidean spaces with which we are so familiar.

The $\mathfrak{L}^2$-norm should not be confused with the *uniform norm* (metric function) introduced in Section 32. We retain the same notation simply because uniform convergence will not be discussed here.

The concept of completeness is extended to the space $\mathfrak{L}^2$ in a natural way. A *Cauchy sequence* of square integrable functions f_n is a sequence such that $\| f_m - f_n \| \to 0$ as $m, n \to \infty$. The sequence $\{ f_n \}$, $f_n \in \mathfrak{L}^2[-\pi, \pi]$ is said to *converge in the mean square*, or simply to *converge* if there is a function $f \in \mathfrak{L}^2[-\pi, \pi]$ such that $\| f_n - f \| \to 0$ as $n \to \infty$. As in the case of $\mathfrak{R}$, the completeness of the space $\mathfrak{L}^2$ asserts that every Cauchy sequence converges. This is stated as the following theorem:

46.8 / The Riesz–Fischer Theorem The space $\mathfrak{L}^2[-\pi, \pi]$ is complete.

Proof We use here the following fact: If $\{ f_n \}$ is a Cauchy sequence with range in $\mathfrak{L}^2$, and a subsequence thereof converges almost everywhere to f, then so does the sequence itself. This is an immediate consequence of Exercise 8-7.

Thus, let $\{ f_n \}$, $f_n \in \mathfrak{L}^2$, be a Cauchy sequence. Then corresponding to the null sequence $\{1/2^k\}$ there is a subsequence $\{n_k\}$ of $\{n\}$ such that

$$\| f_{n_k} - f_{n_{k+1}} \| < \frac{1}{2^k}$$

for each $k \in \mathfrak{N}$. When we apply the Schwarz Inequality 46.3 to the functions $f = f_{n_k} - f_{n_{k+1}}$ and $g = 1$, we find that

$$\int_{-\pi}^{\pi} | f_{n_k}(x) - f_{n_{k+1}}(x) | \, d\mu \leq \left(\int_{-\pi}^{\pi} [f_{n_k}(x) - f_{n_{k+1}}(x)]^2 \, d\mu \right)^{1/2}$$

$$\leq \frac{\sqrt{2\pi}}{2^k}.$$

Beppo Levi's Theorem 43.6 can now be applied to tell us that the series

$$\sum_{k=1}^{\infty} | f_{n_k} - f_{n_{k-1}} |,$$

where $f_{n_0} = 0$ converges almost everywhere on the interval $[-\pi, \pi]$. But absolute convergence implies convergence, and we can therefore conclude that also the series

$$\sum_{k=1}^{\infty} (f_{n_k} - f_{n_{k-1}})$$

converges almost everywhere on $[-\pi, \pi]$ to some function f. However,

$$\sum_{k=1}^{m} (f_{n_k} - f_{n_{k-1}}) = f_{n_m},$$

and it follows that

$$\lim_{m \to \infty} f_{n_m} = f$$

almost everywhere. We know, however, that this implies that also $\lim_{n\to\infty} f_n = f$ (see, e.g. Exercises 8-7 and 32-6). Finally, let us show that $f \in \mathfrak{L}^2$. For this, let a number $\epsilon > 0$ be fixed. We can then find an integer p such that

$$\int_{-\pi}^{\pi} (f_{n_k}(x) - f_{n_m}(x))^2 \, d\mu < \epsilon$$

for all values $k, m > p$. According to Fatou's Theorem (Exercise 43-2) we can take the limit as $m \to \infty$ under the integral sign, yielding the inequality

$$\int_{-\pi}^{\pi} (f_{n_k}(x) - f(x))^2 \, d\mu < \epsilon.$$

But this implies that $f \in \mathfrak{L}^2$.

Returning now to the study of Fourier series, consider an arbitrary function $f \in \mathfrak{L}^2$ and a trigonometric polynomial

$$T_n(x) = \sum_{k=0}^{n} (\alpha_k \cos kx + \beta_k \sin kx).$$

We ask if there is a trigonometric polynomial T_n which approximates f best in the metric of $\mathfrak{L}^2$. That is, we ask if coefficients α_k and β_k can be so determined, that if T_n' is an arbitrary trigonometric polynomial, then $\| f - T_n \| \leq \| f - T_n' \|$. With the notation introduced above,

$$\begin{aligned} \| f - T_n \|^2 &= (f - T_n, f - T_n) \\ &= (f, f) - (T_n, f) - (f, T_n) + (T_n, T_n) \\ &= (f, f) - 2(f, T_n) + (T_n, T_n). \end{aligned}$$

If, as in Definition 45.5, we let a_k and b_k represent the Fourier coefficients of f, then we see with the help of Theorem 46.6 that

$$\begin{aligned} (f, T_n) &= (f, \sum_{k=0}^{n} (\alpha_k \cos kx + \beta_k \sin kx)) \\ &= (f, \sum_{k=0}^{n} \alpha_k \cos kx) + (f, \sum_{k=0}^{n} \beta_k \sin kx) \\ &= \alpha_0 (f, 1) + \sum_{k=1}^{n} (\alpha_k (f, \cos kx) + \beta_k (f, \sin kx)) \\ &= \alpha_0 a_0 + \sum_{k=1}^{n} (\alpha_k a_k + \beta_k b_k). \end{aligned}$$

Likewise, with the orthogonality relations in Theorem 45.1 we show that

$$\begin{aligned}(T_n, T_n) &= \left(\sum_{k=0}^{n} (\alpha_k \cos kx + \beta_k \sin kx), \sum_{k=0}^{n} (\alpha_k \cos kx + \beta_k \sin kx)\right) \\ &= \sum_{k,m=0}^{n} [(\alpha_k \cos kx, \alpha_m \cos mx) + 2(\alpha_k \cos kx, \beta_m \sin mx) \\ &\qquad + (\beta_k \sin kx, \beta_m \sin mx)] \\ &= \sum_{k,m=0}^{n} [\alpha_k\alpha_m(\cos kx, \cos mx) + 2\alpha_k\beta_m(\cos kx, \sin mx) \\ &\qquad + \beta_k\beta_m(\sin kx, \sin mx)] \\ &= 2\alpha_0^2 + \sum_{k=1}^{n} (\alpha_k^2 + \beta_k^2).\end{aligned}$$

Thus, it follows that

$$\begin{aligned}\|f - T_n\|^2 &= \|f\|^2 - 2\alpha_0 a_0 - 2\sum_{k=1}^{n} (\alpha_k a_k + \beta_k b_k) + 2\alpha_0^2 + \sum_{k=1}^{n} (\alpha_k^2 + \beta_k^2) \\ &= \|f\|^2 + (2\alpha_0^2 - 2\alpha_0 a_0) + \sum_{k=1}^{n} ((\alpha_k^2 - 2\alpha_k a_k) + (\beta_k^2 - 2\beta_k b_k)).\end{aligned}$$

To this last expression we add and subtract the quantity

$$\tfrac{1}{2}a_0^2 + \sum_{k=1}^{n} (a_k^2 + b_k^2)$$

to arrive at the equation

$$\begin{aligned}\|f - T_n\|^2 &= \|f\|^2 - \left[\frac{1}{2}a_0^2 + \sum_{k=1}^{n} (a_k^2 + b_k^2)\right] \\ &\quad + \left\{2\left(\alpha_0 - \frac{a_0}{2}\right)^2 + \sum_{k=1}^{n} ((\alpha_k - a_k)^2 + (\beta_k - b_k)^2)\right\}. \qquad \text{(iv)}\end{aligned}$$

Since each entry within the parentheses is nonnegative, it is apparent that a minimum is attained when $\alpha_0 = a_0/2$, $\alpha_k = a_k$ and $\beta_k = b_k$ for all admitted values of k.

Formally we state the result just established as follows:

46.9 / Theorem Let f be an arbitrary function of $\mathfrak{L}^2$ with Fourier coefficients a_k and b_k. If α_k and β_k are arbitrary constants, then for each $n \in \mathfrak{N}$,

$$\left\| f - \left[\tfrac{1}{2} a_0 + \sum_{k=1}^{n} (a_k \cos kx + b_k \sin kx) \right] \right\|$$

$$\leq \left\| f - \left[\tfrac{1}{2} \alpha_0 + \sum_{k=1}^{n} (\alpha_k \cos kx + \beta_k \sin kx) \right] \right\|.$$

From the formula (iv) we obtain the so-called *Bessel's Identity*

$$\left\| f - \left[\tfrac{1}{2} a_0 + \sum_{k=1}^{n} (a_k \cos kx + b_k \sin kx) \right] \right\|^2$$

$$= \| f \|^2 - \left[\tfrac{1}{2} a_0^2 + \sum_{k=1}^{n} (a_k^2 + b_k^2) \right].$$

We note that the left side of this identity is nonnegative. Hence

$$\tfrac{1}{2} a_0^2 + \sum_{k=1}^{n} (a_k^2 + b_k^2) \leq \| f \|^2.$$

Since the right hand side in this inequality is independent of n it follows that the series in the left side converges. This leads us to the following theorem:

46.10 / Bessel's Inequality If f is a function in $\mathfrak{L}^2$ with Fourier coefficients a_k and b_k, then

$$\tfrac{1}{2} a_0^2 + \sum_{k=1}^{\infty} (a_k^2 + b_k^2) \leq \| f \|^2.$$

With the preceding calculations we can now prove:

46.11 / Theorem Each function $f \in \mathfrak{L}^2$ has a Fourier series representation, in the sense that

$$\lim_{n \to \infty} \| f - [\tfrac{1}{2} a_0 + \sum_{k=1}^{n} (a_k \cos kx + b_k \sin kx)] \| = 0,$$

when a_k and b_k are the Fourier coefficients of f.

Proof Let f_n be the nth partial sum of the Fourier series of f:

$$f_n(x) = \tfrac{1}{2} a_0 + \sum_{k=1}^{n} (a_k \cos kx + b_k \sin kx).$$

Then for arbitrary positive integers, $m < n$, Bessel's identity shows that

$$\| f_m - f_n \|^2 = \| \sum_{k=m+1}^{n} (a_k \cos kx + b_k \sin kx) \|^2 = \sum_{k=n+1}^{n} (a_k^2 + b_k^2).$$

Owing to Bessel's inequality the series $\sum (a_k^2 + b_k^2)$ converges, so that

$$\lim_{m,n\to\infty} \| f_m - f_n \| = 0.$$

According to the Riesz–Fischer Theorem 46.8, the sequence $\{ f_n\}$ converges in the mean square to some function $f^* \in \mathfrak{L}^2$, and it remains to show that $f = f^*$ almost everywhere on $[-\pi, \pi]$. In view of the Uniqueness Theorem 45.9 it suffices to show that the Fourier coefficients of f and f^* coincide.†

Let a nonnegative integer k be fixed. The orthogonality relations displayed in Theorem 45.1 admit the following calculation for each integer $n \geq k$:

$$\begin{aligned} \left| \frac{1}{\pi} \int_{-\pi}^{\pi} f^*(x) \cos kx \, d\mu - a_k \right| &= \left| \frac{1}{\pi} \int_{-\pi}^{\pi} f^*(x) \cos kx \, d\mu - \frac{1}{\pi} \int_{-\pi}^{\pi} f_n(x) \cos kx \, d\mu \right| \\ &= \left| \frac{1}{\pi} \int_{-\pi}^{\pi} [f^*(x) - f_n(x)] \cos kx \, d\mu \right| \\ &= |(f^*(x) - f_n(x), \cos kx)| \\ &\leq \| f^* - f_n \| \cdot \| \cos kx \|, \end{aligned}$$

the last step owing to the Schwarz Inequality. But $\| f^* - f_n \| \to 0$ as $n \to \infty$, and so a_k is a Fourier coefficient of f^*. A similar calculation shows each b_k to be a Fourier coefficient of f^* and the theorem is hereby proved.

When we now let $n \to \infty$ in Bessel's identity we obtain:

46.12 / The Parseval Formula If $f \in \mathfrak{L}^2$ and a_k and b_k are its Fourier coefficients, then

$$\| f \|^2 = \tfrac{1}{2} a_0^2 + \sum_{k=1}^{\infty} (a_k^2 + b_k^2).$$

† Actually, to apply Theorem 45.9 we would have to replace the functions f and f^* with new functions g and g^* such that $g(-\pi) = g(\pi)$ and $g^*(-\pi) = g^*(\pi)$, but this, of course, can be done.

As a final result in this section we consider the following converse of Bessel's inequality:

46.13 / Theorem Consider arbitrary sequences of numbers, $\{a_k\}$ and $\{b_k\}$, such that

$$\tfrac{1}{2}a_0^2 + \sum_{k=1}^{\infty} (a_k^2 + b_k^2) < \infty.$$

Then there is a function $f \in \mathfrak{L}^2$ having these numbers as its Fourier coefficients.

The proof of this theorem is left in the hands of the reader.

EXERCISES

46-1 Is the function

$$\sum_{k=2}^{\infty} \frac{\cos kx}{\sqrt{k} \ln k}$$

the Fourier series of a function of $\mathfrak{L}^2$?

46-2 Show that

$$\sum_{k=1}^{\infty} \frac{1}{k^2} = \frac{\pi^2}{6}.$$

(*Hint*: Apply Parseval's Formula 46.12 to the function $f(x) = x$ $(-\pi \leq x \leq \pi)$.)

46-3 Prove or disprove: If $f \in \mathfrak{L}^2$ and

$$f \sim \frac{a_0}{2} + \sum_{k=1}^{\infty} (a_k \cos kx + b_k \sin kx),$$

then

$$\lim_{k\to\infty} a_k = 0 \qquad \text{and} \qquad \lim_{k\to\infty} b_k = 0.$$

What is the status of this problem when $f \in \mathfrak{L}$?

46-4 Prove that if $f \in \mathfrak{L}^2$, then for each number $\epsilon > 0$ there is a bounded measurable function g such that $\| f - g \| < \epsilon$, the norm being the $\mathfrak{L}^2$ norm.

46-5 Prove that if $f \in \mathfrak{L}^2$ and $\epsilon > 0$ is given, then there is a continuous function h such that $\| f - h \| < \epsilon$.
(*Hint*: Consult the proof of Theorem 45.8.)

46-6 If $f \in \mathfrak{L}^2$, show that

$$\lim_{h \to 0} \int_{-\pi}^{\pi} | f(x + h) - f(x) | \, d\mu = 0.$$

Note that this result is immediate when f is continuous.

46-7 Prove or disprove: If $f \in \mathfrak{L}^2$, then the function

$$h(x) = \int_{-\pi}^{\pi} f(\xi + x) f(\xi) \, d\mu$$

is continuous in x.

46-8 Prove Theorem 46.13.

46-9 Consider a sequence $\{ f_n \}$ with range in $\mathfrak{L}^2$. Show that if the sequence converges to f in the mean square, then $\{ f_n \}$ is a Cauchy sequence.

47 / THE QUESTION OF CONVERGENCE

Just as in the case of power series expansions, it is unreasonable to expect even arbitrary continuous periodic functions to have a Fourier series representation. Indeed, an everywhere continuous function of period 2π whose Fourier series diverges at a point was constructed as early as 1876 by DuBois Reymond. Once such an example is known it is easy, of course, to construct continuous periodic functions whose Fourier series diverge on any given finite set.

47.1 / ***Example*** We are going to follow Fejér's construction to obtain a continuous function on $\mathfrak{R}$, of period 2π, whose Fourier series diverges at a point (and hence does not represent the function there). The main components of this construction are the following trigonometric (cosine) polynomials

$$T(n; x) = \frac{\cos nx}{n} + \frac{\cos(n+1)x}{n-1} + \cdots + \frac{\cos(2n-1)x}{1} - \frac{\cos(2n+1)x}{1} - \frac{\cos(2n+2)x}{2} - \cdots - \frac{\cos(3n)x}{n}, \qquad \text{(i)}$$

in which the terms will always be assumed to be listed in the given order. Setting $n_k = 2^{k^3}$ we form the series

$$\sum_{k=1}^{\infty} \frac{1}{k^2} T(n_k; x). \qquad \text{(ii)}$$

We assert that this series converges uniformly on $\mathcal{R}$. This will follow from the Weierstrass M-Test 33.12 once we show that there is a constant c such that $|T(n;x)| < c$ for all $x \in \mathcal{R}$ and $n \in \mathfrak{N}$. To show this, we observe that

$$T(n;x) = \sum_{k=1}^{n} \frac{\cos(2n-k)x - \cos(2n+k)x}{k} = \sum_{k=1}^{n} \frac{2\sin 2nx \cdot \sin kx}{k}$$

$$= 2\sin 2nx \sum_{k=1}^{n} \frac{\sin kx}{k}. \qquad \text{(iii)}$$

Putting

$$s_n(x) = \sum_{k=1}^{n} \frac{\sin kx}{k}$$

let us show that $|s_n(x)| < 5$ for all $x \in \mathcal{R}$. Since, however, $|s_n(x)| = |s_n(x+m)|$ for each integer m, we may restrict x to the interval $0 < x \leq \pi$.

Look at the identity

$$2\sin\frac{x}{2}\sum_{k=1}^{n}\sin kx = \cos\frac{x}{2} - \cos\frac{(2n+1)x}{2}.$$

Since $0 < x \leq \pi$ we find that $\sin x/2 \neq 0$ and hence

$$\left|\sum_{k=1}^{n}\sin kx\right| \leq \frac{\left|\cos\frac{x}{2}\right| + \left|\cos\frac{(2n+1)x}{2}\right|}{2\left|\sin\frac{x}{2}\right|} \leq \frac{1}{\left|\sin\frac{x}{2}\right|}. \qquad \text{(iv)}$$

Now let x be fixed and let m stand for the integral part of $1/x$; then $m \leq 1/x < m+1$. Consequently, when $n \leq m$ then

$$|s_n(x)| = \left|\sum_{k=1}^{n}\frac{\sin kx}{k}\right| \leq \sum_{k=1}^{n}\frac{kx}{k} = nx \leq mx \leq 1.$$

When $n > m$ then

$$|s_n(x)| \leq |s_m(x)| + \left|\sum_{k=m+1}^{n}\frac{\sin kx}{k}\right| \leq 1 + \left|\sum_{k=m+1}^{n}\frac{\sin kx}{k}\right|.$$

From Abel's Lemma 13.7 and (iv) above we deduce that

$$\left|\sum_{k=m+1}^{n}\frac{\sin kx}{k}\right| \leq \frac{1}{(m+1)\sin\frac{x}{2}}.$$

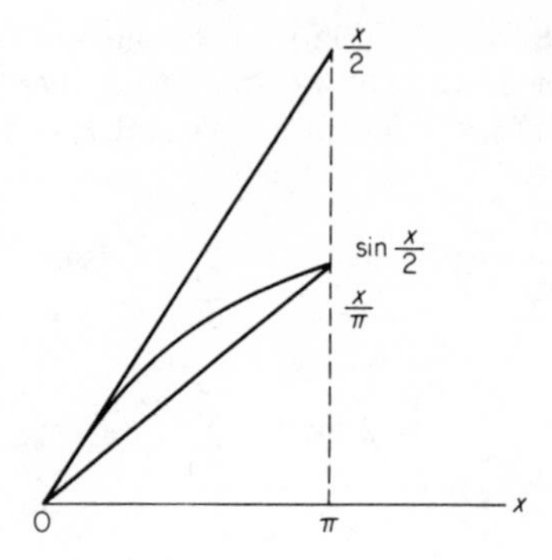

Figure 51

The inequalities $\sin x/2 \geq x/\pi$ (see Figure 51) and $1/x < m + 1$ show that

$$(m+1)\sin\frac{x}{2} \geq \frac{(m+1)x}{\pi} \geq \frac{1}{\pi}$$

and accordingly $|s_n(x)| \leq 1 + \pi < 5$. It now follows from (iii) that

$$|T(n; x)| < 2 \cdot 5 = 10.$$

We have just shown that the series in (ii) converges uniformly. Let its limit function, which is continuous, be f. Because $n_k = 2^{k^3}$ it follows that $3n_k < n_{k+1}$ and an inspection of (i) shows that no two trigonometric polynomials $T(n_j; x)$ and $T(n_k; x)$ have the same cosine factor when $j \neq k$. Theorem 45.7 therefore tells us that if we write out each term $\frac{1}{k^2} T(n_k; x)$ in (ii), then this is the Fourier series of f. However,

$$s_{2n_k}(0) - s_{n_k-1}(0) = \frac{1}{k^2}\left(\frac{1}{n_k} + \frac{1}{n_k - 1} + \cdots + \frac{1}{1}\right) > \frac{1}{k^2}\int_1^{n_k} \frac{dt}{t}$$

$$= \frac{1}{k^2} \cdot \log n_k = \frac{1}{k^2} \cdot \log 2^{k^3} = k \log 2.$$

It follows that the Fourier series of f diverges at $x = 0$.

47.2 / ***Example*** Let us show that the series

$$\sum_{k=1}^{\infty} \frac{\sin kx}{k^\alpha} \qquad (0 < \alpha \leq \tfrac{1}{2}),$$

converges pointwise for each $x \in \mathfrak{R}$, yet it is not the Fourier series of a function of $\mathfrak{L}^2$. The convergence of the series follows again from the Dirichlet Test 13.6. If the numbers $1/k^\alpha$ would be the Fourier coefficients of a function $f \in \mathfrak{L}^2$, then we could appeal to Bessel's Inequality 46.10, according to which

$$\sum_{k=1}^{\infty} \frac{1}{k^{2\alpha}} \leq \frac{1}{\pi} \int_{-\pi}^{\pi} f^2(x)\, d\mu.$$

But the right series diverges and our claim is thus true.

The point of the preceding examples is to warn us that we have to proceed with caution when we deal with Fourier series. Kolmogorov, for example, constructed an integrable (not square integrable) function whose Fourier series diverges everywhere. Steinhaus produced a continuous function whose Fourier series converges pointwise everywhere, but on no interval is the convergence uniform. It is not known if there is a continuous function whose Fourier series diverges on a set of positive measure, but there are examples in which divergence occurs on an uncountable set of measure zero.

Before proceeding to establish some convergence tests, we shall prove the following important result:

47.3 / Riemann–Lebesgue Lemma Let $f \in \mathfrak{L}[-\pi, \pi]$ and let its Fourier coefficients be a_k and b_k. Then

$$\lim_{k\to\infty} a_k = 0 \qquad \text{and} \qquad \lim_{k\to\infty} b_k = 0.$$

Proof Consider the first of these limits. Let an arbitrary number $\epsilon > 0$ be given. Then there is a bounded measurable function g such that

$$\frac{1}{\pi} \int_{-\pi}^{\pi} | f(x) - g(x) |\, d\mu < \frac{\epsilon}{2}.$$

But a bounded measurable function is square integrable. Consequently, if the Fourier coefficients of g are α_k and β_k, then owing to Bessel's Inequality 46.10,

$$\sum \alpha_k^2 + \beta_k^2 < \infty.$$

In particular, $\sum \alpha_k^2$ converges, so that

$$\lim_{k\to\infty} \alpha_k = 0.$$

As a result, there is an integer n_ϵ such that

$$| \alpha_k | < \frac{\epsilon}{2} \qquad (k > n_\epsilon).$$

If we select an arbitrary integer $k \in \mathfrak{N}$, then the following estimate is seen to be true:

$$| a_k - \alpha_k | = \left| \frac{1}{\pi} \int_{-\pi}^{\pi} f(x) \cos kx \, d\mu - \frac{1}{\pi} \int_{-\pi}^{\pi} g(x) \cos kx \, d\mu \right|$$

$$= \frac{1}{\pi} \left| \int_{-\pi}^{\pi} [f(x) - g(x)] \cos kx \, d\mu \right|$$

$$\leq \frac{1}{\pi} \int_{-\pi}^{\pi} | f(x) - g(x) | \, d\mu < \frac{\epsilon}{2}.$$

Hence,

$$| a_k | = | (a_k - \alpha_k) + \alpha_k | \leq | a_k - \alpha_k | + | \alpha_k | < \frac{\epsilon}{2} + \frac{\epsilon}{2} = \epsilon,$$

and the first assertion in the theorem follows. The second assertion is verified in a like manner.

Next, let us introduce the *Dirichlet Kernel*: Consider the periodic function $f \in \mathfrak{L}[-\pi, \pi]$ with Fourier coefficients a_k and b_k and partial sums

$$s_n(x) = \frac{a_0}{2} + \sum_{k=1}^{n} (a_k \cos kx + b_k \sin kx).$$

Then with the customary representation of the Fourier coefficients, we see that

$$s_n(x) = \frac{1}{2\pi} \int_{-\pi}^{\pi} f(t) \, d\mu + \frac{1}{\pi} \sum_{k=1}^{n} \left[\int_{-\pi}^{\pi} f(t) \cos kt \cos kx \, d\mu \right.$$

$$\left. + \int_{-\pi}^{\pi} f(t) \sin kt \sin kx \, d\mu \right]$$

$$= \frac{1}{\pi} \int_{-\pi}^{\pi} f(t) \left[\frac{1}{2} + \sum_{k=1}^{n} \cos k(x - t) \right] d\mu.$$

With the transformation $x - t = u$ we can write the last integral as

$$s_n(x) = \frac{1}{\pi} \int_{x-\pi}^{x+\pi} f(x - u) \left[\frac{1}{2} + \sum_{k=1}^{n} \cos ku \right] d\mu.$$

The function

$$D_n(u) = \frac{1}{2} + \sum_{k=1}^{n} \cos ku \qquad \text{(i)}$$

is called the *Dirichlet Kernel.* The identities

$$\sin\frac{u}{2}\cos ku = \frac{1}{2}\left[\sin\left(k+\frac{1}{2}\right)u - \sin\left(k-\frac{1}{2}\right)u\right] \qquad (k \in \mathfrak{N})$$

show that, for each $n \in \mathfrak{N}$,

$$\sin\frac{u}{2}D_n(u) = \frac{1}{2}\left\{\sin\frac{u}{2} + \left[\sin\left(1+\frac{1}{2}\right)u - \sin\left(1-\frac{1}{2}\right)u\right] + \cdots\right.$$

$$\left. + \left[\sin\left(n+\frac{1}{2}\right)u - \sin\left(n-\frac{1}{2}\right)u\right]\right\} = \frac{1}{2}\sin\left(n+\frac{1}{2}\right)u,$$

and accordingly we have the representation

$$D_n(u) = \frac{\sin\left(n+\frac{1}{2}\right)u}{2\sin\frac{u}{2}}.$$

From (i) it is seen that

$$\frac{2}{\pi}\int_0^{\pi} D_n(u)\, d\mu = 1. \tag{ii}$$

Recognizing that the Dirichlet kernel is an even function, we have the formula

$$s_n(x) = \frac{1}{\pi}\int_0^{\pi} [f(x+u) + f(x-u)]D_n(u)\, d\mu. \tag{iii}$$

The relation in (ii), therefore, permits us to write

$$s_n(x) - f(x) = \frac{1}{\pi}\int_0^{\pi} [f(x+u) + f(x-u) - 2f(x)]D_n(u)\, d\mu. \tag{iv}$$

The last two formulas are known as *Dirichlet's Formulas.*

The stage is now set for the following sufficient condition of Dini for the convergence of a Fourier series at a point.

47.4 / Dini's Test Let

$$\varphi_x(u) = f(x+u) + f(x-u) - 2f(x).$$

If, for some $x \in [-\pi, \pi]$, $\varphi_x(u)/u$ is an integrable function of u in a neighborhood of $u = 0$, then

$$\lim_{n\to\infty} s_n(x) = f(x).$$

Proof Using the formula in (iv), we can put

$$s_n(x) - f(x) = \frac{1}{\pi} \int_0^{\pi} \frac{\varphi_x(u)}{u} \frac{u}{2 \sin \frac{u}{2}} \cdot \sin\left(n + \frac{1}{2}\right) u \, d\mu.$$

If we let

$$g(u) = \begin{cases} \dfrac{u}{2 \sin \frac{u}{2}} & (u \neq 0) \\ 1 & (u = 0), \end{cases}$$

then $g(u)$ is continuous on the interval $[0, \pi]$. Hence, if the function $\frac{\varphi_x(u)}{u}$ is integrable, then so is $\frac{\varphi_x(u)}{u} \cdot g(u) = h(u)$, and we shall show that

$$\lim_{n \to \infty} \int_0^{\pi} h(u) \sin\left(n + \frac{1}{2}\right) u \, d\mu = 0.$$

But

$$\sin\left(n + \frac{1}{2}\right) u = \sin nu \cos \frac{u}{2} + \sin \frac{u}{2} \cos nu,$$

and if we let $h(u) = 0$ for $-\pi \le u < 0$, then $h \in \mathfrak{L}[-\pi, \pi]$,

$$\int_0^{\pi} h(u) \sin\left(n + \frac{1}{2}\right) u \, d\mu = \int_{-\pi}^{\pi} \left[h(u) \sin \frac{u}{2}\right] \cos nu \, d\mu$$

$$+ \int_{-\pi}^{\pi} \left[h(u) \cos \frac{u}{2}\right] \sin nu \, d\mu,$$

and owing to the Riemann–Lebesgue Lemma 47.3 each integral in the right side tends to zero as $n \to \infty$, as was to be shown.

Writing

$$\frac{\varphi_x(u)}{u} = \frac{f(x + u) - f(x)}{u} + \frac{f(x - u) - f(x)}{u}$$

it is apparent that $\frac{\varphi_x(u)}{u}$ is integrable when the limits $f(x+)$ and $f(x-)$, as well as the generalized right and left derivatives $f'(x+)$ and $f'(x-)$, exist (see Definition 23.1 and Exercise 28-5). Hence we have:

47.5 / Corollary Let $f \in \mathfrak{R}[-\pi, \pi]$ be periodic,† and let $x \in [-\pi, \pi]$ be fixed. If

(a) $f(x+)$ and $f(x-)$ exist,

(b) $f'(x+)$ and $f'(x-)$ exist,

then the Fourier series of f will converge to the value

$$\frac{f(x+) + f(x-)}{2}$$

at this point x.

47.6 / *Example* Let us examine the Fourier series of the characteristic function χ of the interval $[0, \pi]$. Clearly, $\chi \in \mathfrak{R}[-\pi, \pi]$, and we note, in particular, that $\chi(0+) = 1$, $\chi(0-) = 0$, $\chi'(0+) = 0$, and $\chi'(0-) = 0$. Owing to the corollary, therefore, the Fourier series of χ should converge to the value $\dfrac{0+1}{2} = \dfrac{1}{2}$ at $x = 0$: Let us see if this is so.

The Fourier coefficients of χ are as follows:

$$a_0 = \frac{1}{\pi}\int_{-\pi}^{\pi} \chi(x)\, d\mu = \frac{1}{\pi}\int_0^{\pi} d\mu = 1;$$

for $k > 0$,

$$a_k = \frac{1}{\pi}\int_{-\pi}^{\pi} \chi(x) \cos kx\, d\mu = \frac{1}{\pi}\int_0^{\pi} \cos kx\, d\mu = 0;$$

$$b_k = \frac{1}{\pi}\int_{-\pi}^{\pi} \chi(x) \sin kx\, d\mu = \frac{1}{\pi}\int_0^{\pi} \sin kx\, dx = \frac{1 - \cos k\pi}{k\pi}$$

$$= \begin{cases} 0 & \text{when } k = 2, 4, 6, \ldots \\ \dfrac{2}{k\pi} & \text{when } k = 1, 3, 5, \ldots . \end{cases}$$

Hence,

$$\chi(x) \sim \frac{1}{2} + \frac{2}{\pi}\left[\sin x + \frac{1}{3}\sin 3x + \frac{1}{5}\sin 5x + \cdots\right],$$

and, in particular, the right side equals $\frac{1}{2}$ at $x = 0$.

† We may assume that f is extended by its periodicity to all of $\mathfrak{R}$.

The student should have no trouble in deducing the following test from Dini's Test:

47.7 / Lipschitz's Test Let $\varphi_x(u)$ be as in Dini's Test 47.4, and let $x \in [-\pi, \pi]$ be fixed. If there are constants $c > 0$ and α, $0 < \alpha \leq 1$, such that

$$| \varphi_x(u) | \leq c | u |^\alpha$$

throughout some neighborhood of $u = 0$, then

$$\lim_{n\to\infty} s_n(x) = f(x),$$

where s_n is the nth partial sum of the Fourier series of f.

Let us now consider the following important problem: How is the convergence behavior at a point x of the Fourier series of a function f affected by changes in f outside a neighborhood of x? The answer to this lies in the following theorem:

47.8 / Riemann's Localization Theorem Let $f \in \mathfrak{L}[-\pi, \pi]$. If $f = 0$ on some interval $(a, b) \subset [-\pi, \pi]$, then its Fourier series converges to zero pointwise on (a, b).

Thus, the convergence behavior at a point x of a Fourier series depends only on the behavior of the function in an arbitrarily small neighborhood of x. This means that if $f \in \mathfrak{L}[-\pi, \pi]$ has a convergent Fourier series at x, and if $g \in \mathfrak{L}[-\pi, \pi]$ is an arbitrary function which agrees with f in an arbitrary neighborhood of x, then also the Fourier series of g converges at x.

Proof It suffices to consider intervals of the form $[a + \epsilon, b - \epsilon]$, where

$$0 < \epsilon < \frac{b - a}{2}.$$

Let a number $\epsilon > 0$ as indicated be fixed. Then

$$f(x + u) + f(x - u) = 0$$

whenever $a + \epsilon \leq x \leq b - \epsilon$ and $0 \leq u < \epsilon$. If $s_n(x)$ stands for the nth partial sum of the Fourier series of f, then by virtue of the formula in (iii), $s_n(x)$ may be expressed as

$$s_n(x) = \frac{1}{\pi}\int_\epsilon^\pi [f(x+u) + f(x-u)] \frac{\sin\left(n+\frac{1}{2}\right)u}{2\sin\frac{u}{2}}\,d\mu$$

$$= \frac{1}{\pi}\int_\epsilon^\pi f(x+u) \frac{\sin\left(n+\frac{1}{2}\right)u}{2\sin\frac{u}{2}}\,d\mu$$

$$+ \frac{1}{\pi}\int_\epsilon^\pi f(x-u) \frac{\sin\left(n+\frac{1}{2}\right)u}{2\sin\frac{u}{2}}\,d\mu.$$

Let us examine the first of these integrals. On the interval $[\epsilon, \pi]$ the function $1/2\sin\frac{u}{2}$ is continuous and thus

$$h_x(u) = \frac{f(x+u)}{2\sin\frac{u}{2}}$$

is integrable over the same interval. According to the last portion of the proof of Theorem 47.4,

$$\lim_{n\to\infty}\int_\epsilon^\pi h_x(u)\sin(n+\tfrac{1}{2})u\,d\mu = 0$$

for each $x \in [a+\epsilon, b-\epsilon]$. The second integral is handled in the same way.

EXERCISES

47-1 Regarding the series $\sum \frac{\sin kx}{k}$, show that it is the Fourier series of the function $\frac{1}{2}(\pi - x)$ on the interval $(0, 2\pi)$. Show further that the Fourier series converges uniformly to $\frac{1}{2}(\pi - x)$ on each closed subinterval of $(0, 2\pi)$.

47-2 Let $f \in \mathcal{R}[-\pi, \pi]$ be periodic, and suppose f to be extended to $\mathcal{R}$ through the specification $f(x + 2\pi) = f(x)$. Show that if x is an arbitrary number, then

$$\int_{-\pi}^{\pi} f(u)\, d\mu = \int_{x-\pi}^{x+\pi} f(u)\, d\mu.$$

47-3 Show that for each fixed number α, $0 < \alpha \leq \pi$,

$$\lim_{n\to\infty} \int_{\alpha}^{\pi} D_n(u)\, d\mu = 0.$$

47-4 Establish the following stronger version of Theorem 47.8: If $f \in \mathfrak{L}[-\pi, \pi]$, and if $f = 0$ on some interval $(a, b) \subset [-\pi, \pi]$, then the Fourier series of f converges to zero uniformly on every closed interval $[\alpha, \beta] \subset (a, b)$.

BIBLIOGRAPHY

GENERAL REFERENCES

[1] Apostol, T. M., Mathematical Analysis. Addison-Wesley, Reading, 1957.

[2] Dieudonné, J., Treatise on Modern Analysis (2nd ed.). Academic Press, New York, 1970.

[3] Gelbaum, B. R., and Olmsted, J. M., Counterexamples in Analysis (2nd ed.). Holden-Day, San Francisco, 1964.

[4] Goffman, C., Real Functions, Holt, New York, 1963.

[5] Natanson, I. P., Theory of Functions of a Real Variable. Frederick Ungar, New York, 1964.

[6] Rudin, W., Principles of Mathematical Analysis (2nd ed.). McGraw-Hill, New York, 1964.

SET THEORY

[7] Fraenkel, A., Abstract Set Theory. North-Holland, Amsterdam, 1961.

[8] Halmos, P., Naive Set Theory. Van Nostrand, Princeton, 1960.

[9] Kamke, E., Theory of Sets. Dover, New York, 1950.

REAL NUMBERS

[10] Cohen, L., and Ehrlich, G., The Structure of The Real Number System. Van Nostrand, Princeton, 1963.

[11] Landau, E., Foundations of Analysis. Chelsea, New York, 1951.

TOPOLOGY

[12] Hocking, J., and Young, G., Topology. Addison-Wesley, Reading, 1961.

[13] Kelley, J., General Topology. Van Nostrand, Princeton, 1955.

MEASURE AND INTEGRATION

[14] Berberian, S. K., Measure and Integration. Macmillan, New York, 1965.

[15] Halmos, P., Measure Theory. Van Nostrand, Princeton, 1950.

[16] Kolmogorov, A. N., and Fomin, S. V., Functional Analysis, Vol. 2. Graylock, Albany, 1961.

FOURIER SERIES

[17] Sz.-Nagy, B., Introduction to Real Functions and Orthogonal Expansions. Oxford Univ. Press, New York, 1965.

[18] Zygmund, A., Trigonometric Series. Cambridge Univ. Press, New York, 1969.

GLOSSARY OF SYMBOLS

$\{a, b, \cdots\}$	set, 4
$\{x \mid p(x)\}$	set of all x having property p, 4
A^c	complement, 7
$P(X)$	power set, 7
$\emptyset$	empty set, 6
$A \times B$	product set, 19
$\langle a, b \rangle$	ordered pair, 18
$\mathfrak{N}$	set of positive integers, 9
$\mathcal{I}$	set of all integers, 27
$\mathcal{R}^*$	rational number system, 29, 37
$\mathcal{R}$	real number system, 20, 56
$\mathcal{R}^E$	extended real number system, 76
$N(a; r)$	neighborhood, 79
N^-	left neighborhood, 152
N^+	right neighborhood, 152
N^*	deleted neighborhood, 152
A'	derived set, 127
$\bar{A}$	closure, 127
$\mathring{A}$	interior, 137
$\mathcal{P}$	Cantor set, 139
F_σ	set of type F-sigma, 171
$\in$	belongs to, 5
$\notin$	does not belong to, 5
$\subset$	contained in, 5
$\not\subset$	not contained in, 5
$\supset$	contains, 5
$\not\supset$	does not contain, 5
$\cup$	union, 10
$\cap$	intersection, 10
$\setminus$	difference (relative complement), 10

$\triangle$	symmetric difference, 15
$\sim$	equivalent, 14, 26
$\nsim$	not equivalent, 13, 26
$\aleph_0$	cardinality of $\mathfrak{N}$, 27
$\aleph$	cardinality of the continuum, 27
$\infty, -\infty$	infinities, 76
lub	least upper bound, 71
glb	greatest lower bound, 71
sup	supremum, 71
inf	infimum, 71
lim sup	upper limit, 88
lim inf	lower limit, 88
f^{-1}	inverse, 17
$f \mid C$	restriction of f to C, 17
$A[f < \alpha]$	set of x for which $f(x) < \alpha$, 160
$A[f > \alpha]$	set of x for which $f(x) > \alpha$, 160
χ_A	characteristic function, 20
d	metric, 21, 22
$\delta(A, B)$	distance between sets, 143
$f(a-)$	left limit, 152
$f(a+)$	right limit, 152
f'	derivative, 184
$f'(a-)$	left derivative, 184
$f'(a+)$	right derivative, 184
$\omega(f; A)$	oscillation of f on A, 172
$\omega(f; \xi)$	oscillation of f at ξ, 172
(f, g)	inner product, 316
$B_n(x; f)$	Bernstein polynomial, 228
$D_n(u)$	Dirichlet kernel, 328
$\Vert\ \Vert$	norm, 23, 204
$m(I)$	length of interval, 72, 247
μ^*	outer measure, 249
μ_*	inner measure, 249
μ	Lebesgue measure, 249
$\int_A f\,d\mu,\ \int_a^b f\,d\mu$	Lebesgue integral, 275, 305
$R\int f\,dx$	Riemann integral, 297
$\mathcal{C}, \mathcal{C}(A)$	space of continuous functions, 22, 204
$\mathcal{L}, \mathcal{L}(A)$	space of Lebesgue integrable functions, 280
$\mathcal{L}^2$	space of square Lebesgue integrable functions, 314
$R(\mathcal{I})$	class of Riemann integrable functions, 297

INDEX

A CATALOG OF SELECTED

DOVER BOOKS

IN ALL FIELDS OF INTEREST

A CATALOG OF SELECTED DOVER BOOKS IN ALL FIELDS OF INTEREST

THE BOOK OF BEASTS: Being a Translation from a Latin Bestiary of the Twelfth Century, T. H. White. Wonderful catalog real and fanciful beasts: manticore, griffin, phoenix, amphivius, jaculus, many more. White's witty erudite commentary on scientific, historical aspects. Fascinating glimpse of medieval mind. Illustrated. 296pp. 5⅜ × 8¼. (Available in U.S. only) 24609-4 Pa. $5.95

FRANK LLOYD WRIGHT: ARCHITECTURE AND NATURE With 160 Illustrations, Donald Hoffmann. Profusely illustrated study of influence of nature—especially prairie—on Wright's designs for Fallingwater, Robie House, Guggenheim Museum, other masterpieces. 96pp. 9¼ × 10¾. 25098-9 Pa. $7.95

FRANK LLOYD WRIGHT'S FALLINGWATER, Donald Hoffmann. Wright's famous waterfall house: planning and construction of organic idea. History of site, owners, Wright's personal involvement. Photographs of various stages of building. Preface by Edgar Kaufmann, Jr. 100 illustrations. 112pp. 9¼ × 10. 23671-4 Pa. $7.95

YEARS WITH FRANK LLOYD WRIGHT: Apprentice to Genius, Edgar Tafel. Insightful memoir by a former apprentice presents a revealing portrait of Wright the man, the inspired teacher, the greatest American architect. 372 black-and-white illustrations. Preface. Index. vi + 228pp. 8¼ × 11. 24801-1 Pa. $9.95

THE STORY OF KING ARTHUR AND HIS KNIGHTS, Howard Pyle. Enchanting version of King Arthur fable has delighted generations with imaginative narratives of exciting adventures and unforgettable illustrations by the author. 41 illustrations. xviii + 313pp. 6⅛ × 9¼. 21445-1 Pa. $5.95

THE GODS OF THE EGYPTIANS, E. A. Wallis Budge. Thorough coverage of numerous gods of ancient Egypt by foremost Egyptologist. Information on evolution of cults, rites and gods; the cult of Osiris; the Book of the Dead and its rites; the sacred animals and birds; Heaven and Hell; and more. 956pp. 6⅛ × 9¼. 22055-9, 22056-7 Pa., Two-vol. set $20.00

A THEOLOGICO-POLITICAL TREATISE, Benedict Spinoza. Also contains unfinished *Political Treatise*. Great classic on religious liberty, theory of government on common consent. R. Elwes translation. Total of 421pp. 5⅜ × 8½. 20249-6 Pa. $6.95

INCIDENTS OF TRAVEL IN CENTRAL AMERICA, CHIAPAS, AND YUCATAN, John L. Stephens. Almost single-handed discovery of Maya culture; exploration of ruined cities, monuments, temples; customs of Indians. 115 drawings. 892pp. 5⅜ × 8½. 22404-X, 22405-8 Pa., Two-vol. set $15.90

LOS CAPRICHOS, Francisco Goya. 80 plates of wild, grotesque monsters and caricatures. Prado manuscript included. 183pp. 6⅝ × 9⅝. 22384-1 Pa. $4.95

AUTOBIOGRAPHY: The Story of My Experiments with Truth, Mohandas K. Gandhi. Not hagiography, but Gandhi in his own words. Boyhood, legal studies, purification, the growth of the Satyagraha (nonviolent protest) movement. Critical, inspiring work of the man who freed India. 480pp. 5⅜ × 8½. (Available in U.S. only) 24593-4 Pa. $6.95

HOW TO WRITE, Gertrude Stein. Gertrude Stein claimed anyone could understand her unconventional writing—here are clues to help. Fascinating improvisations, language experiments, explanations illuminate Stein's craft and the art of writing. Total of 414pp. 4⅜ × 6⅜. 23144-5 Pa. $5.95

ADVENTURES AT SEA IN THE GREAT AGE OF SAIL: Five Firsthand Narratives, edited by Elliot Snow. Rare true accounts of exploration, whaling, shipwreck, fierce natives, trade, shipboard life, more. 33 illustrations. Introduction. 353pp. 5⅜ × 8½. 25177-2 Pa. $7.95

THE HERBAL OR GENERAL HISTORY OF PLANTS, John Gerard. Classic descriptions of about 2,850 plants—with over 2,700 illustrations—includes Latin and English names, physical descriptions, varieties, time and place of growth, more. 2,706 illustrations. xlv + 1,678pp. 8½ × 12¼. 23147-X Cloth. $75.00

DOROTHY AND THE WIZARD IN OZ, L. Frank Baum. Dorothy and the Wizard visit the center of the Earth, where people are vegetables, glass houses grow and Oz characters reappear. Classic sequel to *Wizard of Oz*. 256pp. 5⅜ × 8. 24714-7 Pa. $4.95

SONGS OF EXPERIENCE: Facsimile Reproduction with 26 Plates in Full Color, William Blake. This facsimile of Blake's original "Illuminated Book" reproduces 26 full-color plates from a rare 1826 edition. Includes "The Tyger," "London," "Holy Thursday," and other immortal poems. 26 color plates. Printed text of poems. 48pp. 5¼ × 7. 24636-1 Pa. $3.50

SONGS OF INNOCENCE, William Blake. The first and most popular of Blake's famous "Illuminated Books," in a facsimile edition reproducing all 31 brightly colored plates. Additional printed text of each poem. 64pp. 5¼ × 7. 22764-2 Pa. $3.50

PRECIOUS STONES, Max Bauer. Classic, thorough study of diamonds, rubies, emeralds, garnets, etc.: physical character, occurrence, properties, use, similar topics. 20 plates, 8 in color. 94 figures. 659pp. 6⅛ × 9¼. 21910-0, 21911-9 Pa., Two-vol. set $14.90

ENCYCLOPEDIA OF VICTORIAN NEEDLEWORK, S. F. A. Caulfeild and Blanche Saward. Full, precise descriptions of stitches, techniques for dozens of needlecrafts—most exhaustive reference of its kind. Over 800 figures. Total of 679pp. 8⅛ × 11. Two volumes. Vol. 1 22800-2 Pa. $10.95
Vol. 2 22801-0 Pa. $10.95

THE MARVELOUS LAND OF OZ, L. Frank Baum. Second Oz book, the Scarecrow and Tin Woodman are back with hero named Tip, Oz magic. 136 illustrations. 287pp. 5⅜ × 8½. 20692-0 Pa. $5.95

WILD FOWL DECOYS, Joel Barber. Basic book on the subject, by foremost authority and collector. Reveals history of decoy making and rigging, place in American culture, different kinds of decoys, how to make them, and how to use them. 140 plates. 156pp. 7⅞ × 10¾. 20011-6 Pa. $7.95

HISTORY OF LACE, Mrs. Bury Palliser. Definitive, profusely illustrated chronicle of lace from earliest times to late 19th century. Laces of Italy, Greece, England, France, Belgium, etc. Landmark of needlework scholarship. 266 illustrations. 672pp. 6⅛ × 9¼. 24742-2 Pa. $14.95

ILLUSTRATED GUIDE TO SHAKER FURNITURE, Robert Meader. All furniture and appurtenances, with much on unknown local styles. 235 photos. 146pp. 9 × 12. 22819-3 Pa. $7.95

WHALE SHIPS AND WHALING: A Pictorial Survey, George Francis Dow. Over 200 vintage engravings, drawings, photographs of barks, brigs, cutters, other vessels. Also harpoons, lances, whaling guns, many other artifacts. Comprehensive text by foremost authority. 207 black-and-white illustrations. 288pp. 6 × 9. 24808-9 Pa. $8.95

THE BERTRAMS, Anthony Trollope. Powerful portrayal of blind self-will and thwarted ambition includes one of Trollope's most heartrending love stories. 497pp. 5⅜ × 8½. 25119-5 Pa. $8.95

ADVENTURES WITH A HAND LENS, Richard Headstrom. Clearly written guide to observing and studying flowers and grasses, fish scales, moth and insect wings, egg cases, buds, feathers, seeds, leaf scars, moss, molds, ferns, common crystals, etc.—all with an ordinary, inexpensive magnifying glass. 209 exact line drawings aid in your discoveries. 220pp. 5⅜ × 8½. 23330-8 Pa. $3.95

RODIN ON ART AND ARTISTS, Auguste Rodin. Great sculptor's candid, wide-ranging comments on meaning of art; great artists; relation of sculpture to poetry, painting, music; philosophy of life, more. 76 superb black-and-white illustrations of Rodin's sculpture, drawings and prints. 119pp. 8⅜ × 11¼. 24487-3 Pa. $6.95

FIFTY CLASSIC FRENCH FILMS, 1912–1982: A Pictorial Record, Anthony Slide. Memorable stills from Grand Illusion, Beauty and the Beast, Hiroshima, Mon Amour, many more. Credits, plot synopses, reviews, etc. 160pp. 8¼ × 11. 25256-6 Pa. $11.95

THE PRINCIPLES OF PSYCHOLOGY, William James. Famous long course complete, unabridged. Stream of thought, time perception, memory, experimental methods; great work decades ahead of its time. 94 figures. 1,391pp. 5⅜ × 8½. 20381-6, 20382-4 Pa., Two-vol. set $19.90

BODIES IN A BOOKSHOP, R. T. Campbell. Challenging mystery of blackmail and murder with ingenious plot and superbly drawn characters. In the best tradition of British suspense fiction. 192pp. 5⅜ × 8½. 24720-1 Pa. $3.95

CALLAS: PORTRAIT OF A PRIMA DONNA, George Jellinek. Renowned commentator on the musical scene chronicles incredible career and life of the most controversial, fascinating, influential operatic personality of our time. 64 black-and-white photographs. 416pp. 5⅜ × 8¼. 25047-4 Pa. $7.95

GEOMETRY, RELATIVITY AND THE FOURTH DIMENSION, Rudolph Rucker. Exposition of fourth dimension, concepts of relativity as Flatland characters continue adventures. Popular, easily followed yet accurate, profound. 141 illustrations. 133pp. 5⅜ × 8½. 23400-2 Pa. $3.50

HOUSEHOLD STORIES BY THE BROTHERS GRIMM, with pictures by Walter Crane. 53 classic stories—Rumpelstiltskin, Rapunzel, Hansel and Gretel, the Fisherman and his Wife, Snow White, Tom Thumb, Sleeping Beauty, Cinderella, and so much more—lavishly illustrated with original 19th century drawings. 114 illustrations. x + 269pp. 5⅜ × 8½. 21080-4 Pa. $4.50

SUNDIALS, Albert Waugh. Far and away the best, most thorough coverage of ideas, mathematics concerned, types, construction, adjusting anywhere. Over 100 illustrations. 230pp. 5⅜ × 8½. 22947-5 Pa. $4.00

PICTURE HISTORY OF THE NORMANDIE: With 190 Illustrations, Frank O. Braynard. Full story of legendary French ocean liner: Art Deco interiors, design innovations, furnishings, celebrities, maiden voyage, tragic fire, much more. Extensive text. 144pp. 8⅜ × 11¾. 25257-4 Pa. $9.95

THE FIRST AMERICAN COOKBOOK: A Facsimile of "American Cookery," 1796, Amelia Simmons. Facsimile of the first American-written cookbook published in the United States contains authentic recipes for colonial favorites—pumpkin pudding, winter squash pudding, spruce beer, Indian slapjacks, and more. Introductory Essay and Glossary of colonial cooking terms. 80pp. 5⅜ × 8½. 24710-4 Pa. $3.50

101 PUZZLES IN THOUGHT AND LOGIC, C. R. Wylie, Jr. Solve murders and robberies, find out which fishermen are liars, how a blind man could possibly identify a color—purely by your own reasoning! 107pp. 5⅜ × 8½. 20367-0 Pa. $2.00

THE BOOK OF WORLD-FAMOUS MUSIC—CLASSICAL, POPULAR AND FOLK, James J. Fuld. Revised and enlarged republication of landmark work in musico-bibliography. Full information about nearly 1,000 songs and compositions including first lines of music and lyrics. New supplement. Index. 800pp. 5⅜ × 8¼. 24857-7 Pa. $14.95

ANTHROPOLOGY AND MODERN LIFE, Franz Boas. Great anthropologist's classic treatise on race and culture. Introduction by Ruth Bunzel. Only inexpensive paperback edition. 255pp. 5⅜ × 8½. 25245-0 Pa. $5.95

THE TALE OF PETER RABBIT, Beatrix Potter. The inimitable Peter's terrifying adventure in Mr. McGregor's garden, with all 27 wonderful, full-color Potter illustrations. 55pp. 4¼ × 5½. (Available in U.S. only) 22827-4 Pa. $1.75

THREE PROPHETIC SCIENCE FICTION NOVELS, H. G. Wells. *When the Sleeper Wakes, A Story of the Days to Come* and *The Time Machine* (full version). 335pp. 5⅜ × 8½. (Available in U.S. only) 20605-X Pa. $5.95

APICIUS COOKERY AND DINING IN IMPERIAL ROME, edited and translated by Joseph Dommers Vehling. Oldest known cookbook in existence offers readers a clear picture of what foods Romans ate, how they prepared them, etc. 49 illustrations. 301pp. 6⅛ × 9¼. 23563-7 Pa. $6.00

SHAKESPEARE LEXICON AND QUOTATION DICTIONARY, Alexander Schmidt. Full definitions, locations, shades of meaning of every word in plays and poems. More than 50,000 exact quotations. 1,485pp. 6½ × 9¼. 22726-X, 22727-8 Pa., Two-vol. set $27.90

THE WORLD'S GREAT SPEECHES, edited by Lewis Copeland and Lawrence W. Lamm. Vast collection of 278 speeches from Greeks to 1970. Powerful and effective models; unique look at history. 842pp. 5⅜ × 8½. 20468-5 Pa. $10.95

THE BLUE FAIRY BOOK, Andrew Lang. The first, most famous collection, with many familiar tales: Little Red Riding Hood, Aladdin and the Wonderful Lamp, Puss in Boots, Sleeping Beauty, Hansel and Gretel, Rumpelstiltskin; 37 in all. 138 illustrations. 390pp. 5⅜ × 8½. 21437-0 Pa. $5.95

THE STORY OF THE CHAMPIONS OF THE ROUND TABLE, Howard Pyle. Sir Launcelot, Sir Tristram and Sir Percival in spirited adventures of love and triumph retold in Pyle's inimitable style. 50 drawings, 31 full-page. xviii + 329pp. 6½ × 9¼. 21883-X Pa. $6.95

AUDUBON AND HIS JOURNALS, Maria Audubon. Unmatched two-volume portrait of the great artist, naturalist and author contains his journals, an excellent biography by his granddaughter, expert annotations by the noted ornithologist, Dr. Elliott Coues, and 37 superb illustrations. Total of 1,200pp. 5⅜ × 8.
Vol. I 25143-8 Pa. $8.95
Vol. II 25144-6 Pa. $8.95

GREAT DINOSAUR HUNTERS AND THEIR DISCOVERIES, Edwin H. Colbert. Fascinating, lavishly illustrated chronicle of dinosaur research, 1820's to 1960. Achievements of Cope, Marsh, Brown, Buckland, Mantell, Huxley, many others. 384pp. 5¼ × 8¼. 24701-5 Pa. $6.95

THE TASTEMAKERS, Russell Lynes. Informal, illustrated social history of American taste 1850's–1950's. First popularized categories Highbrow, Lowbrow, Middlebrow. 129 illustrations. New (1979) afterword. 384pp. 6 × 9.
23993-4 Pa. $6.95

DOUBLE CROSS PURPOSES, Ronald A. Knox. A treasure hunt in the Scottish Highlands, an old map, unidentified corpse, surprise discoveries keep reader guessing in this cleverly intricate tale of financial skullduggery. 2 black-and-white maps. 320pp. 5⅜ × 8½. (Available in U.S. only) 25032-6 Pa. $5.95

AUTHENTIC VICTORIAN DECORATION AND ORNAMENTATION IN FULL COLOR: 46 Plates from "Studies in Design," Christopher Dresser. Superb full-color lithographs reproduced from rare original portfolio of a major Victorian designer. 48pp. 9¼ × 12¼. 25083-0 Pa. $7.95

PRIMITIVE ART, Franz Boas. Remains the best text ever prepared on subject, thoroughly discussing Indian, African, Asian, Australian, and, especially, Northern American primitive art. Over 950 illustrations show ceramics, masks, totem poles, weapons, textiles, paintings, much more. 376pp. 5⅜ × 8. 20025-6 Pa. $6.95

SIDELIGHTS ON RELATIVITY, Albert Einstein. Unabridged republication of two lectures delivered by the great physicist in 1920–21. *Ether and Relativity* and *Geometry and Experience*. Elegant ideas in non-mathematical form, accessible to intelligent layman. vi + 56pp. 5⅜ × 8½. 24511-X Pa. $2.95

THE WIT AND HUMOR OF OSCAR WILDE, edited by Alvin Redman. More than 1,000 ripostes, paradoxes, wisecracks: Work is the curse of the drinking classes, I can resist everything except temptation, etc. 258pp. 5⅜ × 8½. 20602-5 Pa. $3.95

ADVENTURES WITH A MICROSCOPE, Richard Headstrom. 59 adventures with clothing fibers, protozoa, ferns and lichens, roots and leaves, much more. 142 illustrations. 232pp. 5⅜ × 8½. 23471-1 Pa. $3.95

PLANTS OF THE BIBLE, Harold N. Moldenke and Alma L. Moldenke. Standard reference to all 230 plants mentioned in Scriptures. Latin name, biblical reference, uses, modern identity, much more. Unsurpassed encyclopedic resource for scholars, botanists, nature lovers, students of Bible. Bibliography. Indexes. 123 black-and-white illustrations. 384pp. 6 × 9. 25069-5 Pa. $8.95

FAMOUS AMERICAN WOMEN: A Biographical Dictionary from Colonial Times to the Present, Robert McHenry, ed. From Pocahontas to Rosa Parks, 1,035 distinguished American women documented in separate biographical entries. Accurate, up-to-date data, numerous categories, spans 400 years. Indices. 493pp. 6½ × 9¼. 24523-3 Pa. $9.95

THE FABULOUS INTERIORS OF THE GREAT OCEAN LINERS IN HISTORIC PHOTOGRAPHS, William H. Miller, Jr. Some 200 superb photographs capture exquisite interiors of world's great "floating palaces"—1890's to 1980's: *Titanic, Ile de France, Queen Elizabeth, United States, Europa,* more. Approx. 200 black-and-white photographs. Captions. Text. Introduction. 160pp. 8⅞ × 11¾. 24756-2 Pa. $9.95

THE GREAT LUXURY LINERS, 1927–1954: A Photographic Record, William H. Miller, Jr. Nostalgic tribute to heyday of ocean liners. 186 photos of Ile de France, Normandie, Leviathan, Queen Elizabeth, United States, many others. Interior and exterior views. Introduction. Captions. 160pp. 9 × 12. 24056-8 Pa. $9.95

A NATURAL HISTORY OF THE DUCKS, John Charles Phillips. Great landmark of ornithology offers complete detailed coverage of nearly 200 species and subspecies of ducks: gadwall, sheldrake, merganser, pintail, many more. 74 full-color plates, 102 black-and-white. Bibliography. Total of 1,920pp. 8⅝ × 11¼. 25141-1, 25142-X Cloth. Two-vol. set $100.00

THE SEAWEED HANDBOOK: An Illustrated Guide to Seaweeds from North Carolina to Canada, Thomas F. Lee. Concise reference covers 78 species. Scientific and common names, habitat, distribution, more. Finding keys for easy identification. 224pp. 5⅜ × 8½. 25215-9 Pa. $5.95

THE TEN BOOKS OF ARCHITECTURE: The 1755 Leoni Edition, Leon Battista Alberti. Rare classic helped introduce the glories of ancient architecture to the Renaissance. 68 black-and-white plates. 336pp. 8⅜ × 11¼. 25239-6 Pa. $14.95

MISS MACKENZIE, Anthony Trollope. Minor masterpieces by Victorian master unmasks many truths about life in 19th-century England. First inexpensive edition in years. 392pp. 5⅜ × 8½. 25201-9 Pa. $7.95

THE RIME OF THE ANCIENT MARINER, Gustave Doré, Samuel Taylor Coleridge. Dramatic engravings considered by many to be his greatest work. The terrifying space of the open sea, the storms and whirlpools of an unknown ocean, the ice of Antarctica, more—all rendered in a powerful, chilling manner. Full text. 38 plates. 77pp. 9¼ × 12. 22305-1 Pa. $4.95

THE EXPEDITIONS OF ZEBULON MONTGOMERY PIKE, Zebulon Montgomery Pike. Fascinating first-hand accounts (1805–6) of exploration of Mississippi River, Indian wars, capture by Spanish dragoons, much more. 1,088pp. 5⅜ × 8½. 25254-X, 25255-8 Pa. Two-vol. set $23.90

KRAUSKOPF LIBRARY
515.8 Sp77
STACKS
Sprecher, David A./Elements of real anal
3 1896 00031 7184

WD